AF199804

Palgrave Studies in Environmental Transformation, Transition and Accountability

Series Editors
Siddharth Sareen⬛, Fridtjof Nansen Institute,
Lysaker, Norway
Tor Håkon Jackson Inderberg, Fridtjof Nansen Institute,
Lysaker, Norway

The monographs and edited collections published in this series boast cutting-edge, interdisciplinary energy and environmental social science, offering compelling insights on the accountable governance of low-carbon transitions. The series features books that combine conceptual rigour and empirical robustness to address urgent yet timeless questions of how to bring about socioecologically equitable and just energy futures. Publications span diverse spatial scales and governance levels, examining the evolution of energy system across increasingly electrifying and digitalising sectors.

Siddharth Sareen · Sirkku Juhola
Editors

Societal Transitions to Sustainability

The Prefigurative Politics of Present Transformation

Editors
Siddharth Sareen
Fridtjof Nansen Institute
Lysaker, Norway

Sirkku Juhola
Ecosystems and Env Research Program
University of Helsinki
Helsinki, Finland

![CC BY NC ND license logo]

ISSN 2523-8183   ISSN 2523-8191 (electronic)
Palgrave Studies in Environmental Transformation, Transition and Accountability
ISBN 978-3-032-07394-5   ISBN 978-3-032-07395-2 (eBook)
https://doi.org/10.1007/978-3-032-07395-2

This work was supported by Forschungsinstitut für biologischen Landbau FiBL.

© The Editor(s) (if applicable) and The Author(s) 2026. This book is an open access publication.

Open Access This book is licensed under the terms of the Creative Commons Attribution-NonCommercial-NoDerivatives 4.0 International License (http://creativecommons.org/licenses/by-nc-nd/4.0/), which permits any noncommercial use, sharing, distribution and reproduction in any medium or format, as long as you give appropriate credit to the original author(s) and the source, provide a link to the Creative Commons license and indicate if you modified the licensed material. You do not have permission under this license to share adapted material derived from this book or parts of it.
The images or other third party material in this book are included in the book's Creative Commons license, unless indicated otherwise in a credit line to the material. If material is not included in the book's Creative Commons license and your intended use is not permitted by statutory regulation or exceeds the permitted use, you will need to obtain permission directly from the copyright holder.
This work is subject to copyright. All commercial rights are reserved by the author(s), whether the whole or part of the material is concerned, specifically the rights of translation, reprinting, reuse of illustrations, recitation, broadcasting, reproduction on microfilms or in any other physical way, and transmission or information storage and retrieval, electronic adaptation, computer software, or by similar or dissimilar methodology now known or hereafter developed. Regarding these commercial rights a non-exclusive license has been granted to the publisher.
The use of general descriptive names, registered names, trademarks, service marks, etc. in this publication does not imply, even in the absence of a specific statement, that such names are exempt from the relevant protective laws and regulations and therefore free for general use.
The publisher, the authors and the editors are safe to assume that the advice and information in this book are believed to be true and accurate at the date of publication. Neither the publisher nor the authors or the editors give a warranty, expressed or implied, with respect to the material contained herein or for any errors or omissions that may have been made. The publisher remains neutral with regard to jurisdictional claims in published maps and institutional affiliations.

Cover credit: © Diephosi/gettyimages

This Palgrave Macmillan imprint is published by the registered company Springer Nature Switzerland AG
The registered company address is: Gewerbestrasse 11, 6330 Cham, Switzerland

If disposing of this product, please recycle the paper.

Foreword: Patterning a Sustainable Future

Imagine a world where life flourishes—where children thrive, and where people feel deeply connected to each other and to nature, treating the Earth and its thin atmosphere as our shared and sacred home. Imagine a world that channels its resources not into weapons and destruction, but into cultivating an ethic of care for life itself and the systems that sustain it.

What if you were to learn that this world is not only imagined, but already being built? Across the globe, countless individuals and groups are working every day to realise an equitable and sustainable world, here and now. This is the essence of prefigurative politics: we are patterning the future through the choices and practices that we make today.

There is a global movement on this planet to pattern a new world, and it is taking shape quietly and steadily, with surprisingly little fanfare and attention. Like a mycelial network beneath the forest floor, it is largely invisible, but vibrantly alive. Paul Hawken (2007) described this phenomenon as "blessed unrest," the largest social movement in human history, encompassing environmental activism, social justice initiatives, and Indigenous resistance to globalisation. As Hawken (2007, 23) writes, "A healthy global civilization cannot be constructed without building blocks of meaning, which are hewn from rights and respect."

What differentiates this prefigurative movement from the polarising politics of these times is its embodiment of universal values such as equity, justice, pluralism, and reciprocal relationships between humans and nature. These values are not selective or partial. As with human rights, they apply to all. These values also underpin the principles of transformative change identified by the Transformative Change Assessment (IPBES, 2024), which highlights the need to address the underlying causes of biodiversity loss and nature's decline. These causes include the disconnection and domination of nature and people; the concentration of wealth and power; and the prioritisation of short-term, individual, and material gains.

Prefigurative politics is repatterning society at all scales. As this edited volume emphasises, prefigurative politics refers to "*embodied strategies to render desirable futures with immediacy*" (Sareen & Juhola, this book). These strategies contribute to what Eric Olin Wright (2010) calls "real utopias"—alternatives that acknowledge tensions between dreams and practices. Drawing on emancipatory social science, real utopias represent viable alternatives to current patterns. They also reflect the radical relationality of Arturo Escobar's *Pluriversal Politics*, which envisions "pathways that may be unthinkable from the perspective of Eurocentric theories" (Escobar, 2020, 70).

The examples presented in this book show that prefigurative politics is not a singular or prescriptive approach—it is diverse, experimental, and context-specific. But what unites these cases is a shared commitment to doing things differently. In contrast to the politics of wishful thinking, where visions and goals are proclaimed but rarely realised, prefigurative initiatives make values tangible through practice. Practitioners do not wait for ideal conditions, but instead put transformations into practice in the current context to fashion recursive change.

The chapters in this volume are organised around four themes that reflect the breadth of prefigurative efforts: (1) urban sites of experimentation and contestation, (2) spaces of conviviality and politics, (3) sectoral movements, and (4) cross-sectoral and transdisciplinary transitions.

Together, they demonstrate how prefigurative politics is contributing to transformative change in diverse ways. Taken together, it becomes clear that the power of prefigurative politics should not be underestimated.

One critique of prefigurative politics is that it lacks scalability and societal impact (Monticelli, 2022). Some argue that in the face of a growing polycrisis, such efforts are an ineffective means of instigating systemic change, especially given the urgency. However, this critique typically relies on linear, hierarchical models of change. Prefigurative approaches, in contrast, follow a fractal approach to scaling transformations by generating self-similar patterns that embody universal values and resonate across scales (O'Brien et al., 2023). When these patterns are embodied, embedded, and repeated, they have the potential to disrupt and eventually replace dominant norms, structures, and institutions.

This edited volume is a timely response to a growing desire to not only imagine transformative change for a just and sustainable world but to experience it. Through more than two dozen examples, it explores how prefigurative politics looks and feels in practice, distilling shared themes and insights that emphasise the diversity of pathways for transformative change. While politics of polarisation and paralysis persist, a willingness to embody values and do things differently may be the most powerful strategy for prefiguring a real utopia.

Karen O'Brien
Department of Sociology
and Human Geography
University of Oslo
Oslo, Norway

References

Escobar, A. (2020). *Pluriversal politics: The real and the possible*. Duke University Press.

Hawken, P. (2008). *Blessed unrest: How the largest social movement in history is restoring grace, justice, and beauty to the world*. Penguin Books.

IPBES. (2024). *Summary for policymakers of the thematic assessment report on the underlying causes of biodiversity loss and the determinants of transformative change and options for achieving the 2050 Vision for Biodiversity*. Intergovernmental Science-Policy Platform on Biodiversity and Ecosystem Services (K. O'Brien, L. Garibaldi, A. Agrawal, E. Bennett, O. Biggs, R. Calderón Contreras, E. Carr, N. Frantzeskaki, H. Gosnell, J. Gurung, S. Lambertucci, J. Leventon, C. Liao, V. Reyes García, L. Shannon, S. Villasante, F. Wickson, Y. Zinngrebe, & L. Perianin, Eds.). IPBES Secretariat.

Monticelli, L. (2022). Prefigurative politics within, despite and beyond contemporary capitalism. In *The future is now: An introduction to prefigurative politics* (pp. 15–31). Bristol University Press. https://doi.org/10.51952/978 1529215687.ch001

O'Brien, K., Carmona, R., Gram-Hanssen, I., Hochachka, G., Sygna, L., & Rosenberg, M. (2023). Fractal approaches to scaling transformations to sustainability. *Ambio, 52*(9), 1448–1461. https://doi.org/10.1007/s13280-023-01873-w

Sareen, S., & Juhola, S. (2025). The prefigurative politics of present transformation. In S. Sareen & S. Juhola (Eds.), *Societal transitions to sustainability: The prefigurative politics of present transformation* (this book). Palgrave Macmillan.

Wright, E. O. (2010). *Envisioning real utopias*. Verso.

Acknowledgements

This publication is based upon work from COST Action CA22156 TransformERS, supported by COST (European Cooperation in Science and Technology). See https://www.cost.eu.

COST (European Cooperation in Science and Technology) is a funding agency for research and innovation networks. Our Actions help connect research initiatives across Europe and enable scientists to grow their ideas by sharing them with their peers. This boosts their research, career and innovation.

The editors are grateful to the COST Association for workshop funding to convene working group 3 (present transformation) book chapter authors in Oslo in April 2025, and for sponsoring open access publication. Sareen acknowledges time on Research Council of Norway projects 349994 (ENERGY4ALL) and 295704 (INCLUDE).

Contents

Notes on Contributors

Juan Manuel Amezcua-Ogáyar (PhD) is a permanent professor at the Higher Polytechnic School of Jaén, University of Jaén (Spain). An industrial engineer with over 25 years of experience, he teaches and researches logistics, production management, supply chain, quality and tourism. He has accumulated over 17 years in one-person management positions at university.

Cheshta Arora is a Senior Researcher at Vestlandsforsking. Her work focuses on unpacking socio-technical systems in the contemporary, traversing fields such as feminist (techno) science studies, science and technology studies, media studies, and work/labour studies. Her analytical toolkit includes approaches such as auto/ethnography, material-semiotic analysis, and speculative design.

Corelia Baibarac-Duignan is at the Department of Technology, Policy, and Society; Knowledge, Transformation & Society (KiTeS), Faculty of Behavioural, Management, and Social Science, University of Twente, The Netherlands.

Samyajit Basu is a postdoctoral researcher at the Free University of Brussels (VUB), at the Research Group Mobilise.

Prof. Simon Batterbury is a Professor of Environmental Studies, University of Melbourne, Australia. He is co-editor of the Journal of Political Ecology and part of WeCycle Melbourne.

Elisa T. Bertuzzo is an urban ethnographer affiliated to Leibniz-Zentrum Moderner Orient (ZMO) in Berlin, Germany, and specialising on migration in/from South Asia. Bridging political ecology, visual anthropology, gender and postcolonial studies, her research investigates the everyday life facets of resistance and self-organization in urban and rural contexts.

Jens Brandt is a researcher at Tampere University, School of Architecture. His work examines the intersections of architecture, urban studies, and performative and visual arts, focusing on spatial practices, political transformation, and experimental inhabitation.

Alberto Calahorro-López is a PhD candidate and Assistant Professor in the Department of Business Organisation, Marketing, and Sociology at the University of Jaén (Spain). His research interests include finance, management, and accounting, with a particular focus on the sports sector. He is also interested in rural development.

Dr. Sabrina Chakori is CERC Postdoctoral Fellow at CSIRO and Honorary Research Fellow at University of Melbourne, Australia. She is co-editor of the Degrowth journal and founder of the Brisbane Tool Library.

Dagmar Diesner is a postdoctoral researcher at the Maastricht Sustainability Institute. She manages SURFIT, which is transforming the food system through collective learning and practice with local food. She is part of the collective thecommoner.org and the International Association for the Study of the Commons at the Elinor Ostrom Foundation.

Aleksandra S. Dragin (PhD) is a Full Professor at the Faculty of Sciences (UNS, Serbia). She also teaches in Croatia and Spain. With over 20 years of experience, she researches local communities, rural transformation, sustainability, entrepreneurship, management and tourism. She is a member of several international networks. She has participated in 41 projects.

Frank Eckerle is at the Department of Psychology at Klagenfurt University.

Giuseppe Feola is at the Copernicus Institute of Sustainable Development, Utrecht University, The Netherlands.

Dr. Turkan Firinci Orman is an independent scholar based in Finland, working at the intersection of children's and youth geographies, citizenship studies, transformative pedagogies, and human behaviour in political contexts. Holding a docentship in Sociology, her interdisciplinary research explores youth environmental citizenship, everyday activism, and sustainable futures.

Zali Fung is a postdoctoral researcher in political ecology at the Institute of Geography and Sustainability at the University of Lausanne, Switzerland. Her research examines uneven development and contested resource extraction in Southeast Asia's transboundary river basins, with a focus on community and civil society resistance to hydropower development.

Juliana E. Gonçalves is an Assistant Professor in the Department of Urbanism at TU Delft. She has an interdisciplinary background with expertise in socio-technical studies, sustainability transitions, and spatial justice. Her most recent research focuses on the potential of collective action and prefigurative politics for transformative climate adaptation.

Bianca Griffani is a social anthropologist at Goldsmiths College, University of London. She focuses on the infrastructuring labour of working-class communities who repurpose institutions to sustain collective life in the face of capitalist restructure, using fieldwork in Terni, where she co-organised a mutual-aid food pantry and co-founded a tenants' union.

Bob Grumiau is a PhD candidate at University College Cork, Ireland, for the DSIS project at the Environmental Research Institute. His PhD addresses the transformation of Gender, Religion and Education. With a transdisciplinary background, he studies social movements and alternative worlding practices in Latin America and the poetics of re-enchantment.

Bård Torvetjønn Haugland is a post-doctoral researcher at the Norwegian University of Science & Technology (NTNU) and researcher at SINTEF Digital. His core research interest revolves around the relationship between innovation and preservation, particularly how these concepts relate to specific notions of stability and change.

John Holmberg is at the Department of Space, Earth and Environment, Chalmers University of Technology, Gothenburg, Sweden.

Johan Holmén is at the Department of Engineering Science, University West, Trollhättan, Sweden, and the Department of Space, Earth and Environment, Chalmers University of Technology, Gothenburg, Sweden.

Sheila Htoo is a PhD researcher in the Faculty of Environmental and Urban Change at York University in Toronto, Canada. Her doctoral research examines the political ecology of war, resource extraction, ceasefire capitalism, and peace-building efforts by Indigenous Karen people in southeastern Burma/Myanmar through the Salween Peace Park movement.

Ian Hughes is Senior Research Fellow at MaREI Centre, Environmental Research Institute, University College Cork, Ireland. He is co-leading the Deep Institutional Innovation for Sustainability and Human Flourishing initiative at UCC, which comprises a transdisciplinary community of over 30 academics and practitioners, and has developed wide institutional collaborations.

Sirkku Juhola is a Professor of urban environmental policy, Ecosystems and Environment Research Program, University of Helsinki and leads the Urban Environmental Policy group. An expert in environmental policy and governance, she is a contributing author of IPCC AR6 and a lead author in AR3 on Climate Change and Cities, UCCRN.

Kristina Košić (PhD) is a Full professor at the Faculty of Sciences (University of Novi Sad, Serbia). She teaches and researches rural tourism, spa tourism and tourism destination management. She is an author and co-author of around 120 scientific papers, monographs and books and a member of several scientific projects.

Ana Kotevska is a Professor at the Institute of Agricultural Economics, Faculty of Agricultural Sciences and Food, University of Ss. Cyril and Methodius in Skopje. She is interested in understanding farmer behaviour, monitoring and evaluation of the agricultural policy in the Western Balkans, and recently also applying Theory U or Neuro-Linguistic Programming.

Jan Lilliendahl Larsen is a freelance researcher at Praxis, Copenhagen. His work explores the possibilities for democratic participation in the interface between urban life, culture, politics and development in the field of tension between high-profile conflicts and the vague spaces of the city.

Julia Leventon is a professor and Department Head at the Global Change Research Institute of the Czech Academy of Sciences, and IPBES Coordinating Lead Author. She engages broadly with inter- and trans-disciplinarily research on sustainability, biodiversity loss and climate change through systems thinking, including understanding place-based social-ecological interactions and multilevel governance.

Maria Luisa Lode is a postdoctoral researcher at the Free University of Brussels (VUB), at the House of Sustainable Transitions.

Cathy Macharis is a Professor at the Free University of Brussels (VUB). She is the host of the House of Sustainable Transitions, and the founder and supporter of the Research Group Mobilise.

Aleksandra Martinovska Stojcheska is a Professor at the Faculty of Agricultural Sciences and Food, Ss. Cyril and Methodius University in Skopje, North Macedonia. Her research interests are in agrifood economics and technical performance, food systems assessment, decision-making and strategic choices behaviour, economic aspects of climate change, agricultural and rural development policy.

Steven R. McGreevy is at the Governance and Technology for Sustainability Section (CSTM), Department of Technology, Policy, and Society, Faculty of Behavioural, Management, and Social Science, University of Twente, The Netherlands.

Peter Mederly is a professor at the Department of Ecology and Environmental Sciences, Constantine the Philosopher University in Nitra, Slovakia. His main expertise is in environmental science, geography and landscape ecology, with particular focus on valuation of ecosystem services, landscape perspective on societal challenges, and environmental politics.

Maja Mijatov Ladičorbić (PhD) is an Assistant Professor in the field of tourism, with the research focused towards business ethics, corporate social responsibility and management. She is a member of international networks, Erasmus+ department coordinator and co-author of sustainable tourism strategy and visitor management plan for protected area in Serbia.

Sarah Milliken is a landscape architect and Research Fellow in the School of Design and Creative Industries at the University of Greenwich, London. Her research interests primarily revolve around the science–policy–society interface in the sustainable cities and food systems agendas.

Kavitha Ravikumar is at Loughborough University, London. Her research focuses on organising, sustainability, design and creative practices. With a background in Anthropology and Economics, Kavitha has worked in the private sector on healthcare technologies, chemicals, and sustainability impact research, with an informed orientation towards collective spaces of hope and transformation.

João Rocha-Gomes is at the Department of Community Medicine, Health Information and Decision, Faculty of Medicine, University of Porto, Porto, Portugal.

Clara Saglietti is at the Department of Space, Earth and Environment, Chalmers University of Technology, Gothenburg, Sweden.

Siddharth Sareen is a Research Professor at the Fridtjof Nansen Institute, Norway and Professor II at the Centre for Climate and Energy Transformation, University of Bergen. He is an award-winning environmental social scientist with grounding in development studies, political ecology, and human geography. He works on the governance of energy transitions.

Zerrin Savaşan is Assoc. Prof. Dr., Department of International Relations, Sub-Department of International Law, Faculty of Economics and Administrative Sciences, Selçuk University, Konya, Türkiye.

Christian Scholl (PhD) is an Assistant Professor, Maastricht Sustainability Institute, Maastricht University, PI of SURFIT (Scaling Urban Regenerative Food Systems in Transition) and Programme Leader of the Master's Programme Sustainability Science, Policy and Society. His research focuses on urban sustainability, participatory and collaborative forms of governance, social movements and social learning.

Tyler Schuenemann is an Assistant Professor of Environmental and Sustainability Studies at Keene State College. His research examines the politicisation of natural disasters in the United States and the

Middle East through ethnographic observations, interviews, and archival research.

Shayan Shokrgozar is at the Centre for Climate and Energy Transformation (CET), University of Bergen, Bergen, Norway.

David Singh is at the Department of Food and Resource Economics (IFRO), University of Copenhagen, Copenhagen, Denmark.

Maximilian Spiegelberg is programme director of the SDG+Dialog at University of Kassel, Germany, and co-director of NPO FEAST in Taipei, Taiwan.

David Stella is a researcher at the Global Change Research Institute of the Czech Academy of Sciences, Prague. His work focuses on ecosystem services, biodiversity research, citizen science, and socio-ecological transformations.

Tamara Surla (PhD) is a Research Assistant at the Faculty of Sciences (UNS, Serbia). She specialises in tourism, with a particular focus on slow movement, stakeholder collaboration, and rural community engagement. She refined her expertise through postdoctoral studies at Breda University of Applied Sciences, enhancing contributions to sustainable tourism development.

Gerald Taylor Aiken is a Senior Fellow at the Research Institute for Sustainability (RIFS), Potsdam.

Aleksandra Tešin (PhD) is a Research Assistant at the Faculty of Sciences, University of Novi Sad, Serbia. She is engaged in the field of tourism, primarily exploring its social and psychological dimensions. Her main areas of interest include tourism experience, tourist behaviour, tourism development, and sustainable development.

Emelj Tuna is Professor at the Institute of Agricultural Economics, Faculty of Agricultural Sciences and Food, University of Ss. Cyril and

Methodius in Skopje. Her research interests include social networks and agricultural cooperatives, socio-economic aspects of rural development, value chain analyses, AKIS, and agricultural advisory.

Dr. Carlos Uxo is a Senior Lecturer, Monash University, Australia, and is also at WeCycle Melbourne.

Simeon Vaňo is a postdoctoral researcher at the Global Change Research Institute of the Czech Academy of Sciences, and at the Department of Ecology and Environmental Sciences, Constantine the Philosopher University in Nitra, Slovakia. His expertise in social-ecological systems research, including policy and governance, focuses on challenges for transformative change.

Timo von Wirth works at the Planetary Futures Studio, Frankfurt University of Applied Sciences, Frankfurt a.M., Germany; the Dutch Research Institute for Transitions, Erasmus University Rotterdam, The Netherlands; and the Center for Sustainability Transitions (CST), Stellenbosch University, South Africa.

Zrinka Zadel (PhD) is a Full Professor and a Head of Tourism Department at the Faculty of Tourism and Hospitality Management, University of Rijeka (Croatia). Her research interests include destination management, cultural tourism and sustainable development. Her work includes over 50 publications and participations in over 20 projects.

List of Figures

Chapter 5

Chapter 6

Chapter 9

Chapter 14

Chapter 15

Chapter 18

Chapter 19

Chapter 21

Chapter 22

Chapter 23

List of Tables

Chapter 23

Chapter 28

1

The Prefigurative Politics of Present Transformation

Siddharth Sareen and Sirkku Juhola

1   Being the Change One Wants to See in the World

Human ways of organising the world are in urgent need of transformation. Since the 'Our common future' report in 1987, too much time has been spent rearranging the metaphorical deck chairs on the Titanic while the ship sinks, or rather, the global sea level rises. The climate has become increasingly unpredictable as a consequence of anthropogenic disruption, and the main certainty in recent decades has become our inability to address its root causes and to rapidly mitigate climate change.

In a predominantly capitalist and extractivist world order that is lurching distinctly towards authoritarianism at the global scale a quarter

S. Sareen (✉)
Fridtjof Nansen Institute, Lysaker, Norway
e-mail: ssareen@fni.no

S. Juhola
Faculty of Biological and Environmental Sciences, University of Helsinki, Helsinki, Finland

© The Author(s) 2026   **1**
S. Sareen and S. Juhola (eds.), *Societal Transitions to Sustainability*, Palgrave Studies in Environmental Transformation, Transition and Accountability,
https://doi.org/10.1007/978-3-032-07395-2_1

of the way into the heady 21st century, having the space to reflect collectively has unfortunately come to feel like a luxury. We find ourselves at global crossroads, confronted by floods, wildfires, genocide, fascism, and any manner of socioecological tensions that are creating rifts between humans and with more-than-human kin. As we have arguably already shot past 1.5°C global warming by 2025, present transformations take on unprecedented importance at scales of human existence and planetary wellbeing. Planning to reach targets by 2030 or 2050, while essential to enable systemic measures, too often stalls adequate action in the present, while silencing demands for urgently-needed change. In recognition of our polycrisis moment (Lawrence et al., 2024), this book responds to the need for rapid and transformative societal change within socially just and safe ecological limits.

Transformation is distinct from transition (Hölscher et al., 2018). It signals fundamental shifts in ways of doing things in society on a systemic scale, and changing overarching relationships such as the ones between economy, energy, nature, and society. Cognate concepts have emerged in the environmental social sciences and humanities, such as sufficiency, circular economy, and degrowth. Literature on institutional change, the spatiotemporality of transitions, socioecological justice, responsible research and innovation, and stemming from place-based approaches, is undergoing a renaissance of interest in interdisciplinary and transdisciplinary research. In this book, we refer to present transformations as the emergence and flourishing of experiments that strive towards fundamentally altering a system towards sustainability.

Attention to present transformations invokes scholarly work on prefigurative politics, which engages and examines the dynamics of putting mechanisms into play to bring about envisioned futures, aptly captured in the age-old quip about being the change that one wants to see in the world. This conceptual hook is of interest to diverse epistemic communities, ranging from solarpunk artists engaging with utopian futures, to modellers informing energy system policies at lower spatial scales based on climate scenarios; and from system design practitioners using techniques like backcasting and foresight, to ethnographic scholars examining change dynamics in specific contexts set in complex histories.

This rich yet somewhat fragmented understanding of prefigurative politics (Yates, 2021) can be productively channelled to address the issue of present transformations. This is the task we undertake collectively through this book. We feature this editorial introduction and a conclusion that bookend 27 case chapters structured into four themes: the urban scale, spaces of conviviality and politics, sectoral movements, and cross-sectoral and transdisciplinary transitions. These short case chapters, averaging 4,000 words each, address the prefigurative politics of present transformations based on specific empirical cases or conceptual cross-case analysis. They span and address societal transitions at local and larger spatial scales, in sectors such as energy, food, the built environment, health, and transport, including explicitly cross-sectoral contributions.

In this editorial introduction, we originally planned to situate the book in cutting-edge scholarship on prefigurative politics and on the politics of present transformations to sustainability. We envisaged this as bringing together these rich yet somewhat separate strands of literature, which matured into an intent to propose four themes that emerged organically and could lead to fruitful engagement and more intertwined discussions towards the substantive enactment of societal transformation in diverse and substantive ways, across sectors, spatial scales, and contexts. Yet in convening and working through the palimpsest of chapters, a process through which we inductively arrived at these themes, it dawned upon us that part of the value of prefigurative politics is its malleability. It is mobilised in a multitude of apt and useful ways in and across various disciplinary traditions as well as interdisciplinary and even transdisciplinary applications. Several definitions and reference works have gained traction, but we discern that these are invoked in a variety of ways that the chapter authors reflect upon, explain, and occasionally reshape into versions of their own making. We wish to recognise this propensity as inherent to the practice of prefigurative politics, also in the explicit act of embracing it as a scholarly approach. We value and wish to highlight this diversity.

We therefore allow ourselves to venture our own choice definition, for the purposes of this introduction as well as for others to borrow, mobilise, and redefine in playful, co-creative ways that they find fit-to-purpose: prefigurative politics refers to *embodied strategies to render desirable futures with immediacy.* By offering our definition, we look towards furthering the debate on prefigurative politics in two ways.

First, we position this volume and its contributions into the literature, acknowledging the previous work that has introduced the concept to begin with. Our definition is easily reconcilable with—and certainly has distinct resonance with—several established and emergent definitions that our book chapters do draw upon, ranging from the influential Jeffrey and Dyson (2021) article to the edited volume introduction by Monticelli (2022) and numerous relevant chapters therein; the classic work by Maeckelbergh (2011) anchored in social movement scholarship with recent echoes in Törnberg (2021); a special issue by Cornish et al. (2016) in the tradition of social psychology and a recent review and synthesis of that literature by Clarke and Drury (2025); an up-to-date review by Avelino et al. (2024) that explicates its linkage to sustainability; the work by Schiller-Merkens (2024) that brings it into conversation with transformative struggle or by Feola et al. (2023) on the politics of imaginary futures in transformation, or a related push to address institutional aspects of prefiguration in relation to orchestration (Lawford and Sareen 2025) or experimentation (Haugland, 2023); and any number of cognate concepts such as the seminal work on the pluriverse and the making of worlds by Escobar (2018) or fractal approaches to scaling transformation by O'Brien et al. (2023), of which the latter served as an inspiring point of common focus during a keynote at the workshop.

Second, our definition, which broadens the existing analytical lens of prefigurative politics, serves the purpose of examining emerging experimentation towards sustainability through it. This allows us to question and identify the explanatory power of the concept and experiment with it. How can a concept which derives from emancipatory social science (Monticelli, 2021) help us to interpret the diversity of cases present in this volume? In adopting this strategy, we align ourselves with Laamanen et al., (2023), who illustrate the power of shifting viewpoints examining prefigurative trajectories.

The acceptance of this broader definition has the consequence that we do not offer a systematic or comprehensive review of the concept. But rather we emphasise the importance of an accessible and relatable book that encompasses a broad, generative sweep of specialised thematic concerns. In the contributors' telling, these concerns take on life through the lens of prefigurative politics, employed in a veritable bouquet of ways. Like an assortment of blossoming flowers, we want the book to evoke a sense of wonder and possibility that can catalyse engagement and action, rather than to carve a narrow theoretical niche of consensus that inevitably runs the risk of alienating those schooled in other traditions than the range we are able to traverse. Phenomenologically, we intend for this book to draw readers in across the tapestry of interests and alternatives (Kothari, 2022) that we seek to build resonance with, and through this act to draw them into the heart of an inclusive and permissive socioecological movement that takes on global polycrisis.

2   Embodied Strategies to Render Desirable Futures with Immediacy

If conciseness and succinctness are markers of quality, then our definition meets the mark. Yet it maintains a certain ambiguity: what are desirable futures? Who strategises and renders? How proximate is immediacy, and relative to what? One might contend that such a lack of clarity is a fatal flaw in any conceptual definition, yet we would argue that a sense of animated suspension is in keeping with the very nature of prefigurative politics. Just as there is no single definition of sustainability, nor of transformation, desirability must be specified in a situated manner, but does signal a general orientation towards a better future. Similarly, immediacy suggests movement and intensity gravitating towards the present and a given place. Strategy imbues proceedings with a sense of purpose, and embodiment relates to actors and agency, while rendering has to do with acting in relation to some structural change, however ephemeral or institutionalised. This lack of commitment to a particular temporality is a hallmark of prefigurative politics, because it relates to something emergent, something actively being negotiated through contestation and confrontation, but with scope for compromise in agonistic and antagonistic practices. There is a temporal horizon being drawn into closer view, but whether it will stay there or lead to a particular reaction or response is an unfolding process.

This definition, in its flexibility, bestows us with a dynamic that fluctuates between fluidity and tautness, between softness and hardness, between initiation and closure, between embrace and rejection, between unison and alterity. Prefigurative politics emerge and exist in a space that breathes, where there is a sense of possibility being navigated from different standpoints, testing the limits of combativeness and alignment, idealism and pragmatism, and in that movement, enlarging spaces of possibility.

We organise the book around four such spaces of possibility, recognised in four parts. Over the year of conceptualising and bringing this collective endeavour into being, our efforts clustered around (A) urban sites of experimentation and contestation, (B) spaces of conviviality and politics, (C) sectoral movements, and (D) cross-sectoral and transdisciplinary transitions. As editors, we have led a working group of the large European Cooperation on Science and Technology (COST) Action network 'Transformations International Experience and Research Network for Sustainable Futures' (TransformERS: https://www.cost.eu/actions/CA22156/) since 2023, within which we anchor interdisciplinary and transdisciplinary collaboration on 'present transformation' across over 200 working group members based in more than 30 countries. We selected the 27 case chapters featured in this book from over thrice as many abstracts submitted in response to a competitive open call. In so doing, we intentionally did not foreclose fertile possibilities, but rather continued to engage with them in the process of curating the book workshop and the final result. In the remainder of this introduction, we gently insert the reader into these conversations, complementing our broad point of departure above with a brief motivation of the four themes that constitute these parts of the book, each of which features five to eight chapters. In a short conclusion, we pull together synthesis reflections across the parts.

2.1   Urban Sites of Experimentation and Contestation

Since prefigurative politics is about immediacy, the urban scale where humans congregate, attention is focused, and spectacle highlighted, is a naturally important type of site for it to emerge and play out (Minuchin, 2021). Strategies are necessarily complex given the multitude of actors entangled in complex relation to each other and with enhanced expression of aspirations and desires concentrated at these nexuses of accelerated societal metabolism. The urban scale is where momentum gain becomes prominently visible, be it in counter-hegemonic protests or demonstrations of hegemonic control. Ideas can travel fast and burgeon into material manifestations with startling rapidity through the translation mechanism of cities at the juncture of human density and capital flow in consequential ways for prefigurative politics.

Yet there are also expressions of agency at the urban scale that escape notice, forms of organising whose time in the limelight has not yet come, within the grander vicissitudes of national policies and geopolitical trends. As Krznaric (2024) explains using the triangle of the 'disruption nexus', society moves between crises, ideas, and social movements; these ideas are produced and percolate in and through cities to enliven social movements that help them take form to meet crises in hope of a novel resolution. Thus, the urban scale features some of the key sites where futures are being contested and formed in the present.

2.2   Spaces of Conviviality and Politics

It is impossible to understand prefigurative politics without dwelling on what politics itself entails, namely, the art and craft of decision-making through various forms of institutionalisation, both formal and informal. While institutions imply some structural rigidity to give decision-making both firmer shape and continuity, the inclusion of 'spaces of conviviality' alongside politics recognises and emphasises the relational nature of decision-making, whether seen as emergent (Kennedy et al., 2018) or a governmental strategy (De Wilde & Duyvendak, 2016). Richter (2022) zooms in on Alexander Hamilton's keenness to be in 'the room where it happens', an embodiment of this relational and place-based character of politics. While one only has selective purchase to the goings-on in the corridors of power, it is often possible to collectively fathom the nature of transactions that have occurred in hidden convivial spaces. Likewise, these intrigues are discussed in contrasting convivial spaces such as community gardens, local community spaces, and places of worship where agreement or discontent are expressed, far from formal authority but with a claim to power that rests at the heart of democratic possibility to express dissent and resist abuse of authority.

Prefigurative politics does not simply consist of capturing the ebbs and flows of decision-making; rather, it is constitutive of moments of change, of emergent spaces of enhanced possibility (Dinerstein, 2015), that may flicker briefly as impulses that fade, or may be nurtured into more enduring being through popular support or a broadly shared sense of social resistance that diffuses in numerous forms of empowered agency imbued with collective meaning.

2.3   Sectoral Movements

While sectoral action is often critiqued for its siloed ontology, it holds important advantages of specialisation and coherence that can increase the reach of prefigurative politics. Actors who are keenly aware of how a sector functions, and who often cultivate particular relationships with other central actors in a given sector, become positioned to enact prefigurative politics by virtue of their embedded engagement, social networks, and immediate relation to specific stakes (Schiller-Merkens, 2022). A claim deeply coated in someone's sense of everyday involvement has strong social resonance and thus carrying power to challenge the legitimacy of particular developments or to build social momentum towards demands for a given type of change.

This thematic focus emphasises the complex and specialised ways in which prefigurative politics need to intervene to bring about real and lasting change in complex, already existing systems that are often organised and governed by sector. It is important to understand and engage with such ontologies due to their structuring effect over key infrastructures and practices, without taking for granted everything the status quo rests upon. It is in engaging at a level of nuance that it becomes possible to perceive and intuit scope and appetite for change. Combined with the recognition of the necessity of transformation, sectoral movements can perform vital prefigurative politics of advancing cognitive and discursive shifts, to bring about conditions for reconfigured financial and infrastructural arrangements.

2.4 Cross-Sectoral and Transdisciplinary Transitions

As society evolves through new entrenchments in infrastructural innovation—such as digitalisation in its various manifestations—sectors are increasingly coupled and integrated, bringing the rhythms of previously separate practices into closer relation. For instance, low-carbon electrification of sectors like transport and the built environment makes it desirable to coordinate their rhythms in relation to desired energy demand patterns, to avoid undue pressure on existing energy systems with a given tradition of balancing supply from an evolving mix of energy sources with real-time energy demand. Yet such sociomaterial changes hold significance for how prefigurative politics can engage with and reconfigure decision-making, as they necessitate engagement across sectors and competencies, especially given the concern that prefigurative political efforts can become inward-looking and exclusive (Jeffrey & Dyson, 2021). For instance, someone wishing to exert influence over future low-carbon energy systems would have to consider novel integration across sectors in ways that qualitatively alter our basic assumptions about sectors from before, e.g., the landscape possibilities to combine solar energy and agriculture.

Prefigurative politics at such a conjuncture entail setting in place the mechanisms for desired future states of being, whether of an energy system or a farm, or indeed of the way these work together going forward. At base, this requires an openness of spirit and keenness to learn about how other sectors work and what stakes actors have across sectors. Taking stock of concerns and competencies, while devising and institutionalising ways to coordinate action to steer collectively towards commonly desired futures, constitutes prefigurative politics in highly transdisciplinary action.

3   The Prefigurative Politics of Present Transformation

What, then, is the prefigurative politics of present transformation? It is the convergence of *embodied strategies to render desirable futures with immediacy* with the recognition of the urgent necessity to act in order to create meaningful systemic change at scale and along an accelerated timeline. The gravity of present crises requires nothing of lesser ambition.

But this prefigurative politics has to be steadfast in its resolve to set ideas in motion as mechanisms that have the potential to endure and spread, not stuck in a fixed vision of the form that change will take. There is no such certainty in the space of contingent possibility that constitutes the condition for the emergence of prefigurative politics. A predefined future target can work as an anti-politics machine (Ferguson, 1994). This is not to say that prefigurative politics cannot strive towards a definite end expressed in general terms, such as lower consumerism and greenhouse gas emissions, more convivial spaces and an embrace of heterogeneity, or other desired futures. But the emphasis remains squarely on the *doing* of change, even if this change aims to *unmake* something (Feola et al., 2021), to end, to discontinue. This processual, ontological focus does not obviate structural aspects to prefigurative politics. To succeed over time, prefigurative politics must manage to institutionalise novel mechanisms to exert lasting influence in a systemic way. By such a point, however, one exits the space of prefigurative politics and enters a state of consolidation beyond emergence. Contingency remains, as the present and future are both characterised by uncertainty. But prefigurative politics arises in a liminal space that, once settled into firmer rhythms, shrinks scope for a politics of novel possibility.

Unfortunately for the world at large, but perhaps somewhat fortunately for the actors embracing prefigurative politics to enable change, we remain far from a state of affairs where the dust has settled on embodied strategies to render desirable futures with immediacy. Entering the second quarter of the 21st century, there appears no dearth of opportunity to advance the prefigurative politics of present transformation. Every polycrisis holds the seeds of societal change; they can open at the close.

Acknowledgements We are grateful to the COST Association for workshop funding through COST Action 22156 TransformERS to convene working group 3 (present transformation) book chapter authors in Oslo in April 2025. Sareen acknowledges time on Research Council of Norway projects 349994 (ENERGY4ALL) and 295704 (INCLUDE).

Competing Interests The authors have no conflicts of interest to declare that are relevant to the content of this chapter.

References

Avelino, F., Wijsman, K., van Steenbergen, F., Jhagroe, S., Wittmayer, J., Akerboom, S., Bogner, K., Jansen, E. F., Frantzeskaki, N., & Kalfagianni, A. (2024). Just sustainability transitions: Politics, power, and prefiguration in transformative change toward justice and sustainability. *Annual Review of Environment and Resources, 49*. https://doi.org/10.1146/annurev-environ-112321-081722

Clarke, D., & Drury, J. (2025). Emergent prefigurative politics and social psychological processes: A systematic review and research agenda. *Journal of Community & Applied Social Psychology, 35*(1), e70040. https://doi.org/10.1002/casp.70040

Cornish, F., Haaken, J., Moskovitz, L., & Jackson, S. (2016). Rethinking prefigurative politics: Introduction to the special thematic section. *Journal of Social and Political Psychology, 4*(1), 114–127. https://doi.org/10.5964/jspp.v4i1.640

De Wilde, M., & Duyvendak, J. W. (2016). Engineering community spirit: The pre-figurative politics of affective citizenship in Dutch local governance. *Citizenship Studies, 20*(8), 973–993. https://doi.org/10.1080/13621025.2016.1229194

Dinerstein, A. C. (2015). Confronting value with Hope: Towards a prefigurative critique of political economy. In *The politics of autonomy in Latin America. Non-Governmental Public Action*. Palgrave Macmillan. https://doi.org/10.1057/9781137316011_8

Escobar, A. (2018). *Designs for the pluriverse: Radical interdependence, autonomy, and the making of worlds*. Duke University Press.

Feola, G., Goodman, M. K., Suzunaga, J., & Soler, J. (2023). Collective memories, place-framing and the politics of imaginary futures in sustainability transitions and transformation. *Geoforum, 138*, 103668. https://doi.org/10.1016/j.geoforum.2022.103668

Feola, G., Vincent, O., & Moore, D. (2021). (un) making in sustainability transformation beyond capitalism. *Global Environmental Change, 69*, 102290. https://doi.org/10.1016/j.gloenvcha.2021.102290

Ferguson, J. (1994). *Anti-politics machine: Development, depoliticization, and bureaucratic power in Lesotho*. University of Minnesota Press.

Haugland, B. T. (2023). The future is present: Prefiguration in policy and technology experimentation. *Environmental Innovation and Societal Transitions, 48*, 100750. https://doi.org/10.1016/j.eist.2023.100750

Hölscher, K., Wittmayer, J. M., & Loorbach, D. (2018). Transition versus transformation: What's the difference? *Environmental Innovation and Societal Transitions, 27*, 1–3. https://doi.org/10.1016/j.eist.2017.10.007

Jeffrey, C., & Dyson, J. (2021). Geographies of the future: Prefigurative politics. *Progress in Human Geography, 45*(4), 641–658. https://doi.org/10.1177/0309132520926569

Kennedy, E. H., Johnston, J., & Parkins, J. R. (2018). Small-p politics: How pleasurable, convivial and pragmatic political ideals influence engagement in eat-local initiatives. *The British Journal of Sociology, 69*(3), 670–690. https://doi.org/10.1111/1468-4446.12298

Kothari, A. (2022). Earth Vikalp Sangam: Proposal for a global tapestry of alternatives. In *Challenging authoritarian capitalism* (pp. 84–88). Routledge. https://doi.org/10.4324/9781003325871-10

Krznaric, R. (2024). *History for tomorrow: Inspiration from the past for the future of humanity*. Random House.

Laamanen, M., Forno, F., & Wahlen, S. (2023). Neo-materialist movement organisations and the matter of scale: Scaling through institutions as prefigurative politics? *Journal of Marketing Management, 39*(9–10), 857–878. https://doi.org/10.1080/0267257X.2022.2045342

Lawford, H. L., & Sareen, S. (2025). Institutional prefiguration: Community energy development through spaces of orchestration. *Sustainability Science, 1-14.* https://doi.org/10.1007/s11625-025-01652-4

Lawrence, M., Homer-Dixon, T., Janzwood, S., Rockstöm, J., Renn, O., & Donges, J. F. (2024). Global polycrisis: The causal mechanisms of crisis entanglement. *Global Sustainability, 7,* e6. https://doi.org/10.1017/sus.2024.1

Maeckelbergh, M. (2011). Doing is believing: Prefiguration as strategic practice in the alterglobalization movement. *Social Movement Studies, 10*(01), 1–20. https://doi.org/10.1080/14742837.2011.545223

Minuchin, L. (2021). Prefigurative urbanization: Politics through infrastructural repertoires in Guayaquil. *Political Geography, 85,* 102316. https://doi.org/10.1016/j.polgeo.2020.102316

Monticelli, L. (2021). On the necessity of prefigurative politics. *Thesis Eleven, 167*(1), 99–118. https://doi.org/10.1177/0725513621105699

Monticelli, L. (2022). Prefigurative politics within, despite and beyond contemporary capitalism. In *The future is now: An introduction to prefigurative politics* (pp. 15–31). Bristol University Press. https://doi.org/10.51952/9781529215687.ch001

O'Brien, K., Carmona, R., Gram-Hanssen, I., Hochachka, G., Sygna, L., & Rosenberg, M. (2023). Fractal approaches to scaling transformations to sustainability. *Ambio, 52*(9), 1448–1461. https://doi.org/10.1007/s13280-023-01873-w

Richter, J. (2022). Thinking about "the room where it happens": Using place to teach about Alexander Hamilton and Early America. In *(ed. C. Northrop) the Hamilton phenomenon* (pp. 185–202). Vernon Press.

Schiller-Merkens, S. (2022). Social transformation through prefiguration? A multi-political approach of prefiguring alternative infrastructures. *Historical Social Research/Historische Sozialforschung, 47*(4), 66–90. https://doi.org/10.12759/hsr.47.2022.39

Schiller-Merkens, S. (2024). Prefiguring an alternative economy: Understanding prefigurative organizing and its struggles. *Organization, 31*(3), 458–476. https://doi.org/10.1177/13505084221124189

Törnberg, A. (2021). "prefigurative politics and social change: A typology drawing on transition studies." *Distinktion. Journal of Social Theory, 22*(1), 83–107. https://doi.org/10.1080/1600910X.2020.1856161

Yates, L. (2021). Prefigurative politics and social movement strategy: The roles of prefiguration in the reproduction, mobilisation and coordination of movements. *Political Studies,* 69(4), 1033–1052. https://doi.org/10.1177/0032321720936046

Open Access This chapter is licensed under the terms of the Creative Commons Attribution-NonCommercial-NoDerivatives 4.0 International License (http://creativecommons.org/licenses/by-nc-nd/4.0/), which permits any noncommercial use, sharing, distribution and reproduction in any medium or format, as long as you give appropriate credit to the original author(s) and the source, provide a link to the Creative Commons license and indicate if you modified the licensed material. You do not have permission under this license to share adapted material derived from this chapter or parts of it.

The images or other third party material in this chapter are included in the chapter's Creative Commons license, unless indicated otherwise in a credit line to the material. If material is not included in the chapter's Creative Commons license and your intended use is not permitted by statutory regulation or exceeds the permitted use, you will need to obtain permission directly from the copyright holder.

Part I

Urban Sites of Experimentation and Contestation

2

Acclimatising to the Future: Prefiguration and Urban Experimentation in Public Policy

Bård Torvetjønn Haugland⑩ and Timo von Wirth

1   Introduction

The threat of climate change has imposed a specific logic on all societal sectors: rather than reacting to events that have already transpired, societies are forced to act pre-emptively. Simultaneously, there is no one way to act in order to avoid the catastrophic consequences of global warming; rather, a wide range of strategies are available. Hence, strategies seldom converge around a common direction, at least not beyond "away-from-here" (Aiken, this volume). On one hand, addressing climate change means developing workable arrangements within existing social, institutional, and material structures; on the other, it requires challenging, transforming, and/or deconstructing the same structures (Feola, this volume). Prefiguration and experimentation are two perspectives that

B. T. Haugland (✉)
Department of Interdisciplinary Studies of Culture, Norwegian University of Science and Technology (NTNU), Trondheim, Norway
e-mail: bard.t.haugland@ntnu.no

Department of Technology Management, SINTEF Digital, Trondheim, Norway

© The Author(s) 2026
S. Sareen and S. Juhola (eds.), *Societal Transitions to Sustainability*, Palgrave Studies in Environmental Transformation, Transition and Accountability,
https://doi.org/10.1007/978-3-032-07395-2_2

in distinct, though partially overlapping ways take on the challenge of manifesting future states within present socio-material configurations. In this chapter, we explore the overlaps and complementarities between the two perspectives through an empirical case study of the Car-Free City Life project.

Car-Free City Life was initiated by the City Council of Oslo in 2015, and attempted to establish a car-free city centre in Oslo, Norway's capital. In terms of strategy, the project had clear prefigurative elements, though the organisers never explicitly used the term prefiguration. Simultaneously, the project was enacted through urban experimentation in the form of temporary socio-material interventions in the urban fabric. These aspects make Car-Free City Life an exemplary case for illustrating how prefigurative approaches may be operationalised in institutional settings, and for discussing the relationship between strategic orientation, subsequent enactment through experimentation, and the resulting politics.

T. von Wirth
Planetary Futures Studio, Frankfurt University of Applied Sciences, Frankfurt am Main, Germany

Dutch Research Institute for Transitions, Erasmus University Rotterdam, Rotterdam, The Netherlands

Center for Sustainability Transitions (CST), Stellenbosch University, Stellenbosch, South Africa

2   Prefiguration and Urban Experimentation

2.1   Prefiguration and Prolepsis

Prefiguration is a political strategy with roots in anarchist, feminist, and (to a lesser extent) Marxist thinking (Boggs, 1977; Rowbotham, 1979). In the narrow sense, prefiguration refers to the present-day embodiment of specific deliberative structures in preparation for a post-capitalist society (Raekstad, 2018). In the broader sense, prefiguration refers to "the embodiment within the ongoing political practice of a movement, of those forms of social relations, decision-making, culture, and human experience that are the ultimate goal" (Boggs, 1977: 100). Furthermore, one can distinguish between two main approaches to prefiguration: within or outside governance institutions. The former sees prefiguration in relation to institutional politics (e.g., Haugland, 2023; Lawford & Sareen, 2025); the latter focuses on prefigurative practices within social movements (e.g., Wittmayer et al., 2022; Firinci, this volume). We use Boggs' definition as a starting point for exploring prefiguration in institutional politics.

If prefiguration is the present-day embodiment of a movement's ultimate goal, prefigurative strategy might appear a contradiction in terms. Strategy usually refers to the means for realising a future end; if an end can readily be embodied in the present, there is no need for developing means for realising it. Consequently, Dan Swain (2019) argues, the means-ends relationship cannot be abolished. Rather, drawing on De Smet (2015), Swain suggests understanding prefiguration in terms of *prolepsis*—that is, "an action or context in which a developing subject projects themselves to a later stage of development, and acts as if they have already reached that stage" (Swain, 2019: 58). In this sense, prolepsis is a rehearsal *and* a learning process. By trying to imitate some future stage of development, one may learn what is currently possible as well as what *might* be possible, "since it is only from within a given stage of development that it is possible to recognise what is possible to achieve" (Swain, 2019: 59). Simultaneously, understanding prefiguration in terms of prolepsis places certain limitations on prefigurative practice: it leaves no room for means that do not reflect ends (e.g., building an enormous state apparatus to, eventually, abolish the state); rather, a proleptic approach to prefiguration requires seeing "ends and means as distinct but still *of a kind*" (Swain, 2019: 59, emphasis in original). Simultaneously, the generative aspect of prolepsis points to a kinship with a principle commonly practised in (urban) experimentation: to act and see what happens (Torrens & von Wirth, 2021: 4; see also Schön, 1991).

2.2   Experimentation in Urban Climate Governance

Experimentation has been a long-standing focus in the fields of sustainability transitions research (e.g., Sengers et al., 2019) and climate governance (e.g., Bernstein & Hoffmann, 2018; Voß & Simons, 2018), including urban climate governance (Bulkeley & Castán Broto, 2013). Harriet Bulkeley (2023) has suggested that the increasing prevalence of experimentation in urban climate governance results from a global shift in the framing of climate policy. In the decades since *Our Common Future* (World Commission on Environment and Development, 1987), focus has shifted from addressing global greenhouse gas emissions to decarbonising all societal domains. This shift, Bulkeley argues, has given rise to a paradigm of experimentation. In short, experimentation

> *… represents a significant and potentially paradigm-shifting break with established norms and practices concerning the nature of the climate problem which revolve around who is authorized to govern, the relation between knowledge and policy, the indeterminacy of climate futures, and what it means to improve the condition of society.* (Bulkeley, 2023: 3)

On this background, Bulkeley (2023) identifies four key tensions in urban climate governance: "the shifting dynamics of governing authority, the relation between knowledge and policy, how to address indeterminacy, and what progress or improvement looks like in the condition of a climate-changed socio-natural world" (Bulkeley, 2023: 7). Below, we elaborate each point:

1. *Shifting governance constellations.* The condition of urban experimentation entails a reconfiguration of governance. Experimentation—and thus also governance—brings together diverse sets of actors, including experts, practitioners, policymakers, and the citizenry. Scholarship interprets this development in different ways: some consider this a fragmentation of political institutions' governance capability (or authority), others as a necessary re-development in the face of changing empirical realities.

2. *Action and knowledge.* Conventionally, policies were supposed to build on evidence and analysis. More recently, experimentation has been used to also inform policy development. However, under the condition of urban experimentation, action and knowledge are combined: experiments have become a means for better identifying the nature of a problem, rather than assessing prospective solutions (Bulkeley, 2023: 9). However, when enacting urban experiments diverse actors still conflate these, that is, some actors expect experiments to deliver short-term solutions for long-term, complex challenges. Experiments have become means for assessing the viability of specific interventions, and thus also of new socio-material realities (see, e.g., Lösch & Schneider, 2016; Voß & Simons, 2018).

3. *Indeterminacy of problems.* Experimentation is often employed to counter problems characterised by high uncertainty and complexity. In such instances, traditional methods for long-term planning cannot be applied (e.g., Sharp & Raven, 2021; Wanzenböck et al., 2020), and experimentation becomes a means for suspending indeterminacy through action. Simultaneously, the experimental nature of the action helps sustain indeterminacy (Bulkeley, 2023: 10).

4. *The measure of progress.* In the paradigm of ecological modernisation, the measures of progress were clear. For example, economic growth remained an invariable condition. In the paradigm of experimentation, they are not (yet). This raises the question of what normative standard or standards urban experimentation is held to.

Though each tension might not be present in every instance of urban experimentation, Bulkeley suggests that when aggregating such experiments, the four tensions indicate "the contours of a mode of governing by and through experimentation" (Bulkeley, 2023: 11). In conclusion, Bulkeley emphasises the importance of "[examining] which kinds of socio-material orders are being produced and excluded through experimentation, by and for whom, and with which consequences" (Bulkeley, 2023: 12).

For the purposes of this chapter, Bulkeley's (2023) emphasis on the production of socio-material orders is key. Boggs (1977) defined prefiguration as the present-day embodiment of a movement's ultimate goals. However, the emphasis on the embodiment of such goals *within* a movement's ongoing political practice means that the mechanisms for fostering change *beyond* the movement remains unclear (e.g., Yates, 2015; Wittmayer et al., 2022). Following Bruno Latour's (1990) contention that social relations are made durable by embedding them in materiality, urban experimentation may be one tool for enacting change beyond a movement: through such experimentation, one may facilitate the emergence of new social relations and even make them durable by materialising them in urban environments (for a relevant perspective, see Hommels, 2005). This suggests that prefigurative strategies and urban experiments may be complementary approaches for testing and producing new socio-material orders.

3   Prefiguring a Low-Carbon Future

The following analysis builds upon ten qualitative interviews focusing on the Car-Free City Life project in Oslo, Norway. Barring one interviewee representing business interests in the inner city, all interviewees were politicians or public employees who were involved in planning and/ or implementing Car-Free City Life (for more information on methods and analytical strategy, see Haugland 2023: 5–6).

3.1   The Means and Ends of Prefiguration

Car-Free City Life was initiated by a broad coalition of centre-left parties after Oslo's 2015 local elections. Starting in 2016, Car-Free City

Life entailed establishing a 1.5 square-kilometre car-restricted area in Oslo's city centre. The early phase of the project entailed the implementation of temporary measures. As temporary and initially reversible interventions, these measures fit with the definition of urban experimentation. The measures served a dual purpose: first, materialising the project faster than a "traditional" planning process would allow; second, to learn which measures worked and not (e.g., regarding acceptability, unintended effects). The project employed two interrelated measures: transport-reducing measures (push-factors) and measures improving the urban environment (pull-factors). The measures relating to transport primarily focused on reducing and redirecting traffic, in particular private cars. The measures ranged from removing 700 kerbside parking spaces to changing driving patterns. These measures were intended to make car use less attractive. The measures focusing on the urban environment included adding benches, parklets, and playgrounds, improving lighting, and shutting down streets which already experienced little traffic. In short, these measures aimed to make walking or cycling more attractive. In terms of means and ends, one interviewee argued that "removing cars is a means towards something else, it is not an end in itself"; the car restrictions were a means for changing the urban environment in a way that would improve urban quality of life.

The goal of improving the urban environment was implicated in another means-ends linkage, a linkage which illustrates Car-Free City Life's prefigurative strategy. While traffic reductions were a means towards a present-day end, the resulting improvement of the urban environment was also linked to the future. All interviewees who took part in planning and/or implementing Car-Free City Life emphasised this aspect. For example, one interviewee argued that "no one has been in a car-free city centre before. Hence, it is necessary that we display it". Another interviewee noted that the project was "a symbol of another way of living, and another way of developing cities". The combined attention to materiality and symbolism was also reflected in the emphasis on quality even when implementing temporary measures: "temporary measures are no fun if they don't appear to be of high quality".

Most interviewees acknowledged that Car-Free City Life's effect on CO_2 emissions was negligible. For example, a representative from the

Urban Development Committee noted that the project was "not a climate measure per se"; rather, it was "a spearhead in the broader work of changing urban transport". Similarly, another interviewee noted that "if larger and larger areas in the city become attractive to live and reside in without a car, that matters for the city's climate footprint over time". Hence, even if the project was marginal in terms of present-day emission reductions, it would prefigure a specific future where the same kind of measures made a difference. It would show that another city is possible.

3.2   Adaptation, Controversy, and a Resolution of Sorts

Car-Free City Life created considerable backlash from when it was first presented. One interviewee traced the controversy to the initiative's first title: Car-Free City *Centre*. The slight difference in name, the interviewee believed, indicated that the removal of cars was an end, rather than a means for improving urban life. This misunderstanding—or miscommunication—plagued the project, especially in its first years. The interviewees sketched up two key challenges: unintended consequences and "culture war" issues.

Store owners operating in the Car-Free City Life area expressed concerns regarding the project's effect on their businesses. While the project organisers considered their concerns legitimate, they simultaneously found it untenable to refrain from changing the city centre simply because businesses got part of their revenue from customers who got there by car. The project sought to improve the city centre for all of Oslo's citizenry. This required trade-offs: improving urban quality of life meant giving lower priority to business owners and car-driving customers. Simultaneously, the organisers chose a different strategy when it turned out that some measures—for example, removing parking spaces—negatively impacted people with disabilities. When this issue came to light, the organisers sought to rectify it as fast as possible. As expressed by one interviewee, "these are the people you want to attend to in every step". In other words, some concerns were considered more legitimate than others within the project's priorities. Together, these examples illustrate how the project was a normatively guided learning process prioritising action over knowledge.

The second line of conflict shaped around so-called culture war issues, that is, deeper underlying value contestations that surfaced due to the measures taken. During the project's first years, there was a prominent Facebook group, "Ja til bilen i Oslo" ("Yes to the car in Oslo"), dedicated to criticising Car-Free City Life. At its height, the group counted approximately 22,500 members. The group regularly featured threats of violence toward specific politicians as well as more generic fantasies of car-based violence towards cyclists (see also Michael, 2001, on road rage). Despite Car-Free City Life being initiated by a broad centre-left coalition, Green Party politicians were often targeted. In particular, Green Party representative Lan Marie Berg Nguyen received a series of threats and was at one point given police protection.

The culture war issues did not only play out in Facebook groups. One interviewee referred to a press conference where Siv Jensen (then-leader of the right-wing Progress Party and then-Minister of Finance in Norway's Conservative Government) spoke out against Oslo's City Council with specific reference to the car-reducing measures (see, e.g., Berge, 2018; on the performative relationship between Norwegian politicians and automobility, see Anfinsen, 2024: 146). The interviewee found the séance curious, with "the Government's second most powerful member pretending to be the little woman speaking truth to power". The two examples show how Car-Free City Life's challenging of car-users' privileges opened a culture war fought along two fronts: on Facebook and in national politics.

Though Car-Free City Life was contested, the organisers experienced that the contestation waned over time. The interviewees explained this in similar ways. One interviewee argued that "when people see what we are trying to do, the resistance disappears". Another saw the fact "that the critical voices die down and we see people starting to use the areas we made available" to be "the proof of the pudding". Yet another interviewee found that "after a while, the business owners understood that this was not too bad, that it was a bit forward-looking". In other words, they observed how time appeared to soften measures that had initially been considered (too) radical.

The strategic choices underpinning Car-Free City Life can be described as prefigurative and proleptic. The organisers pursued present-day means whose nature reflected future ends. Simultaneously, these means helped gauge what futures appeared possible from the present. On one hand, the organisers saw how the public grew to like—or at least tolerate—a car-reduced city centre. On the other, the organisers learnt that some things were not possible: for example, many interviewees noted that Oslo's city centre *could not* become entirely car-free. Furthermore, approximately 1000 people live within the Car-Free City Life area, and most of the area was established in the 1600s—long before the era of mass motorisation. As such, the project organisers recognised that expanding the project beyond the inner city would require other strategies, both in terms of finding appropriate means and participatory approaches. These lessons support Swain's contention that "it is only from within a given stage of development that it is possible to recognise what is possible to achieve" (Swain, 2019: 59).

4   Getting Used to the Future?

Car-Free City Life's prefigurative strategy was organised within traditional governance structures with a legitimated democratic mandate. As such, the project did not operate within radically altered governance structures. However, throughout the implementation process, the organisers remained attentive to feedback from the public. Some concerns were prioritised over others, a reflection of the organisers' conscious choice to challenge car-use as the self-evident starting point for urban organisation. Though this choice triggered controversies and value contestations, it also allowed the organisers to gauge the acceptability of specific interventions. In other words, the experiments allowed for "[testing] the nature and political salience of the problem itself" (Bulkeley, 2023: 9). The same relationship between action and knowledge is reflected in the project's use of temporary measures. Though the project was organised with specific short-term *and* long-term ends in mind, the project organisers did not know the exact configuration of a future low-carbon society. By developing a series of means that reflected the project's ends, the use of temporary measures allowed the organisers to navigate indeterminacy. This process may be understood in terms of prolepsis (De Smet, 2015; Swain, 2019): the temporary measures served as an inquiry into what was currently possible to realise, practically *and* politically. This observation also relates to the project's conception of "good": the project built on the assumption that urban life without cars can be good and sought to exemplify this by improving the urban environment in a way that would also serve as a small-scale demonstration of a low-carbon future.

With this case study, we have heeded Bulkeley's call for "[examining] which kinds of socio-material orders are being produced and excluded through experimentation, by and for whom, and with which consequences" (Bulkeley, 2023: 12). Car-Free City Life successfully challenged a well-established hierarchy in urban planning, thus establishing a new socio-material order in Oslo's city centre. Though the organisers found the resulting controversies challenging, they felt redeemed by how the public perception changed over time. This points to a notable temporal dynamic associated with this instance of prefigurative strategy and the accompanying experimentation. The organisers of Car-Free City Life experienced something resembling a paradox: though their strategy brought the future into the present, this present only became acceptable—or at least tolerable—in the future. Still, to the organisers, this suggested that the prefigurative strategy pursued in Car-Free City life did not only reconfigure urban landscapes: it had reshaped mental landscapes, too.

This chapter builds almost exclusively upon interviews with project organisers and implementers. Hence, we cannot reliably gauge how or to which extent Car-Free City Life *actually* changed the public's perceptions and normative intuitions, though a 2018 survey indicated that public perceptions of the project improved between 2016 and 2018 (Polle, 2018). However, future studies on prefigurative strategies must inquire further into this seemingly asynchronous reconfiguration of socio-material, mental, and normative landscapes, including the time-frame(s) of reconfiguration and how the effects and perceptions differ between audiences (on the latter point, see Eckerle, this volume).

There is clearly a resonance between prefiguration and urban experimentation. Simultaneously, our study design does not permit us to explore how prefigurative strategies, experimental practices, and value contestations interact over time. Similarly, we are unable to say anything about the longer-term (intended and unintended) consequences of temporary interventions—for example, how they can be replicated, scaled, and/or adapted to new contexts, or whether they deepen political conflict lines. Future research may address these questions through longitudinal studies, for example, through long-term urban observatories. Finally, we have focused on a prefigurative strategy that succeeded. However, it is equally important to understand why prefigurative strategies fail, as to better understand the conditions for success and failure. Though we have noted how time appeared to soften measures that were initially considered (too) radical, we do not expect this dynamic to be generalisable. Hence, it is crucial to develop a more fine-grained understanding of the kind of prefigurative strategies and experimental measures that, over time, successfully enable and materialise urban transformations.

Competing Interests The authors have no conflicts of interest to declare that are relevant to the content of this chapter.

References

Anfinsen, M. (2024). *A vehicle reframed? Culture, materiality and making of meaning on the road towards electric automobility* [Ph.D. thesis, Norwegian University of Science and Technology (NTNU)]. Trondheim.

Berge, J. (2018, 23/09). Siv Jensen om Oslo-byrådet: –Jeg begriper ikke hva de holder på med. *Nettavisen.* https://www.nettavisen.no/nyheter/innenriks/siv-jensen-om-oslo-byradet-jeg-begriper-ikke-hva-de-holder-pa-med/s/12-95-3423538862

Bernstein, S., & Hoffmann, M. (2018). The politics of decarbonization and the catalytic impact of subnational climate experiments. *Policy Sciences, 51*(2), 189–211. https://doi.org/10.1007/s11077-018-9314-8

Boggs, C. (1977). Marxism, prefigurative communism, and the problem of workers control. *Radical America, 11*(6), 99–122.

Bulkeley, H. (2023). The condition of urban climate experimentation. *Sustainability: Science, Practice and Policy, 19*(1), 2188726. https://doi.org/10.1080/15487733.2023.2188726

Bulkeley, H., & Castán Broto, V. (2013). Government by experiment? Global cities and the governing of climate change. *Transactions of the Institute of British Geographers, 38*(3), 361–375. http://www.jstor.org/stable/24582453

De Smet, B. (2015). *A Dialectical Pedagogy of Revolt: Gramsci, Vygotsky, and the Egyptian Revolution.* Brill.

Haugland, B. T. (2023). The future is present: Prefiguration in policy and technology experimentation. *Environmental Innovation and Societal Transitions, 48*, 100750. https://doi.org/10.1016/j.eist.2023.100750

Hommels, A. (2005). Studying obduracy in the City: Toward a productive fusion between technology studies and urban studies. *Science, Technology, & Human Values, 30*(3), 323–351. https://doi.org/10.1177/0162243904271759

Latour, B. (1990). Technology is society made durable. *The Sociological Review, 38*(1_suppl), 103–131. https://doi.org/10.1111/j.1467-954X.1990.tb03350.x

Lawford, H. L., & Sareen, S. (2025). Institutional prefiguration: Community energy development through spaces of orchestration. *Sustainability Science, advance online publication.* https://doi.org/10.1007/s11625-025-01652-4

Lösch, A., & Schneider, C. (2016). Transforming power/knowledge apparatuses: the smart grid in the German energy transition. *Innovation: The European Journal of Social Science Research, 29*(3), 262–284. doi:https://doi.org/10.1080/13511610.2016.1154783.

Michael, M. (2001). The invisible car: The cultural purification of road rage. In D. Miller (Ed.), *Car Cultures* (pp. 59–80). Berg Publishers.

Polle, S. (2018). *Bilfritt byliv. Statusrapport 2018—Midtveisevaluering.* Sweco. [Retrieved from 11.03.2025]. https://www.oslo.kommune.no/getfile.php/13303828-1542623637/Tjenester%20og%20tilbud/Politikk%20og%20administrasjon/Byutvikling/Bilfritt%20byliv/0_Bilfritt%20byliv_Samlerapport%202018%20-%20midtveisevaluering%202018_16.11.2018.pdf

Raekstad, P. (2018). Revolutionary practice and prefigurative politics: A clarification and defense. *Constellations, 25*(3), 359–372. https://doi.org/10.1111/1467-8675.12319

Rowbotham, S. (1979). *Beyond the fragments: Feminism and the making of socialism* (2nd ed.). Merlin Press.

Schön, D. A. (1991). *The reflective practitioner: How professionals think in action.* Avebury.

Sengers, F., Wieczorek, A. J., & Raven, R. (2019). Experimenting for sustainability transitions: A systematic literature review. *Technological Forecasting and Social Change, 145*, 153–164. https://doi.org/10.1016/j.techfore.2016.08.031

Sharp, D., & Raven, R. (2021). Urban planning by experiment at precinct scale: Embracing complexity, ambiguity, and multiplicity. *Urban Planning, 6*(1), 195–207. https://doi.org/10.17645/up.v6i1.3525

Swain, D. (2019). Not not but not yet: Present and future in Prefigurative politics. *Political Studies, 67*(1), 47–62. https://doi.org/10.1177/0032321717741233

Torrens, J., & von Wirth, T. (2021). Experimentation or projectification of urban change? A critical appraisal and three steps forward. *Urban Transformations, 3*(1), 8. https://doi.org/10.1186/s42854-021-00025-1

Voß, J.P., & Simons, A. (2018). A novel understanding of experimentation in governance: co-producing innovations between lab and field. *Policy Sciences, 51*(2), 213–229. doi:https://doi.org/10.1007/s11077-018-9313-9.

Wanzenböck, I., Wesseling, J. H., Frenken, K., Hekkert, M. P., & Weber, K. M. (2020). A framework for mission-oriented innovation policy: Alternative pathways through the problem–solution space. *Science and Public Policy, 47*(4), 474–489. https://doi.org/10.1093/scipol/scaa027

Wittmayer, J. M., Campos, I., Avelino, F., Brown, D., Doračić, B., Fraaije, M., Gährs, S., Hinsch, A., Assalini, S., Becker, T., Marín-González, E., Holstenkamp, L., Bedoić, R., Duić, N., Oxenaar, S., & Pukšec, T. (2022). Thinking, doing, organising: Prefiguring just and sustainable energy systems via collective prosumer ecosystems in Europe. *Energy Research & Social Science, 86*, 102425. https://doi.org/10.1016/j.erss.2021.102425

World Commission on Environment and Development. (1987). *Our Common Future*. Oxford University Press.

Open Access This chapter is licensed under the terms of the Creative Commons Attribution-NonCommercial-NoDerivatives 4.0 International License (http://creativecommons.org/licenses/by-nc-nd/4.0/), which permits any noncommercial use, sharing, distribution and reproduction in any medium or format, as long as you give appropriate credit to the original author(s) and the source, provide a link to the Creative Commons license and indicate if you modified the licensed material. You do not have permission under this license to share adapted material derived from this chapter or parts of it.

The images or other third party material in this chapter are included in the chapter's Creative Commons license, unless indicated otherwise in a credit line to the material. If material is not included in the chapter's Creative Commons license and your intended use is not permitted by statutory regulation or exceeds the permitted use, you will need to obtain permission directly from the copyright holder.

3

Prefigurative Politics in Citizen Science: The Golf Vinoř Project as a Case of Present Transformations

David Stella ⓘ

1 Introduction

1.1 Context and Rationale

Golf courses are often perceived as elite spaces, associated with exclusivity, high resource consumption, and large-scale land use. In Czechia, this view is intensified by concerns over environmental sustainability, particularly in terms of water management and biodiversity (Duží et al., 2020). However, emerging approaches to golf course design and management increasingly emphasise ecological considerations and social inclusivity. The Golf Vinoř Citizen Science Project shows how science and practice can challenge land-use paradigms and explore alternative governance models.

D. Stella (✉)
Department of Human Dimensions of Global Change, CzechGlobe - Global Change Research Institute of the Czech Academy of Sciences, Prague, Czechia
e-mail: stella.d@czechglobe.cz

© The Author(s) 2026
S. Sareen and S. Juhola (eds.), *Societal Transitions to Sustainability*, Palgrave Studies in Environmental Transformation, Transition and Accountability,
https://doi.org/10.1007/978-3-032-07395-2_3

The case of Golf Vinoř illustrates how abandoned or underutilised landscapes can transform into ecologically valuable spaces that serve both human and non-human communities. Between 2017 and 2022, an unfinished golf course in the outskirts of Prague lay dormant, becoming a dynamic new wilderness. During this period, local residents organically repurposed the site for recreation, while nature reclaimed the space, fostering a significant increase in biodiversity. Figure 1 shows the site. The project, launched in 2022 by the CzechGlobe, sought to document and facilitate this transformation by integrating citizen science into decision-making processes.

Beyond its ecological and social dimensions, the project embodies prefigurative politics—a concept that describes actions and practices that model desired futures in the present (Raekstad & Gradin, 2020). Rather than waiting for top-down institutional changes, the Golf Vinoř Citizen Science Project actively shapes immediate transformations. By involving citizens, scientists, authorities, and stakeholders, it prefigures a more participatory, adaptive, and environmentally conscious governance model for contested urban landscapes (Mattijssen et al., 2024). This chapter explores how the project aligns with prefigurative politics in

Fig. 1 Golf Vinoř 100 ha size golf course in the outskirts of Prague (*Source* Author)

its attempt to bridge the gap between expert-driven urban planning and grassroots engagement, demonstrating the power of citizen-led initiatives in shaping socio-ecological transformations.

1.2   Research Aims and Methodological Overview

This chapter examines how the project challenges conventional land-use governance, fosters participatory decision-making, contributes to present transformations by prefiguring alternative socio-ecological futures, reveals institutional and behavioural shifts from stakeholder engagement, and informs broader debates on sustainable urban governance. By analysing the project through the lens of prefigurative politics, the chapter contributes to a growing body of literature that examines how grassroots initiatives actively shape present transformations rather than just advocating future policy shifts (Seyfang & Haxeltine, 2012).

This qualitative case study draws on multiple data sources:

- Participatory research: Engagement with local stakeholders, scientists, policymakers, and residents.
- Survey data: Perceptions of ecological and social value from 500+ respondents.
- Spatial and ecological assessments: Mapping of land use and biodiversity changes.
- Stakeholder workshops and expert consultations: Input from hydrologists, ecologists, planners, and golf course representatives.

This interdisciplinary and transdisciplinary methodology aligns with contemporary research on present transformations, allowing for a nuanced understanding of how citizen-driven initiatives contribute to immediate, tangible change (Zapata Campos & Zapata, 2017).

2   Conceptual Framework

2.1   Prefigurative Politics and Citizen Science

In socio-ecological contexts, prefigurative politics can be understood as enacting governance models, environmental management strategies, or social arrangements that reflect a more sustainable and participatory future (Monticelli, 2021). Unlike traditional governance, often slow and bureaucratic, prefigurative politics operates in the present, showing alternative ways to organise human-nature relationships through immediate action.

Citizen science is a key tool in prefigurative politics, as it empowers local actors to co-produce knowledge, influence governance, and challenge established decision-making processes. It is particularly relevant in land-use debates, where expert-driven planning often marginalises local voices. The *Golf Vinoř* exemplifies this dynamic by prefiguring a model of land governance that is more inclusive, participatory, and adaptive. Instead of passively accepting golf development, residents and scientists co-created a vision that balances recreation, biodiversity, and ecological care.

This contrasts sharply with traditional urban planning, which, while structured and systematic, is often rigid, exclusionary, and slow to adapt. Such planning often prioritises economic growth and investment over citizen concerns. In contrast, prefigurative politics transforms space in real time, offering evidence for futures often dismissed in conventional planning (Davoudi, 2023).

2.2   Transformation Versus Transition: A Bottom-Up Approach

Transition is a planned, phased process in which policy and institutional shifts guide long-term change. Transformation is a more immediate, systemic shift, often driven by grassroots action.

The *Golf Vinoř Project* aligns with transformation, having emerged organically from local inhabitants, not from a structured long-term policy process. Residents and community members, witnessing the abandoned golf course evolving into an ecologically rich landscape, claimed agency over its future. They mobilised science and public participation to influence decisions beyond the scope of formal institutions.

This bottom-up approach challenges top-down governance, showing how local knowledge, lived experience, and participatory science drive immediate socio-ecological shifts. Such transformations can serve as precursors to broader institutional change, offering working models that show the feasibility of alternative governance (Monticelli, 2021).

2.3   Ecosystem Services as a Bridge Between Nature and Society

To make socio-ecological transformations more tangible and relatable, the *Golf Vinoř* project employs the ecosystem services framework, which links nature's contributions to human well-being. Rather than viewing conservation as purely ecological, ecosystem services highlight benefits like water retention, biodiversity, recreation, and local climate regulation (Díaz et al., 2015).

In *Golf Vinoř*, the ecosystem services approach helped in shifting the narrative around golf course development. By incorporating citizen science-driven biodiversity monitoring and participatory land-use discussions, the project showed that such landscapes can be reimagined as multifunctional spaces (Dahl Jensen et al., 2017).

Key services in this transformation include:

- Regulating (biodiversity, climate resilience, water retention)
- Cultural (recreation, engagement, sense of place)
- Supporting (habitats for pollinators, birds, other wildlife)

Rather than placing nature in opposition to human use, the framework supports hybrid landscapes where social and ecological needs balance in real time.

2.4   Land-Use Conflict and Governance Tensions

Despite its successes, *Golf Vinoř* remains a contested space, revealing deep-seated tensions in urban land governance. These conflicts are rooted in competing visions for the site. Developers prioritise commercial recreation. Residents emphasise open access, ecology, and diverse uses. Scientists advocate biodiversity and sustainable management.

The project has illuminated three major areas of land-use conflict:

- Access: Residents used the site freely during its abandonment; new golf restrictions raise questions about fair access.
- Biodiversity vs. development: The site's ecological richness challenges the assumption that formal golf use is optimal.
- Regulatory complexities: Municipally owned but privately managed land creates fragmented decision-making.

Citizen science gave new legitimacy to public involvement in land governance. It generated biodiversity and public data that local actors used to negotiate with developers and policymakers.

These conflicts illustrate the broader challenges of implementing prefigurative politics in contested urban spaces. While grassroots initiatives can drive immediate transformation, their long-term sustainability often depends on institutional recognition and legal frameworks (Seyfang & Haxeltine, 2012).

3   The Golf Vinoř Citizen Science Project

3.1   Project Genesis and Objectives

The *Golf Vinoř Citizen* emerged as a response to an unresolved debate surrounding the future of a contested landscape. Planned as a golf resort, the site underwent significant land transformation, but financial constraints halted construction in 2017, leaving the area in limbo. Over the following years, the landscape evolved into a *new wilderness*, fostering a surge in biodiversity and attracting local residents for recreational use.

While discussions about the golf course continued, public debate lacked scientific evidence. Many residents and stakeholders had strong opinions, but these were based largely on personal perceptions rather than empirical data. This evidence gap motivated the citizen science initiative.

Led by the author, a small research team launched the project to integrate science into local land-use debate.

The project had three aims:

- Promote social science's role in sustainable land-use planning
- Co-produce knowledge with residents, scientists, and policymakers
- Explore land-use options that balance biodiversity and human activity

By structuring the project around citizen science and participatory methods, the initiative sought to prefigure a governance model where local knowledge, scientific insights, and multi-stakeholder discussions actively shape present transformations, rather than waiting for top-down interventions (Kimura & Kinchy, 2016).

3.2   Key Activities and Methods

To engage the public and collect data, the project employed a combination of participatory research methods, digital tools, and workshops. The primary methods included:

- Participatory mapping and surveys: The project utilised the Maptionnaire platform, a participatory mapping tool, to collect spatial data on how residents perceive and use the landscape (Stella, b.r.).
- Biodiversity monitoring: Biodiversity monitoring: Schools and residents used the PlantNet app to contribute citizen science data on species (*PlantNet*, b.r.).
- Public opinion and sensory perception: A survey of 500+ people captured views on the golf course, biodiversity, and land-use options.
- Stakeholder workshops and deliberation: A final workshop, attended by over 50 participants, provided a platform for discussion among local residents, scientists, and policymakers.

The scientific and participatory approach helped elevate the quality of debate about the golf course. Instead of an emotionally charged dispute, discussions began incorporating quantitative and qualitative data, shifting the discourse toward evidence-based decision-making.

However, governance gaps became evident, particularly in the inconsistent communication from local authorities. The Prague City Council, which owns the land, demonstrated fragmented decision-making. One department approved a zoning plan change that prioritised golf use, effectively limiting alternative land-use discussions. Meanwhile, the nature protection division of the same institution supported the citizen science project, recognising the ecological value of the new wilderness. This contradiction illustrated a key challenge of land governance—different institutional priorities competing within the same bureaucratic system, creating uncertainty for local actors.

3.3   Stakeholder Involvement and Governance

The *Golf Vinoř Citizen Science Project* engaged a diverse group of stakeholders with often competing interests:

- Local residents: The most consistent group in the project, residents had been using the land for recreation and wanted continued access, fostering interest in multifunctional land use beyond golf.
- Scientists: The research team saw the project as an opportunity to increase awareness of both social and environmental sciences, ensuring that data-driven perspectives influenced the debate.
- Municipality (local government): The municipal government struggled to navigate the tensions between local demands and the broader governance structure controlled by Prague City Council.
- Golf Course Developer & Operator: Initially absent from discussions, the new golf course operator gradually took interest in the project but later ceased cooperation, making future negotiations uncertain.
- Prague City Council: As landowners, they played a critical but ambiguous role. Their internal contradictions reflected the complexity

of land-use governance, where conservation priorities often clash with urban development goals.

Stakeholder engagement evolved over time, shifting from an open-ended community-led discussion toward a more structured negotiation with formal land managers. However, with the golf course operator withdrawing from the project, future implementation of proposed solutions remains uncertain. The project sought openness, not all actors had equal capacity to participate, nor were all views equally recognised by decision-makers. These dynamics underscore the importance of embedding participatory practices within a critical framework of inclusion and justice.

3.4   Key Findings and Outcomes

The project's findings, exemplified in Figs. 2 and 3, provided new empirical insights into biodiversity, public perception, and land-use conflicts:

Fig. 2 An example of the result from the participatory mapping. Spatially based results on the question: *How would you like to move across this area?* (*Source* Author)

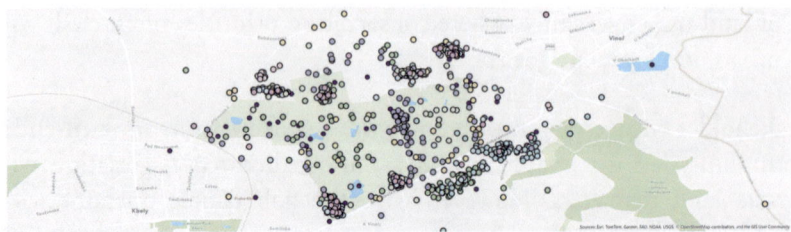

Fig. 3 Spatially given points from one of the participatory mapping activities in the Maptionnaire environment (*Source* Author)

- Biodiversity awareness: Survey results confirmed that residents perceived higher levels of biodiversity during the *new wilderness* phase compared to the finished golf course. This shift in biodiversity awareness strengthened the argument for designated ecological protection zones within the landscape.

- Consistent spatial preferences: Participatory mapping revealed that residents were remarkably consistent in identifying areas that could accommodate alternative land uses (e.g., walking trails, ecological preserves). This demonstrates a collective spatial awareness—people intuitively understand how landscapes function and can contribute meaningfully to planning discussions.

- Shifting public perception: Initially, public opinion was polarised, with strong opposition to the golf course. However, as the project facilitated structured discussions, attitudes evolved from "We don't want golf" to "We appreciate the discussion, even if we don't fully agree". This highlights how prefigurative engagement—where citizens actively shape governance—can increase public acceptance of complex land-use decisions.

- Policy proposals: The project proposed small ecological reserves and a nature trail that would not interfere with golf operations. These micro-transformations could serve as bridging solutions between conservation and recreation. However, without cooperation from the golf course manager, their implementation remains uncertain.

3.5   Challenges and Lessons Learned

The project faced several unexpected challenges, reinforcing key lessons about participatory governance and prefigurative politics:

- Unclear future of the site: The project's initial phase envisioned future "new wilderness" planning, but the situation shifted dramatically when the golf course operator resumed construction.
- Shifting project objectives: Initially focused on exploring land-use alternatives, the project drifted into a more formal negotiation process, engaging with institutional actors. While this transition was necessary, it limited the purely community-driven aspects of prefigurative politics.
- Governance complexity: The fragmentation between local governance (municipality), land ownership (Prague City Council), and land management (golf operator) created bureaucratic roadblocks that made participatory outcomes difficult to implement.
- The role of science in conflict resolution: Science was most powerful when it was not afraid to engage with real-world conflicts. By using data to structure discussions, the project provided legitimacy to public concerns and helped bridge ideological gaps.

4   Prefigurative Politics in Action

4.1   Reshaping Economy, Nature, and Society

One of the most profound contributions of the *Golf Vinoř* is how it challenges the traditional economic model of land use. Typically, large land parcels, especially those located near urban centres, undergo predictable transformations: conversion from agriculture to recreational use (e.g., golf courses, parks, or sports complexes), and potential rezoning for residential or commercial development, leading to permanent urbanisation.

In this trajectory, economic interests often take precedence over ecological and social considerations, with land-use decisions dominated by developers and municipal authorities rather than participatory deliberation. The *Golf Vinoř*, however, intervened in this process at a critical moment, raising a fundamental question: *Can large recreational areas serve multiple functions beyond a single-use designation?*

By engaging citizens in biodiversity monitoring and participatory mapping, the project elevated alternative perspectives—highlighting the potential for land to function both as a space for recreation and as a biodiverse ecosystem. This shifted the discourse from a purely economic valuation toward a multifunctional, ecosystem-based perspective, advocating for a balance between land-use profitability and ecological sustainability.

Furthermore, the project transformed the relationship between nature and society at the site. The period of new wilderness (2017–2022) demonstrated that, when left undisturbed, even formerly constructed landscapes can rapidly evolve into ecologically rich environments. Residents not only experienced this transformation firsthand but also contributed to scientific knowledge about its ecological value. This generated a new cultural relationship with the landscape—one that moved beyond the traditional framing of golf courses as elite spaces and instead positioned the land as a shared ecological and recreational asset.

By integrating these scientific, social, and ecological narratives, the project functioned as a prefigurative model of land-use governance—demonstrating, in the present, a more inclusive and participatory approach to landscape management.

4.2   Institutional and Behavioural Change

While prefigurative politics often operates outside traditional institutions, the *Golf Vinoř* directly influenced local governance and stakeholder attitudes.

One of the most striking transformations occurred at the level of public perception. Initially, local residents viewed the golf course primarily as an imposed, exclusionary land use. However, as the project facilitated discussions and provided data-driven insights, community perspectives evolved from outright rejection ("We don't want golf") to a more nuanced stance ("We appreciate that discussions are taking place").

This shift illustrates the role of prefigurative engagement in shaping institutional attitudes. By allowing residents to actively co-produce knowledge, the project increased public acceptance of negotiated solutions, even in a contested space.

At the municipal level, the project further demonstrated the value of integrating science into policy discussions. Local city authorities, often limited in their capacity to engage in scientific land-use assessments, welcomed the project as a resource that could inform planning decisions. This expanded their perspective on the role of science in governance, making them more receptive to interdisciplinary approaches that combine ecology, urban planning, and citizen participation.

Even private stakeholders—such as the golf course operator and land managers—initially showed interest in the project's insights. Although their eventual withdrawal from cooperation limited the long-term implementation of recommendations, their early engagement demonstrated an openness to interdisciplinary dialogue. This is significant because commercial landowners are often resistant to participatory governance models. The fact that they acknowledged and explored project findings, even if temporarily, suggests that scientifically informed citizen engagement can act as a bridge between corporate landowners and public interests.

Despite these successes, institutional contradictions persisted. The Prague City Council, as the landowner, remained internally divided— one department approving zoning changes to prioritise golf, while another (focused on nature conservation) fully supported the citizen science initiative. This reflects a broader challenge of prefigurative politics interfacing with bureaucratic structures: while grassroots initiatives can drive immediate transformation, their long-term influence often depends on institutional alignment (Seyfang & Haxeltine, 2012).

Rather than seeking full institutional consensus (which may be unrealistic in contested urban governance), grassroots initiatives can benefit from aligning with sympathetic actors within the system to gain legitimacy, resources, or visibility. In the case of Golf Vinoř, collaboration with nature conservation officials created opportunities to elevate biodiversity concerns and sustain momentum, even in the face of bureaucratic pushback elsewhere. This suggests that transformative potential need not be lost when engaging with formal institutions.

4.3   Cross-Sectoral Implications

Beyond its immediate impact, the *Golf Vinoř* offers valuable lessons for other sectors, including urban planning, biodiversity conservation, and participatory governance.

Multidisciplinary approaches foster new governance pathways: One of the most important insights from the project is that multidisciplinary collaboration opens new governance possibilities. The initiative brought together natural scientists (assessing biodiversity changes), social scientists (facilitating participatory engagement), urban planners (interpreting land-use potential), local authorities (balancing regulatory responsibilities), and community members (contributing experiential knowledge).

This combination of expertise was critical in broadening the scope of land-use debates beyond rigid, single-discipline approaches. If sustained, such interdisciplinary cooperation could be institutionalised, serving as a template for more adaptive governance models in contested urban spaces.

The Golf Vinoř model as a replicable case for contested landscapes: While *Golf Vinoř* has unique circumstances—particularly the abandonment phase that enabled the emergence of *new wilderness*—its core principles are transferable to other land-use conflicts.

Key elements that could be replicated elsewhere include using citizen science to generate empirical data on local ecosystem services, facilitating participatory mapping to integrate diverse land-use perspectives, and bridging the gap between institutional governance and grassroots initiatives.

The project also highlights the limitations of replication. In cases where abandonment periods do not exist, direct negotiation between land managers, developers, and local communities is essential. This requires stronger institutional engagement from the outset, ensuring that participatory science is not dismissed once commercial interests regain control.

4.4   Challenges and Tensions

Conflicts with Established Power Structures: One of the most significant tensions in the *Golf Vinoř* arose from the disengagement of the golf course operator, which ultimately limited the project's ability to implement its proposed land-use recommendations. Initially, the operator cited a lack of expert capacity to engage in the discussions, but later the withdrawal was attributed to unresolved property relations between the Prague City Council (the landowner) and the golf course operator (the manager). This bureaucratic ambiguity over land tenure and decision-making power created an environment where the project could diagnose issues but struggled to implement solutions.

Beyond the withdrawal of the golf course operator, the project did not face direct institutional pushback or resistance from local government authorities, as municipal representatives generally welcomed the scientific contributions and participatory approach. However, the fragmented governance within the Prague City Council—where one department supported nature conservation while another prioritised golf course development—created a paralysis in decision-making. The project also encountered political influences that were not explicitly revealed, suggesting that behind-the-scenes interests may have shaped the willingness (or reluctance) of key decision-makers to act on the findings.

Despite these structural constraints, the main limitation was not political opposition but rather the finite capacity of project members. With limited personnel and resources, the team had to balance its research and engagement efforts with other responsibilities, making it difficult to fully institutionalise the project's findings into formal governance structures.

Scalability and Sustainability: While the project successfully engaged over 500 survey participants and facilitated public deliberation, ensuring long-term sustainability remained a significant challenge. Several factors limited the scalability and continuity of the project's impact:

- Lack of dedicated funding: The project operated within an academic framework rather than as an institutionally embedded initiative, making it difficult to sustain engagement beyond its immediate research cycle.

- No long-term citizen engagement mechanism: While participatory mapping and workshops were effective in collecting data and shaping discussions, there was no formal mechanism to continue integrating citizen input into decision-making.
- Fragmented governance structures: Without coherent decision-making within the Prague City Council, even well-supported scientific findings struggled to translate into policy action.

To scale or sustain such an initiative in the future, stronger cooperation among stakeholders would be essential. This would include more structured collaboration with all land-use stakeholders, ensuring continued dialogue between citizens, scientists, municipal authorities, and land managers; sustainable communication strategies that maintain engagement beyond the formal research phase; and a clearer governance framework within the Prague City Council, preventing bureaucratic deadlocks that stall implementation.

Reflections on Citizen Engagement and Long-Term Impact: Despite its successes, the project revealed gaps in citizen engagement that, if addressed, could have strengthened its overall impact. Specifically, local schools were underutilised as potential partners, and the use of the PlantNet biodiversity monitoring app could have been expanded with greater contributions from local residents. In hindsight, a stronger focus on outreach—including additional workshops, stakeholder meetings, and targeted efforts to engage more local actors—could have broadened participation and increased public pressure for more concrete action.

A key lesson from the project is that participatory initiatives must be designed with long-term sustainability in mind. While the scientific and deliberative components were effective in shaping public perception and generating data, the absence of a mechanism to institutionalise findings into governance structures remains a barrier. If repeated, the project would benefit from earlier and deeper engagement with stakeholders who expressed interest, ensuring that momentum is not lost when key decision-makers disengage.

Competing Interests   The author has no conflicts of interest to declare that are relevant to the content of this chapter.

References

Dahl Jensen, A. M., Caspersen, O. H., Jensen, F. S., & Strandberg, M. (2017). Multifunctional golf facilities as a resource of important ecosystem services in a changing urban environment: Nordic case studies. *International Turfgrass Society Research Journal, 13*(1), 236–239. https://doi.org/10.2134/itsrj2 016.05.0424

Davoudi, S. (2023). Prefigurative planning: Performing concrete utopias in the here and now. *European Planning Studies, 31*(11), 2277–2290. https://doi. org/10.1080/09654313.2023.2217853

Díaz, S., Demissew, S., Carabias, J., Joly, C., Lonsdale, M., Ash, N., Larigauderie, A., Adhikari, J. R., Arico, S., Báldi, A., Bartuska, A., Baste, I. A., Bilgin, A., Brondizio, E., Chan, K. M., Figueroa, V. E., Duraiappah, A., Fischer, M., Hill, R., … Zlatanova, D. (2015). The IPBES conceptual framework—Connecting nature and people. *Current Opinion in Environmental Sustainability, 14*, 1–16. https://doi.org/10.1016/j.cosust.2014.11.002

Duží, B., Osman, R., Lehejček, J., Nováková, E., Taraba, P., & Trojan, J. (2020). Exploring citizen science in post-socialist space: Uncovering its hidden character in the Czech Republic. *Moravian Geographical Reports, 27*(4), 241–253. https://doi.org/10.2478/mgr-2019-0019

Kimura, A. H., & Kinchy, A. (2016). Citizen science: Probing the virtues and contexts of participatory research. *Engaging Science, Technology, and Society, 2*, 331–361. https://doi.org/10.17351/ests2016.99

Mattijssen, T. J. M., Hennen, W., Buijs, A. E., De Dooij, P., Van Lammeren, R., & Walet, L. (2024). Urban greening co-creation: Participatory spatial modelling to bridge data-driven and citizen-centred approaches. *Urban Forestry & Urban Greening, 94*, 128257. https://doi.org/10.1016/j.ufug. 2024.128257

Monticelli, L. (2021). On the necessity of prefigurative politics. *Thesis Eleven, 167*(1), 99–118. https://doi.org/10.1177/07255136211056992

PlantNet. (b.r.). Pl@ntNet. Received 26 May 2025, https://plantnet.org/en/

Raekstad, P., & Gradin, S. S. (2020). *Prefigurative politics: Building tomorrow today*. Polity Press. https://doi.org/10.1080/09644016.2020.1741768

Seyfang, G., & Haxeltine, A. (2012). Growing grassroots innovations: Exploring the role of community-based initiatives in governing sustainable energy transitions. *Environment and Planning C: Government and Policy, 30*(3), 381–400. https://doi.org/10.1068/c10222

Stella, D. (b.r.). *Maptionnaire*. Golf Vinoř Citizen Science Projekt. Received 26 May 2025, https://new.maptionnaire.com/p/7hk4c3m9erg8

Zapata Campos, M. J., & Zapata, P. (2017). Infiltrating citizen-driven initiatives for sustainability. *Environmental Politics, 26*(6), 1055–1078. https://doi.org/10.1080/09644016.2017.1352592

Open Access This chapter is licensed under the terms of the Creative Commons Attribution-NonCommercial-NoDerivatives 4.0 International License (http://creativecommons.org/licenses/by-nc-nd/4.0/), which permits any noncommercial use, sharing, distribution and reproduction in any medium or format, as long as you give appropriate credit to the original author(s) and the source, provide a link to the Creative Commons license and indicate if you modified the licensed material. You do not have permission under this license to share adapted material derived from this chapter or parts of it.

The images or other third party material in this chapter are included in the chapter's Creative Commons license, unless indicated otherwise in a credit line to the material. If material is not included in the chapter's Creative Commons license and your intended use is not permitted by statutory regulation or exceeds the permitted use, you will need to obtain permission directly from the copyright holder.

4

Prefiguring Consumption Reduction in Shadow Networks: Recirculating Resources, Knowledge, and Skills in Urban Tool Libraries and Community Bike Workshops

Simon Batterbury, Sabrina Chakori, and Carlos Uxo

1   Introduction

To me, the future already exists. For some people, maybe for everyone. It's just a matter of tuning yourself to it. Scarlet Riviera, 1975, in *Rolling Thunder: A Bob Dylan Story*, dir. Martin Scorsese. (2019).

The impacts of the Capitalocene (Moore, 2017) are felt across the world and include increasing resource consumption and its effects (UNEP, 2024). As this volume shows, there are transformative actions and projects that form the basis for active and passive resistance to current trends (Sareen & Juhola, 2025). In this chapter, we provide

S. Batterbury (✉)
SGEAS & Melbourne Climate Futures Academy, University of Melbourne, Melbourne, Australia
e-mail: simonpjb@unimelb.edu.au
URL: http://www.simonbatterbury.net

S. Chakori
The Commonwealth Scientific and Industrial Research Organisation (CSIRO), Canberra, ACT, Australia

Brisbane Tool Library Inc, Brisbane, Australia

© The Author(s) 2026
S. Sareen and S. Juhola (eds.), *Societal Transitions to Sustainability*, Palgrave Studies in Environmental Transformation, Transition and Accountability,
https://doi.org/10.1007/978-3-032-07395-2_4

51

examples of practical ways in which communities have *already prefigured* a low-carbon future that is also socially-oriented and convivial. In these small and agile initiatives, George Monbiot reminds us to:

Start with this principle: don't face your fears alone. Make friends, meet your neighbours, set up support networks, help those who are struggling. Since the dawn of humankind, those with robust social networks have been more resilient than those without. (Monbiot, 2025)

After exploring the key elements of success in transformational processes, we introduce two movements that challenge the productivist and consumerist growth-driven system. Both are poorly represented in literature on transformations and degrowth and are part of what Sareen et al., (2024, 251) call the 'marginalised, subversive, dominated, silenced, and left-aside spaces of transition studies'.

Community bike workshops and a tool library (in Brisbane) are prefiguring the future through fostering socio-ecological wellbeing, remaining resilient against various internal and external constraints. The form of resistance found in tool lending and bike repair initiatives is in their anti- and post-capitalist nature. We will return to the sometimes-muted political nature of these initiatives, which are not, or cannot always be expressed explicitly.

The authors combine academic and lived experience. Two, Simon and Carlos, have been involved in bike repair workshops for many years, in various roles, and have also conducted research on them (Batterbury, 2024; Batterbury et al., 2025). Sabrina founded the Brisbane Tool Library in 2017 and has managed various aspects of it since, while contributing to other areas of degrowth scholarship.

Department of Management and Marketing, University of Melbourne, Melbourne, Australia

S. Chakori
e-mail: sabrina.chakori@csiro.au

C. Uxo
European Languages, Monash University, Melbourne, Australia
e-mail: carlos.uxo@monash.edu

2   Prefiguring Transformations in the Shadows

In a book produced from Melbourne, Australia during the global pandemic, Sam Alexander and Brendan Gleeson refer to the CERES environmental centre in the city as 'a lived, real-world example that you can touch and see and feel, and therein lies its transformative potential', part of a range of projects that are the 'seeds of the new within the shell of the old' (Alexander & Gleeson, 2020, 207). They argue that the 'incumbent regime' (Ehnert et al., 2021) driving an affluent capitalist city can be challenged by such locally based and progressive initiatives, sometimes termed 'niches' (Geels, 2002; Vandeventer et al., 2019).

CERES, and the niche projects we describe below, are part of shadow networks, outside the dominant economic system (Boonstra & Joose, 2013). They provide a focus for prefigurative political work, through their pedagogy, activities, and environmental contributions. They generate new imaginative responses and ideas, but only if they can maintain *sustained human interest and investment* across diverse participants and actors. It is important that participants in such organisations, even if there are frequent challenges, keep them going, as CERES has done. Many initiatives or niches fail or run into significant problems over time, as we will show (Vadeventer et al., 2019).

Many alternative networks, and the two that we explore below, have sometimes struggled to maintain operations, or to thrive without access to premises and sufficient financial reserves. To summarise, the problems they face are (1) the incommensurability between 'growth' and their purpose as alternative organisations and movements and (2) 'prefiguring' usually exists within a regime (here, capitalism), which can squeeze out competition (a clear example we give is the struggle to find premises). Of course, there are considerable differences between the different incumbent political economies in which prefiguring organisations reside. But as a vast literature about the running of cognate charitable organisations and nonprofits shows (Helmig et al., 2004), a certain agility is needed to balance anti- and post-capitalist values with available finance, personnel, skills, and material requirements (Törnberg, 2021). We return to this point in the discussion.

We now introduce the cases presented, and then we discuss their role in prefiguring alternatives.

3   Community Bike Workshops

Community bike workshops (CBWs) or 'bike kitchens' are a predominantly urban phenomenon dating back to the 1980s, prefiguring sustainable mobilities and active travel in many parts of the world. They offer a convivial environment for the social practice of the restoration and maintenance of bicycles, and are located in a variety of premises, some less suitable for purpose than others. Unlike commercial bike shops, anybody can come and fix their own bikes with salvaged components, learning a range of practical repair skills. In some workshops, bikes are repaired by volunteers and then offered freely or cheaply to people who need them (Batterbury et al., 2025). CBWs are also social entities, niches in which mutual aid, helping others, and convivial social networks operate. The major goal is an extension of the useful life of bicycles, forming part of reuse and circularity, given that major bike components can be kept in use for decades.

In societies where urban rhythms, transportation systems, and infrastructures are deeply tied to car-centric mobility, the presence of CBWs is quite a rare challenge to automobility and consumerism, because they offer low-cost alternatives to both. Most bikes they fix are second-hand, and by learning repair skills, participants challenge the modern division of labour through *vélonomie,* which means developing confidence among riders to repair a bicycle and to ride it safely (Abord de Chatillon & Eskenazi, 2022). CBWs offer an innovative pathway to increasing bike ridership and acceptance of cycling. Workshops link users and a variety of volunteers, and occasionally paid staff, organically (Durkheim, 1933). They demonstrate a passive refusal of automobility rather than active protest against car culture.

Whether they are volunteer-only or social enterprises reinvesting any profit, the focus is not on broad social transformation, but on skill acquisition and a change in mindset—for example, demystification of bike technologies, aiding autonomy, and undertaking more journeys with confidence. Participant-observation and interviews of how this transformation is brought about have been conducted by authors 1 and 3 in around sixty workshops in Australia and across Europe (Batterbury, 2024; Batterbury et al., 2025). We found that transforming urban mobility is not only about expensive 'supply-side' 'hard' intervention like new bike lanes and redesigning streetscapes, but also what CBWs do—'soft' improvements in 'demand' for cycling, by encouraging confidence with mechanical tasks, learning, and convivial sharing of tools and knowledge. Workshops prefigure cultural and social change in how we travel, while recognising social diversity in terms of age, race, gender, and ability (termed 'bike equity') (Batterbury et al., 2025).

WeCycle in Melbourne, Australia is one example (Fig. 1). Formed in 2017, it occupies a cheaply leased building in a public park owned by a municipal council, and distributes over 250 bicycles per year to refugees, asylum seekers, and other people who have difficulty affording a bike. There are from 5 to 15 active volunteers of different ages, genders, and backgrounds including authors 1 and 3, and the budget is quite lean. Bikes and parts are donated by the citywide community, and occasionally from clear-outs of apartment complexes and other businesses. Volunteers, primarily living locally, undertake different tasks according to their skill sets, particularly learning how to swap and substitute broken for re-used parts, and restoring wheels, brakes, and drivetrains. In this they form a 'task group' (Richards, 2010) that co-evolves skills from practices, while developing shared, embodied capacity. But members also develop friendships and new social networks (Illich, 1973). WeCycle is open just one day a week, plus time taken to deliver bikes to households in need, but the city also has a few larger bike recycling operations, most of them linked through an informal recyclers network. These are open longer hours and have larger budgets than WeCycle (Batterbury et al., 2025). WeCycle has elected not to 'grow', fighting back 'regime pressure' to expand (Henriksson & Göransson Scalzotto, 2023) given its constraints on time and human capital.

Fig. 1 A community repair event at WeCycle in Melbourne, Australia. 2023. *Source* Authors at WeCycle. Alt text: WeCycle volunteers repairing bicycles

4   Tool Libraries or 'Libraries of Things'

The Brisbane Tool Library (BTL) (Fig. 2) was founded in 2017 in Australia as a practical response to enable a degrowth urban transition (Chakori & Batterbury, 2025). Degrowth can be defined as 'an equitable downscaling of production and consumption that increases human well-being and enhances ecological conditions at the local and global level, in the short and long-term' (Schneider et al., 2010, 512). Organisations can have political origins (e.g. interest in moving beyond growth, or just reducing waste), but communities can also initiate them for other reasons, such as a passion for 'Do it Yourself (DIY)' activities. Tool libraries (or 'libraries of things') across the world do this by sharing and improving access to tools and resources (Chakori & Hopkinson, 2024; Claudelin et al., 2022; Lynch, 2023). While BTL was founded on principles of degrowth, not everyone who joins and contributes to the project is aware of the founding values or explicitly aligns with them. Concepts such as sharing and the circular economy, or 'war on waste', are more familiar in Australia than degrowth. Equally, Australia has a strong DIY culture, with a strong Men's Sheds movement (Golding et al., 2007; Southcombe et al., 2015), which is supported with various resources and funding (e.g. the National Shed Development Program of the Australian Government Department of Health and Aged Care, see https://www.health.gov.au/our-work/national-shed-development-pro gramme-nsdp [accessed 05.05.2025]). BTL is a space not limited to men, where everyone is welcomed (in fact, BTL has always been led by women). The increased equity in providing access and opportunities to everyone can also be considered a prefiguration of a future society that thrives on justice and fairness.

BTL enables the community to access and borrow everyday items and specialist equipment (e.g. circular saws, drills, lawnmowers, camping gear, party appliances) for short periods. By encouraging the reuse

Fig. 2 Brisbane tool library, Brisbane, Australia. 2019. *Source* Author Sabrina Chakori. Alt text: Electrical tools for loaning at the BTL

and circulation of resources, the goal is to reduce the economic metabolism. By prioritising access over private ownership, BTL provides an alternative circular economic model that challenges productivism and consumerism, reducing the private ownership of un- or under-utilised items (Chakori & Hopkinson, 2021, 2024).

The inventory of the BTL has been built by rescuing second-hand items from recovery centres (i.e. landfill stations) and by private donations that are maintained and repaired by the volunteers. Borrowers have to pay a membership fee to access the inventory. Prices are kept accessible to encourage more people from diverse socio-economic backgrounds to borrow. This revenue covers the maintenance costs of the items (e.g. repair) and other ongoing costs (e.g. rent, insurance), however, it does not yet allow for paid positions.

By prioritising access over ownership, this sharing model helps households save money and space, reducing ecological footprints (Chakori & Hopkinson, 2021, 2024). As of December 2024, BTL had enabled more than 15,000 loans, which corresponds to saving approximately AUD\$1.8M (€1.04M) across the community (with an average value of AUD\$120 (€70) per tool as estimated by BTL's committee).

Further research is required to quantify the impacts of the tool library and to compare them across different localities. Nevertheless, communities show some dematerialisation while increasing their wellbeing (e.g. by providing more access and experiences to everyone), even if degrowth incurs resistance and inertia at a macro level. By advancing alternative ownership and sharing models (i.e. *commoning*), tool libraries help to reduce inequity in material access and related experiences, such as DIY hobbies, house renovations, and camping (Chakori & Batterbury, 2025; Chakori & Hopkinson, 2021, 2024). Yet, BTL also faces considerable constraints and barriers, as presented below.

5   Discussion

We can 'touch and feel' projects like these that prefigure low-carbon and convivial solutions to meet particular urban needs, even if they remain in the shadows of capitalist societies. Existing alternatives can be established and can endure. Nevertheless, we also need to highlight the significant challenges facing such organisations.

1. Funding. A drop-off in funding can end some prefigurative experiments. Workshops with rental and staff costs were affected by a post-pandemic drop in bike use, and rising pressures on household budgets. France has the most workshops, but some failed. In 2024 several including Recup'R in Bordeaux launched emergency crowdfunding appeals, due to declining revenues from repairs and bike sales, or the rising cost of premises. The same happened to the larger Bike For Good in Glasgow, Scotland in 2023, and in Recyke y'Bike in Newcastle England in 2024 (Batterbury, 2024).

2. The transition to a paid workforce in this sector can be difficult. Paying labour is socially just, but realistically impossible without adequate funds and financial stability. The organisations presented here have not secured those resources, and the issue is extensively debated in the 'for-purpose' sector, particularly when state revenues or subsidies are unavailable. Meeting national employment regulations is also complex.

3. In competitive real estate markets, many not-for-profits and small businesses struggle to find and access free or affordable premises. BTL, for example, is financially self-sufficient (through its membership scheme and some smaller grants) and has enough volunteers, but its location has shifted several times. From 2018–2022, it was housed within the public State Library of Queensland. This enabled a collaboration between the tool library and the 'maker space' of the library, a first in Australia. It had to evacuate in 2022, when floods hit the city. After temporary locations, BTL had to rent a warehouse from government in 2025. Chakori and Batterbury (2025) and Chakori and Hopkinson (2021, 2024) expand on these internal and external challenges.

Despite these obstacles, shadow networks persist and continue to model alternatives to the status quo. Their prefiguration embodies mutual aid, manifested beyond commodified capitalist market relations. Mutual aid takes different forms. Durkheim counterposed 'mechanical solidarity' in forms of collective action driven by shared values—visible today in climate movements and protests against the far right for example, with 'organic solidarity', where participants have wider-ranging values (Durkheim, 1933). Many small prefigurative organisations work 'organically', with quite diverse individuals bringing their own skills and attributes to the initiative. Kropotkin's support for 'mutual' aid counterposed it to a Darwinist 'struggle for the means of existence', arguing it plays a leading part in the progress of humankind (Kropotkin, 1902, 293). Diverse volunteers in bike workshops and tool libraries certainly practice 'mutual aid'.

Both organisational forms also operate with 'convivial tools', a term coined by Ivan Illich. Tools should be easy to use and accessible: 'A convivial society should be designed to allow all its members the most autonomous action by means of tools least controlled by others' (Illich, 1973, see 23, 30, 86). Workshops feature '…technologies that are simple to understand, communally owned and democratically developed based on the active involvement of the users' (Linnea-Møller, 2023, 139), with 'Do It Yourself ethics' (Furness, 2010, 141) and convivial friendships and social relationships built around tools also present in tool libraries.

Although we stress the positive aspects of both types of organisations, prefiguring also involves agreement on what they are against. In both domains, participants are against waste and non-circular economies, as well as non-active polluting travel in the case of bike workshops. Publicity and campaigning on these issues brings in new participants who are sympathetic, but also people just seeking an outlet for their talents and time. In *organic* movements where key participants hold diverse values, political messaging is often depoliticised, for practical reasons. Local council funding, for example, depends on stressing the *healthiness* of bike riding and providing *mobility choices* to diverse and sometimes disadvantaged local people. Anti-car sentiments will not get much traction, at least in Australia. The BTL, for example, publicises its waste reduction and circular economy contributions, an exercise in 'degrowth diplomacy'.

In addition, prefiguring organisations cannot totally delink from the economy that surrounds them. WeCycle needs cars and busy roads to deliver bikes to refugees; and BTL still relies on capitalist social media to communicate. Tool libraries (and repair cafes and other similar initiatives) cannot truly move beyond a growth-driven system if they use its discards (i.e. second-hand items). BTL is aware that it is 'one step behind' capitalism, despite supporting change and being observable and real (Im, 1991). A collaboration between various counter-hegemonic niches does build where degrowth becomes a common representative framework (Demaria et al., 2013).

Lastly, the question of scale and impact. BTL has reach across a medium-sized city, while most bike workshops remain localised. They have no need to expand in terms of conventional 'growth' criteria (Törnberg, 2021). However, to lead to systemic change, these initiatives need to amplify their impact (Lam et al., 2020). Expanding or amplifying can be risky for vulnerable organisations. Core values can be challenged. Financial and human sustainability can be compromised. Workshops that have expanded to serve city regions, like Bikes for Good in Glasgow, need income for salaries and premises (Batterbury, 2024). Even mutual aid can prove scarce in financial downturns. Here, 'networking' is important. There is a membership network of several hundred community bike workshops in France (where there are around 400) which lobbies nationally, seeking benefit from national regulations on waste and recycling (including some payments to small organisations keeping bikes out of the waste stream, negotiated in 2025 with Ecologic), and providing open-source software and training for management tasks.

We recognise that bike workshops and tool libraries are *only intervening in specific circuits of material goods and knowledge,* but they are also socialising the process of salvaging a subset of tools and items. As practitioners in this space, we are still happy to be making our small contributions. And as Riviera noted in the introductory quote, the future already exists. Maybe only some can see it.

6 Conclusion

Prefigurative actions based on solidarity and mutual aid challenge existing power structures, and they advocate for more equitable ways of living together while consuming less. The organisations we describe here are organic in nature, rather than manifesting tighter-knit 'mechanical' solidarity or forming activist social movements (Durkheim, 1933). They are 'shadow' networks, distanced from mainstream economic thinking and as Sareen et al. (2024) postulate, they are also 'marginal' and counter-hegemonic (Robra & Hinton, 2024).

Nonetheless, their operations are convivial and vitally, they are sustained through enormous effort. This is a key success of certain prefiguration initiatives, and their sociality extends beyond the material acts of bike repair, and the lending of tools and goods, that anchor them in particular premises or locales. The transformative shifts that they are sympathetic to, and are part of, highlight the possibility of reconfiguring the ways in which capitalist systems organise social and ecological relations. But as we have shown, they are still a speck of space dust in a huge capitalist universe.

7   Competing Interests

The authors have no conflicts of interest to declare that are relevant to the content of this chapter.

Acknowledgements Thank you to the reviewers Frank Eckerle and Gerald Aiken, and numerous volunteer collaborators in our projects. SB acknowledges a British Academy Visiting Fellowship, 2024. Generative AI was not used.

References

Abord de Chatillon, M., & Eskenazi, M. (2022). Devenir cycliste, s'engager en cycliste: Communautés de pratiques et apprentissage de la vélonomie. *Sociologies*, 18924 https://doi.org/10.4000/sociologies.18924

Alexander, S., & Gleeson, B. (2020). Rewilding the suburbs: CERES as a site of enchantment. In Alexander, S & Gleeson, B. (Eds.), *Urban awakenings*. Palgrave Macmillan. https://doi.org/10.1007/978-981-15-7861-8_19

Batterbury, S. P. J. (2024). *Community bike workshops and the culture of sustainable mobility: British cases*. Report to the British Academy. https://doi.org/10.5281/zenodo.14048990

Batterbury, S. P. J., Uxo, C., Abord de Chatillon, M., & Nurse, S. (2025). Community bike workshops in Australia: Increasing demand for cycling through mutual aid. *Transactions on Transport Sciences, 16*, 15–22. https://doi.org/10.5507/tots.2024.023

Boonstra, W. J., & Joose, S. (2013). The social dynamics of degrowth. *Environmental Values, 22*(2), 171–189. https://doi.org/10.3197/096327113X13581561725158

Chakori, S., & Batterbury, S. P. J. (2025) Degrowth urban transitions: A tool library case study. *Journal of City Climate Policy and Economy. 4*(2), 428–455. https://doi.org/10.3138/jccpe-2024-0056

Chakori, S., & Hopkinson, S. (2021). The role of tool libraries in the new economy: Sharing in an economic degrowth society. In T. Sigler & J. Corcoran (Eds.), *A guide to the urban sharing economy* (pp. 237–253). Edward Elgar. https://doi.org/10.4337/9781789909562.00026

Chakori, S., & Hopkinson, S. (2024). Living in abundance: Tool Libraries for convivial degrowth. In L. Eastwood & K. Heron, (Eds.), *De Gruyter handbook of degrowth*. de Gruyter. https://doi.org/10.1515/9783110778359-012

Claudelin, A., Tuominen, K., & Vanhamäki, S. (2022). Sustainability perspectives of the sharing economy: Process of creating a library of things in Finland. *Sustainability, 14*(11), 6627. https://doi.org/10.3390/su14116627

Demaria, F., Schneider, F., Sekulova, F., & Martinez-Alier, J. (2013). What is Degrowth? From an activist slogan to a social movement. *Environmental Values, 22*(2), 191–215. https://doi.org/10.3197/096327113x13581561725194

Durkheim, E. (1933). *The division of labor in society*. The Free Press.

Ehnert, F., Egermann, M., & Betsch, A. (2021). The role of niche and regime intermediaries in building partnerships for urban transitions towards sustainability. *Journal of Environmental Policy & Planning, 24*(2), 137–159. https://doi.org/10.1080/1523908X.2021.1981266

Furness, Z. (2010). *One less car: Bicycling and the politics of automobility*. Temple University Press.

Geels, F. W. (2002). Technological transitions as evolutionary reconfiguration processes: A multi-level perspective and a case-study. *Research Policy, 31*(8–9), 1257–1274. https://doi.org/10.1016/S0048-7333(02)00062-8

Golding, B., Brown, M., Foley, A., Harvey, J., & Gleeson, L. (2007). *Men's sheds in Australia: Learning through community contexts*. National Centre for Vocational Education Research.

Helmig, B., Jegers, M., & Lapsley, I. (2004). Challenges in managing nonprofit organizations: A research overview. *VOLUNTAS: International Journal of Voluntary and Nonprofit Organizations, 15*, 101–116. https://doi.org/10.1023/B:VOLU.0000033176.34018.75

Henriksson, M., & Göransson Scalzotto, J. (2023). Bike-sharing under pressure: The role of cycling in building circular cycling futures. *Journal of Cleaner Production, 395*, Article 136368. https://doi.org/10.1016/j.jclepro.2023.136368

Illich, I. (1973). *Tools for conviviality*. Calder & Boyars.

Im, H. B. (1991). Hegemony and counter-hegemony in Gramsci. *Asian Perspective, 15*(1), 123–156.

Kropotkin, P. (1902). *Mutual aid: a factor of evolution*. MacClure Phillips & Co. https://dn790008.ca.archive.org/0/items/cu3192403024 3640/cu31924030243640.pdf

Lam, D. P., Martín-López, B., Wiek, A., Bennett, E. M., Frantzeskaki, N., Horcea-Milcu, A. I., & Lang, D. J. (2020). Scaling the impact of sustainability initiatives: A typology of amplification processes. *Urban Transformations, 2*(1), 1–24. https://doi.org/10.1186/s42854-020-00007-9

Linnea-Møller, J. (2023). Degrowth and the slow travel movement: Opportunity for engagement or Consumer Fad? *Tvergastein, 2*, 134–150. https://www.tvergastein.com/s/Degrowth-Volume-2.pdf

Lynch, N. (2023). Borrowing spaces: The geographies of 'libraries of things' in the Canadian sharing economy. *Tijdschrift Voor Economische En Sociale Geografie, 114*(2), 157–173.

Monbiot, G. (2025). There are many ways Trump could trigger a global collapse. Here's how to survive if that happens. *The Guardian*, 18 Feb. https://www.theguardian.com/commentisfree/2025/feb/18/donald-trump-global-collapse-wildfires-pandemic-financial-crisis

Moore, J. W. (2017). The Capitalocene, Part I: On the nature and origins of our ecological crisis. *The Journal of Peasant Studies, 44*(3), 594–630. https://doi.org/10.1080/03066150.2016.1235036

Richards, P. (2010). *A green revolution from below? Science and technology for global food security and poverty alleviation*. Farewell retirement address, Wageningen University. https://edepot.wur.nl/165231

Robra, B., & Hinton, J. B. (2024). Economic organizations and the transformation towards degrowth. In E. Weik, C. Land & R. Hartz (Eds.),

The handbook of organizing economic, ecological and societal transformation (p. 209). De Gruyter.

Sareen, S., & Juhola, S. (2025). The prefigurative politics of present transformation. In S. Sareen & S. Juhola (Eds.), *Societal transitions to sustainability: The prefigurative politics of present transformation.* Palgrave Macmillan.

Sareen, S., Ortar, N., & Lis-Plesińska, A. (2024). Framing spatiotemporalities of sustainability in new mobility practices and imaginaries. *Norsk Geografisk Tidsskrift - Norwegian Journal of Geography, 78*(5), 251–257. https://doi.org/10.1080/00291951.2024.2434051

Schneider, F., Kallis, G., & Martinez-Alier, J. (2010). Crisis or opportunity? Economic degrowth for social equity and ecological sustainability. Introduction to this special issue. *Journal of Cleaner Production, 18*(6), 511–518. https://doi.org/10.1016/j.jclepro.2010.01.014

Scorsese, M. (2019). *Rolling Thunder: A Bob Dylan story* [Film]. Netflix.

Southcombe, A., Cavanagh, J., & Bartram, T. (2015). Retired men and Men's Sheds in Australia. *Leadership & Organization Development Journal, 36*(8), 972–989. https://doi.org/10.1108/LODJ-03-2014-0065

Törnberg, A. (2021). Prefigurative politics and social change: A typology drawing on transition studies. *Distinktion: Journal of Social Theory, 22*(1), 83–107. https://doi.org/10.1080/1600910X.2020.1856161

United Nations Environment Programme. (2024). *Global Resources Outlook 2024.* UNEP International Resource Panel. https://wedocs.unep.org/20.500.11822/44901

Vandeventer, J. S., Cattaneo, C., & Zografos, C. (2019). A degrowth transition: Pathways for the degrowth niche to replace the capitalist-growth regime. *Ecological Economics, 156,* 272–286. https://doi.org/10.1016/j.ecolecon.2018.10.002

Open Access This chapter is licensed under the terms of the Creative Commons Attribution-NonCommercial-NoDerivatives 4.0 International License (http://creativecommons.org/licenses/by-nc-nd/4.0/), which permits any noncommercial use, sharing, distribution and reproduction in any medium or format, as long as you give appropriate credit to the original author(s) and the source, provide a link to the Creative Commons license and indicate if you modified the licensed material. You do not have permission under this license to share adapted material derived from this chapter or parts of it.

The images or other third party material in this chapter are included in the chapter's Creative Commons license, unless indicated otherwise in a credit line to the material. If material is not included in the chapter's Creative Commons license and your intended use is not permitted by statutory regulation or exceeds the permitted use, you will need to obtain permission directly from the copyright holder.

5

Starting by Doing: Ethical Orientations for Open-Ended Prefigurative Action

Bob Grumiau and Ian Hughes

1   Introduction

"*Empezar con hacer, después pensar.*" Starting by doing, then thinking. The person speaking is *Andrés,* one of the regular participants in the meetings and activities of one of the local citizen assemblies in Santiago de Chile. Barely two months hae gone by since the start of severe protests against the neoliberal regime of President Sebastian Piñera, and during those days, the assembly has been meeting at least twice a week to discuss what is going on and to organise those services that are considered essential for the human dignity of the inhabitants of the *barrio.* Up till now, as *Andrés* puts it well, the assembly has been so occupied with doing—organising a whole range of services from community schools to health posts and street theatres—that they haven't had the chance to reflect on

B. Grumiau (✉) · I. Hughes
MaREI Centre—Deep Societal Innovation for Sustainability and Human Flourishing (DSIS), Sustainability Institute, University College Cork, Cork, Ireland
e-mail: Bob.Grumiau@ucc.ie

© The Author(s) 2026
S. Sareen and S. Juhola (eds.), *Societal Transitions to Sustainability,* Palgrave Studies in Environmental Transformation, Transition and Accountability,
https://doi.org/10.1007/978-3-032-07395-2_5

how they define themselves, on the purpose of their meetings. However, given the need to communicate with outside elements as other local and supra-local assemblies, the moment has come in which it has become important to reach a certain consensus about the objectives and the nature of this small political body.

Let us take a moment to focus on the fact that it took a citizen assembly almost two months since the start of the social outburst in Chile, in October 2018, to feel the need to ask themselves what they were doing and where they wanted to go. Or even, simply what they themselves represented. Starting by doing, then thinking. These five words, referred to here at the beginning, point towards an antecedence of doing to being—of action before representation. As Sareen and Juhola (2025) mention in the introduction to this book, prefigurative politics refers to embodied strategies that render desirable futures with immediacy. By emphasising the *doing* of change, rather than the thinking, the example of the Chilean revolutionary assemblies shows how—conceptual—silence can be productive for, and crucial to, immediate creative, and collective action. If prefigurative politics is about natality and new beginnings here-and-now, predetermined goals are anti-political and counterproductive when it comes to the enacting of worlds-to-come (Arendt, 2005). Through the analysis of our case study, we argue that focusing on ethical directionality, rather than on fixed objectives, is important if we want prefigurative politics to be pluralist and democratic.

By emphasising the importance of ethical directionality and different ways of seeing, we take inspiration from the Deep Societal Innovation for Sustainability and Human Flourishing (DSIS) model for transformation. The DSIS model holds that climate change must be considered within the context of a wider metacrisis (Rowson, 2023). According to DSIS, moving beyond the current unsustainable paradigm, requires a whole-of-society approach that transforms the way in which our social institutions are enacted and challenges the way in which we understand the political. We will show that the idea of prefigurative politics implies a transformation of what we commonly understand as politics. Seen from this perspective, the political, rather than being the prerogative of the state and hierarchical institutions, is specific to horizontal forms of being-in-common, is of ontological importance to the human being, concerns the here and now of every-day decisions and, for it to be truly democratically and plural, must be rooted in common action.

2 The Deep Societal Innovation for Sustainability and Human Flourishing (DSIS) Model

A growing body of literature views the crises that are currently facing humanity as deeply interlinked and a signal of the necessity for deep systemic and cultural change (Dussel, 2012; Escobar, 2018; Homer-Dixon et al., 2021; Hughes et al., 2021; Kanger & Schot, 2019; Morin & Kern, 1999). Nevertheless, existing narratives and models of transformation continue to focus primarily on climate change as a problem in itself, rather than as a symptom of a broader and deeper polycrisis and of an unsustainable societal construct. As a consequence, dominant narratives limit themselves to emphasising isolated innovations in technology, economy, and (to a lesser extent), politics, without stressing the mutual interdependence of the forenamed nor the need for whole-of-society transformation. The model for Deep Societal Innovation for Sustainability and Human Flourishing (DSIS) aims to reframe the dominant narrative that focuses mainly on socio-technical transition by stressing the need for a deep global cultural transformation that acknowledges the interdependency of the climate crisis with other challenges such as economic inequality, the crisis of democratic systems, continued gender, cultural and racial hierarchies.

Figure 1 shows the DSIS conceptual model. At the centre of the Venn diagram, we find the "reimagining of social institutions," which constitutes the core of societal transformation. The DSIS model focuses on six primary and intertwined social institutions: politics, technology, economy, religion, education, and gender. The deep re-imagining of these institutions is informed by four further elements of the model. As evidenced by the diagram, these four elements present four ways of approaching the imagining of social institutions that overlap and transgress into each other. First, the "evolving narratives" component of the model emphasises the need to go beyond both socio-technical transitions and deep transitions framings to a whole-of-society transformation framing. As we shall see, the latter entails a move beyond the paradigm of modernity and introduces us to the second component of the model: "ways of seeing" (Dussel, 2012).

Acknowledging the idea, advanced by Escobar (2004), that we currently confront modern problems for which there are no modern solutions, the evolving narratives component acknowledges that we are faced with problems that transcend the problem-solving capabilities of modern culture alone. To move beyond the unsustainable and exploitative practices associated with modernity, one has to think and act from an exteriority to the dominant culture and structures (Dussel, 2012, 48–49). The "ways of seeing" element of the DSIS model aims to put into

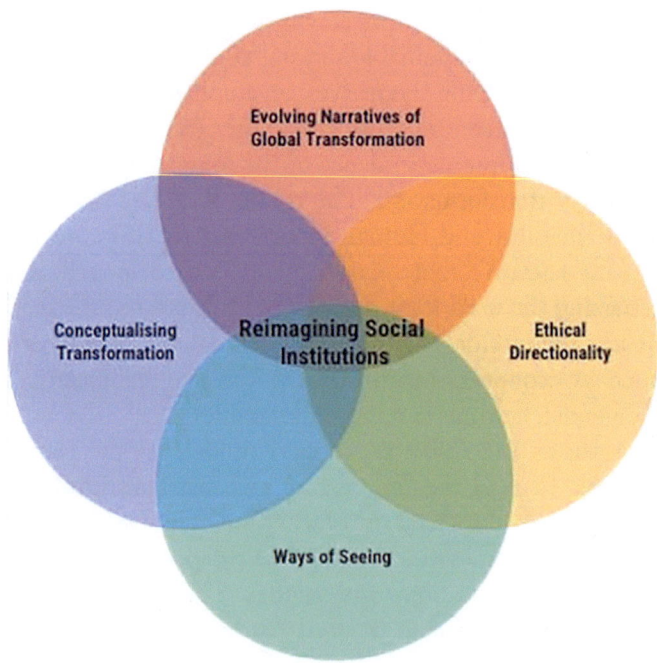

Fig. 1 The DSIS model for deep institutional and global transformation (*Source* Deep Societal Innovation for Sustainability and Human Flourishing (DSIS))

perspective the dominant epistemics by focusing on onto-epistemological plurality, transdisciplinarity, stakeholder engagement, and the inclusion of perspectives from the arts, music, and literature. In other words, the model stresses the importance of exterior and subaltern perspectives for a transmodern project that aims to transcend the unsustainable and hierarchic systematics of modernity, all the while preserving its positive moments and transforming them into pluralist, horizontal, and democratic cultures (Dussel, 2012, 48).

Thirdly, the model holds that "ethical directionality" is essential to ensure transformation towards sustainability and greater human flourishing. Inspired by Sen's (2001) capabilities approach, the DSIS model underscores the need for the new narratives to be grounded in a series of capabilities/values that are considered central for human flourishing—for example the democratic values of cooperation and equality. As we shall see, the case of the revolutionary assemblies shows that a radical democratic and pluralist politics entails the constant rebuttal of final objectives. Making sure to not reproduce the errors from the past, a pluralist and democratic society has to reject symbolic unification under a common denominator because, to use the words of the French philosopher Jean-Luc Nancy (1991, 4), "the association of community with a being that is already known precludes the becoming of new and as-yet unthought ways of being." The refusal to define a final *telos*, and close the totality, increases the relevance of common ethical values that, without imposing any fixed objective or destination, orient action in shared directions. Finally, the "conceptualising transformation" component of the DSIS model, focuses on the dynamics of transformation. Identifying qualities such as the speed, the scale and the tipping points in our search for transformative change.

3   The Case of the Revolutionary Assemblies in Chile (2019–2020)

On 18 October 2019, what started out as protests against the increase in price of a public transport ticket rocked a country accustomed to daily *choques* and earthquakes. Student-led evasions of the public transport fare in Santiago proved to be the spark that ignited a general, long dormant, discontent with a system and a political class that protestors traced back to the tyranny of general Pinochet and the authoritarian implementation of a neoliberal regime in Chile during the last decades of the twentieth century (Spyer Dulci & Sadivia, 2021, 45). For a timespan of several weeks, Santiago and beyond, formed the scene for unrelenting and violent protests and confrontations between citizens and the forces of the order. However, in the wake of the violence and the noise, people also started meeting on more quiet places, such as secluded squares or in university buildings that had been abandoned because of the student *tomas*—literally *taken*, referring to the buildings that were occupied by striking students. At these gatherings, neighbours came together to discuss the ongoing events, but also to organise basic services that were considered vital for the dignified life that they believed the neoliberal government had denied them. Depending on the assembly, services ranged from *ollas communes*, communal pots, a practice dating back to the dictatorship and in which food was collectively cooked and shared with those less fortunate, to small schools, health posts, and artistic activities such as theatres and musical performances (Han, 2012, 73).

Besides the many young students, who have a strong record of organisation and protest in Chile and whose movement formed the background of the protests, the participants at these meetings came from different contexts and backgrounds. They were brought together by their desire for change, and the conviction that, true transformation would have to be enacted there and then. For example, Rosa, a regular attendee of the weekly plenaries formulated it as follows:

> *"Participando es que si… me digo eso tenemos que hacer, esos huevones no van a hacer nada, es otra forma de hacer política"* - Participating means

that yes… I tell myself it is what we have to do, those assholes are not going to do anything, it's another way of doing politics. (Grumiau, 2021)

Or Maria who claims that:

"*La nueva constituciónno va a cambiar nada, aquí ya estamos construyendo y constituyendo*" - "The new constitution is not going to change anything, here we are already building and constituting." (Grumiau, 2021)

Echoing the words of *Andrès*, with which we started our intervention, the two excerpts here-above stress the importance of "another way of doing politics." For many participants of the assemblies, it is plain and clear that the government institutions cannot be relied upon when it comes to social and economic improvements, and that any significant changes will have to be created from the bottom-up. The gatherings and the activities organised in the streets are all about "*building and constituting in the here and now*" (Tomás, member of the asamblea, Grumiau, 2021), but what does that mean?

At the turn of the century, in an article titled "Beyond the Third World: imperial globality, global coloniality and anti-globalisation social movements," Arturo Escobar (2004) observed the emergence of social movement networks operating under a new logic, fostering forms of counter-hegemonic globalisation (Escobar). In the wake of what the New York times referred to as the first post-modern revolution, when the neo-Zapatistas revolted in Chiapas to put an end to the exploitation of the local communities—*Ya Basta!*—these grassroots organisations did not aim to conquer power or impose a fixed program, but rather to foster local autonomy and create democratic spaces for discussion (Graeber, 2013).

Escobar describes how, instead of working with hierarchical structures, these movements operate as self-organising meshworks relying largely on non-hierarchical structures that set them apart from the hierarchical institutions of the state. They tend to show emergent properties and complex adaptive behaviour that movements of the past, with their penchant for centralisation and hierarchy, were never able to manifest (Escobar, 2004, 210). While "hierarchies entail a high degree of centralised control, ranks, overt planning, homogenisation, goals and rules of behaviour, meshworks", Escobar (2004, 222) argues, "function like ants." They exhibit a form of "complex adaptive behaviour" in which many singular entities interact and combine to form complex entities without any need of central planning (Escobar, 2004, 222). Closer to forms of philosophical anarchism and anarcho-socialism the meshwork model of self-organisation presupposes bottom-up processes where agents interacting on lower scales produce behaviour, forms and regularities at higher levels (Escobar, 2004).

The first weeks of the assembly were marked by the need to *attender a la urgencia*—to deal with the urgency of the situation. In other words: one of the main reasons that there had been no time to discuss common objectives was that since the start of the protests, the barrio had suffered particularly under the repression of the state. The continued presence, and harassment, of the police had added itself to the previous insecurity—mostly related to drugs and alcohol consumption—experienced in the neighbourhood. During the first weeks, almost all the assembly's efforts were consequently absorbed by those more pressing issues. However, at moments like this, when there was time and space to define common objectives, the participants in the assembly almost invariably mentioned neighbourhood unity and the creation of social ties as the main objective of their being-there. In what can be considered as an alternative reformulation of the idea of starting by doing then thinking, participants stressed the social ties and networks, the action itself, as their main objective and refused to identify final objectives. For a majority, having the discussion about a common interest was more important than the definition of the interest itself. As if there was a silent, implicit agreement that the establishment of a unique and final objective for the assembly also entailed the possibility of compromising the inclusiveness and diversity that was cherished.

4   The Political Condition of the Human Being

As evidenced from earlier statements, what was at stake for the participants of the assembly was another way of doing politics. One that, contrary to the conventional politics of the state, concerned them at the very core of their being. To say it in the words of *Julio*, a professor in philosophy at one of Santiago's universities, whom I met at one of the Sunday plenaries:

> We are here to see what happened. We are here to try and understand why people have become so cold with each other. (Grumiau, 2021)

For many of those partaking in the plenaries and activities, the assembly formed a way to reconnect to the other inhabitants of their neighbourhood. In a neoliberal society that had grown "cold" due to an imaginary that stressed the importance of cold negotiations and individualised actors in a market-economy (Gibson-Graham, 2006), people felt disconnected and isolated. In the context of the protest movement, the assembly was seen an oasis in a desert landscape, where people could reconnect to fellow citizens through a "new way of doing politics."

Rosa, member of the assembly: "No conocer a la señora del almacén eso no se tiene que cambiar desde casa, hacer micro política porque eso si se puede cambiar." "Not knowing the lady of the corner-store, that does not have to be changed from home, but by doing micro politics because that can be changed" (Grumiau, 2021).

Andrès, member of the assembly: "Before all this, I knew none of the people here while some of them lived just around the corner. We lived two "*cuadras*" away but never saw each other. Now things are different" (quoted in Grumiau, 2021).

One of the main, and few, objectives that the assembly agreed upon was the reconstruction of social tissue—*la reconstrucción del tejido social*. The weekly meetings were seen as an opportunity to rebuild and reaffirm the social relations between the inhabitants of the same neighbourhood. This meant that from the point of view of the members, politics did not merely refer to the participation in the administration of society, of the goods and services considered necessary for—biological—survival but was first an essential element of a dignified and human life. In other words, like for example the conception of the ancient Greeks, politics was seen as an end to be pursued for itself, independent from the economics that managed biological survival (Arendt, 1958).

This resonates with how Hannah Arendt approaches the political. In her book on "the Human Condition," Arendt (1958) identified three fundamental human activities that each corresponds to the basic conditions under which life on earth has been given to the human being: labour, work, and action (Arendt, 7). Of the three activities, action is the only one that goes on directly between human beings without the intermediary of artificial things or matter. Action corresponds to the human condition of plurality, that is, to the fact that that human beings, plural, equal by their difference from each other, inhabit the world. For Arendt, this human plurality is the condition of political life, of the stage on which people can see and be seen, hear and be heard, by others. According to Arendt, any permanent change to these activities will profoundly transform the human condition. In other words, from Arendt's point of view, action—and consequently politics—defines the human being on an ontological level. Human beings exist in and through the networks of social relations that they weave with one another. For the members of the citizen assemblies, it seems that the political concerned first and foremost this seeing and being seen, hearing and being heard by other participants without which they would not feel human.

5   Ethical Orientations for Prefigurative Politics

Monticelli (2014, 3) emphasises how the notion of prefigurative politics focuses on alternative world making as a political process: "Unlike conventional or contentious politics, prefigurative politics focuses on the creation of alternative ontologies: alternative ways of being in the world and, one might even dare to say, 'alternative worlds'." If prefigurative politics is about identifying the mechanics and dynamics to bring about these envisioned futures, here and now, then our discussion of the case of the Chilean assemblies in combination with the framework of the DSIS model points towards several crucial takeaways.

First, there is a need to clarify a theoretical ambiguity that is characteristic of the narratives surrounding the political in the twenty-first century. As evidenced by Monticelli's definition above, the concept of the political hovers between a "conventional" idea of politics, in which the political is still very much associated to the state, and on the other hand, what seems to be a contemporary experience of the political that approaches it in a broader sense as characteristic of grassroots action and as of ontological importance to the human being. We argue that the idea of prefigurative politics falls under the latter as its focus on the processual and emergent elements of politics precedes the consolidation and order that come with more hierarchical forms of political organisation which limit the scope for a politics of novel possibility (Sareen & Juhola, 2025).

The assembly's decision to focus on the action in itself and the refusal to define common and final objectives should be considered off the back of these ideas. Jean-Luc Nancy (2009, 78–85), argues that insofar as democracy promotes and promises the freedom of all human beings, it engages human beings ontologically and is necessarily founded on the absence of a human nature. Consequently, as mentioned earlier, any attempt at a democratic politics of difference must be wary of forms of common being that compromise authentic being-in-common. For Nancy (2000, 47) "being with" lacks symbolic identification and any attempt to define the common being installs hierarchies and closes of the opportunity to cultivate open-ended ethical praxis (Gibson-Graham, 2006, 98).

By refusing to define a commonality, and focusing on the action and the relation itself, the participants of the assemblies maintained an openness to new possibilities in their praxis. Imagining and enacting alternative practices that transcend contemporary unsustainable and exploitative practices towards alternatives that foster human and non-human flourishing inevitably requires a respect for the indeterminacy that defines authentic common action. Counterhegemonic politics are based on the outlining of a positive ideal, but one that respects the indeterminacy and creativity of being in common by describing ethical directions rather than fixing them with ultimate, final and common objectives. In line with the DSIS model, we consequently emphasise the importance of ethical values for the orientation of the process of deep transformation and the significance of identifying the conditions and tipping points that foster human qualities that are favourable to sustainability and—collective—human flourishing.

Finally, Monticelli (2014) emphasises the ontological dimension of prefigurative politics and its embeddedness in the practices of worlding and the creation of alternative ontologies. The outline of an alternative approach of—prefigurative—politics as presented in this chapter, focuses on action as a defining element of the human condition. The example of the Chilean citizen assemblies shows how common action, and the network of interhuman social relations it presupposes, does not merely concern the administration of goods and services or the meeting of human social and economic rights, but touches on the very (in)essence of the human being. The participants of the assemblies did not feel human without the acknowledgement of the relationality of their lived reality, a tissue of social relations through which they could participate in the human play of seeing and being seen and hearing and being heard. Prefigurative politics thus possesses an existential dimension and concerns, beyond the human condition, also the ontologies of the—relational—worlds that human beings shape around them. For prefigurative politics to be truly democratic and pluralist, these ontologies must be founded on a constant rebuttal of any form of ultimate symbolic unification so to continue enabling the becoming of new and as-yet unthought ways of being that respect the plurality that defines our human condition.

Competing Interests   The authors have no conflicts of interest to declare that are relevant to the content of this chapter.

References

Arendt, H. (1958). *The human condition*. The University of Chicago Press.
Arendt, H. (2005). *The promise of politics*. Schocken Books.
Dussel, E. (2012). *Filosofía de la cultura y transmodernidad*. Editorial Docencia.

Escobar, A. (2004). Beyond the Third World: Imperial globality, global coloniality and anti-globalisation social movements. *Third World Quarterly, 25*(1), 207–230. https://doi.org/10.1080/0143659042000185417

Escobar, A. (2018). *Designs for the pluriverse: Radical interdependence, autonomy and the making of worlds.* Duke University Press.

Homer-Dixon, T., Renn, O., Rockstrom, J., Donges, J. F., & Janzwood, S. (2021). *A call for an international research program on the risk of a global polycrisis.* https://doi.org/10.2139/ssrn.4058592

Hughes, I., Byrne, E., Glatz-Schmallegger, M., Harris, C., Hynes, W., Keohane, K., & Gallachóir, B. Ó. (2021). Deep institutional innovation for sustainability and human development. *World Futures, 77*(5), 371–394. https://doi.org/10.1080/02604027.2021.1929013

Gibson-Graham, J. K. (2006). *Postcapitalist politics.* University of Minnesota Press.

Graeber, D. (2013). *The democracy project.* Penguin Books.

Grumiau, B. (2021). *Nos Hemos Escondido: The 21st century commons: Subaltern movements and social resistance* (Master Thesis). Université Libre de Bruxelles.

Han, C. (2012). *Life in debt: Times of care and violence in neoliberal Chile.* University of California Press.

Kanger, L., & Schot, J. (2019). Deep transitions: Theorizing the long-term patterns of socio-technical change. *Environmental Innovation and Societal Transitions, 32*, 7–21. https://doi.org/10.1016/j.eist.2018.07.006

Monticelli, L. (2014). *The future is now: An introduction to prefigurative politics.* Bristol University Press.

Morin, E., & Kern, A. B. (1999). *Homeland earth: A manifesto for the new millennium. Advances in systems theory, complexity, and the human sciences.* Hampton Press.

Nancy, J.-L. (1991). Of being in common. In The Miami Theory Collective (Ed.), *Community at loose ends* (pp. 1–12). University of Minnesota Press.

Nancy, J.-L. (2000). *Being singular plural.* Stanford University Press.

Nancy, J.-L. (2009). Démocratie finie et infinite. In G. Agamben, A. Badiou, D. Bensaïd, W. Brown, J.-L. Nancy, J. Rancière, K. Ross, & S. Zizek, *Démocratie dans quel état?* La fabrique éditions.

Rowson, J. (2023). *Prefixing the world: Why the polycrisis is a permacrisis, which is actually a metacrisis, which is not really a crisis at all.* Perspectiva. https://per specteeva.substack.com/p/prefixing-the-world. Last visited: 07 March 2025.

Sareen, S., & Juhola, S. (2025). The prefigurative politics of present transformation (this volume).

Sen, A. (2001). *Development as freedom*. Oxford University Press.
Spyer Dulci, T. M., & Sadivia, V. A. (2021). El estallido social en Chile: ¿Rumbo a un Nuevo Constitucionalismo? *Revista Katálysis, 24*(1), 43–52. https://doi.org/10.1590/1982-0259.2021.e73555

Open Access This chapter is licensed under the terms of the Creative Commons Attribution-NonCommercial-NoDerivatives 4.0 International License (http://creativecommons.org/licenses/by-nc-nd/4.0/), which permits any noncommercial use, sharing, distribution and reproduction in any medium or format, as long as you give appropriate credit to the original author(s) and the source, provide a link to the Creative Commons license and indicate if you modified the licensed material. You do not have permission under this license to share adapted material derived from this chapter or parts of it.

The images or other third party material in this chapter are included in the chapter's Creative Commons license, unless indicated otherwise in a credit line to the material. If material is not included in the chapter's Creative Commons license and your intended use is not permitted by statutory regulation or exceeds the permitted use, you will need to obtain permission directly from the copyright holder.

6

Prefigurative Politics of Post-growth Futures in the Making: Learning from Transformative Practices that Leverage (Temporary) Convivial Space

Steven R. McGreevy, Corelia Baibarac-Duignan, and Maximilian Spiegelberg

1   Introduction

Despite the dire need for wide-spread and deep transformations to move towards a post-growth society, our capacity to imagine radically different ways of organising our economies, governing ourselves, designing our settlements, and living sustainably is limited by conventional logics and path dependencies. Activating the collective imagination requires fostering capacities within a community to anticipate potential futures and empathising with the lives and experiences of others (Finn & Wylie, 2021). When public institutions, practitioners, researchers, and cultural workers activate space into convivial learning environments through transdisciplinary real-world experimentation, these can be used

S. R. McGreevy (✉) · C. Baibarac-Duignan
Department of Technology, Policy, and Society, University of Twente, Enschede, The Netherlands
e-mail: s.r.mcgreevy@utwente.nl

M. Spiegelberg
SDG+ Lab, University of Kassel, Kassel, Germany

© The Author(s) 2026   **81**
S. Sareen and S. Juhola (eds.), *Societal Transitions to Sustainability*, Palgrave Studies in Environmental Transformation, Transition and Accountability,
https://doi.org/10.1007/978-3-032-07395-2_6

to catalyse the collective imagination, host democratic debates about the future, and share local knowledge to impact change and root transformations in place (Hajer & Pelzer, 2018). An engaged and active citizenry with "critical, collaborative, and experimental skill-sets" and a public discussion and vision of what living well within limits actually means are essential components of a prefigurative post-growth transformation (Jackson, 2011; Steinberger et al., 2024, p. 4).

Space—particularly land—in cities is becoming increasingly threatened by developments that privatise or transform it into spaces for consumption (Miles, 2012), while "lingering" or appropriating public green space for food production are often perceived as undesirable and not as expressions of democratic reaction to the status quo. Rather than promoting meaningful social and ecological connections, such transformations reflect and enhance dominant capitalist logics that separate places of production from places of consumption, humans from nature, professional expertise from lay knowledge. However, in between sites of mainstream development lie glimpses into other ways of inhabiting public space, often through temporary uses of municipal land.

In this chapter, we will discuss examples of grassroots (community-driven) initiatives from three European contexts where land is transformed through everyday practices, like growing food, cooking, building, and repairing, into convivial spaces for learning, experiencing and shaping alternative, post-growth futures, see Fig. 1. Following Ivan Illich, we address conviviality as a form of "individual freedom realized in personal interdependence" (1973), centred on mutually enhancing relationships between humans and nature that counter the damaging effects of industrial productivity and mass consumption. Interviews with team members from each initiative are used to ground online documentation and literature review. We discuss the three initiatives focusing on the conviviality of the spaces where they are situated and the relationships they foster, reflecting on their prefigurative potential and on what is needed to realise it.

Fig. 1 Top: Foodpark Amsterdam protest over business park sign (*Source* Tamalone van den Eijnden). Left bottom: aerial photo of AgroCité Colombes (before demolition) (*Source* Corelia Baibarac-Duignan). Right bottom: the SDGLab public space in Kassel (*Source* Sascha Mannel)

2   Three Case Vignettes

2.1   Foodpark Amsterdam/Voedselpark Amsterdam

FPA is a grassroots alliance of various civic initiatives proposing an alternative vision of the Lutkemeerpolder area in the southwest edge of Amsterdam, which adjoins one of the city's poorest neighbourhoods, Nieuw West. In 2018, a campaign was launched for the preservation of the Lutkemeerpolder and against the construction of distribution centres on what is said to be the last piece of fertile soil of the city. This led to the emergence of FPA in 2022 as a citizen initiative to develop the area into an agro-ecological landscape park, with crowdfunding and support from various citizen-led organisations and, more recently, research collaborations. The aim of FPA is to protect a 42-hectare plot of agricultural land by reclaiming it as an agro-ecological commons through exploring alternative land ownership models (i.e. community land trust). The initiative emerged at a time when this plot of land was being rezoned by the Municipality of Amsterdam from agricultural to industrial use to build logistical buildings and warehouses contributing to the creation of Business Park Ostdorp.

Farming is still practised today on a fraction of the land in the form of the Boterbloem, a care farm, two community-supported agriculture (CSA) farms and a horticulture enterprise. The rest of the land is being prepared for the business park and a new storage building has already been under construction since 2024 at the edge of the CSAs, with potentially devastating consequences for the local biodiversity as well as the social relations supported through the care farm.

Foodpark Amsterdam and Business Park Osdorp are prefigurative visions of what could be on the land, projecting different conflicting values because of the paradigms and worldviews they encapsulate—one related to dominant capitalist values of land financialisation, and the other centred on commons, conviviality, and collective well-being. How the future of the polder is imagined is highly contested by the two groups of actors that promote the two visions, which emerge from distinct understandings of sustainable land use, social justice, and economic success. The two groups propose divergent institutional approaches to land governance while mobilising different values in attempting to establish legitimacy (Savini, 2025).

The vision of a food park managed as a commons has been put forward as an alternative plan for the area by the activists supporting it, which opposes capitalist and extractivist relationships with land (Foodpark Amsterdam, 2023). After months of on-again, off-again negotiations with the municipality, the activist group succeeded in winning the tender for an adjacent plot of land, where a food park can now be realised. While parcel will not encapsulate the entire land area of the original struggle, its proximity can serve as a living example of how transformative practices can leverage the networks and communities (human and non-human) they build through convivial everyday engagements.

2.2   R-Urban/AgroCité Paris

The R-Urban project was conceived as a bottom-up strategy for urban resilience based on a network of collective hubs run by residents as urban commons for living, producing, and consuming. The project was initiated by the activist architecture practice Atelier d'Architecture Autogérée (aaa) in collaboration with public and civic actors, including a local municipality. It was established in 2009 in the Northern Paris suburb Colombes, a multicultural municipality with low-income residents and accommodation consisting primarily of social housing tower blocks. The project was funded with a grant from the EU Life+ Programme of Environmental Governance and local and regional funds provided by the Municipality of Colombes which, at that time, was led by Social Democrats and Greens (Petrescu et al., 2016).

From the proposed pilot units, two were built on unused land in collaboration with the Municipality and local residents, starting in 2012: AgroCité—an urban agriculture unit consisting of a micro-experimental farm, community gardens, educational and cultural spaces and devices for energy production, composting, and rainwater recycling; and RecycLab, a repair cafe and green building constructed around equipment for the recycling of urban waste into materials for eco-construction. These units were originally planned with a 10–15-year time horizon and the goal to expand them into a network of civic hubs for local resilience. Nevertheless, after the 2014 municipal elections, a new conservative right-wing Municipal government took power, which pressured AgroCité to close and be replaced by a temporary car park as a precursor to an area-wide renovation project (including a public square, public facilities and housing) (Tribillon, 2015).

AgroCité continued until early 2017, supported by plans to move the unit to a different suburb in the Northwest of the city, Gennevilliers. Despite a short period of two months that would have been necessary to translocate the hub, including its architecture prize awarded building, the site was flattened by bulldozers through an unexpected eviction by the police met by residents' protest (Par, 2017). Due to the eviction, the community and possibilities for learning and being with each other that many of the women who volunteered at the hub could benefit from disappeared, together with the plants in the garden and the chickens at the farm. In 2018, AgroCité returned in an identical set-up at Gennevilliers, but recreating the feeling of belonging and the social bonds lost due to the relocation takes time and sustained joint efforts by the R-Urban team, volunteers, and organisations involved.

The AgroCité example in Colombes prefigured alternative (post-growth) values through everyday practices and activities centred on conviviality and community well-being, while struggles over land use were eventually brought to a halt through coercive actions (eviction) and re-immersed the land into the mainstream growth-oriented paradigm. Nevertheless, it illustrates an important attempt to propose and build an alternative, as well as cultivate a new ideology centred on post-growth ideas for using land, and convivial modes of living and working in cities based on collaborative ecological practices.

2.3   SDG+ Lab Kassel

SDG+ Lab, located in the industrial city of Kassel, Germany, is an ongoing project now in the second of a four-year lifespan (2023–2027). Kassel is mostly known for the city-wide Documenta art festival that takes place every five years, which most recently featured an exploration of alternative economic models shaped by collectivity, communal resource sharing, and equity under the concept and practice of *lumbung* (Documenta Fifteen, 2022). Documenta Fifteen embraced an approach to community involvement by sharing and renting spaces for public discussions of art and sustainability, such as the "future village", a temporary structure carried out and curated completely by civil society actors, many critical of dominant capitalist dynamics.

The SDG+ Lab project was inspired by the success of these community-based approaches and other future-oriented interventions (University of Kassel, 2022). Riding the positive public response, the City of Kassel and the University of Kassel partnered to trial a government funded effort to envision and transform Kassel into "the Sustainable

Valley". The SDG+ Lab is headquartered in a rented storefront in downtown Kassel with space for exhibitions, workshops, a cafe, and office for a core team to work on small projects and initiatives.

In 2023–2024, four real world labs were set up in neighbourhoods around the city, each one running for a two-month period. The approach allowed the neighbourhoods a high degree of agency—the Lab facilitated team building workshops, rented a space in the neighbourhood decided on by the participants, provided a budget (~15K Euros) with very few restrictions on use. The only stipulation was that the real-world lab do things connected with energy, the environment, and the SDGs in the broadest sense. A "Future Committee" composed of a council member, one Lab representative, and members of the local neighbourhood self-organised into working groups, initiated networking, composed rules on how to use the space, and began working in their community. Not all the real-world labs performed equally, but some were successful in triggering latent networks and projects that just needed a space, small fund, or the right connection to get initiated. For example, the appeal of a weekly night market selling fresh local food from regional farmers in the "post-office hours" was shared by many. The living lab brought organisational capacity, a direct link to the city to acquire necessary permits, and the network to find space to make it happen. In addition, with the help of the lab, a Cargo Bike sharing service in-limbo found the network and funding for bicycle garages and an open-sourced digital booking system enabled the sharing service to expand across the city. A total of 25 projects were initiated from the four living labs.

The 2025 "City & Country" theme explores ways to redevelop areas of Kassel that are suffering from neglect and economic decline through a multi-stakeholder pilot experiment to redesign a neighbourhood area. Like many cities, the area is already included in the municipality's traditional long-term planning process, but this is seen as a continuation of commercial 'development'. The experiment is designed to initiate a discussion of how to reimagine the closed-off space and provide a means to build a bridge between the two very different adjacent neighbourhoods. Although the project is ongoing, multiple challenges have emerged. Structures shall be built to allow for different use of the space and provide an alternative future vision that could be picked up by the planning process, but these must ultimately be temporary by design and a 'piloting fatigue" has set in among some civil society actors. Acquiring the right permit to initiate temporary structures is also difficult as a protocol for such an intervention doesn't exist despite the decades of Documenta experience. No city department is willing to take on responsibility for having given permission. Lastly, after years of austerity, public administrations are short on staff to handle the many challenges of the desired transition and the willingness to take on this 'extra-work' only comes from very motivated individuals or needs to be otherwise ordered from top-down.

There is little to no acknowledgement of post-growth values within official discourse of the project. Since the project is publicly funded, involving the city, university leadership, and regional company associations, explicit questioning of growth is just not an acceptable direction. That said, prefiguring a post-growth world does begin to surface from a closer look at the activities on ground and the involved civic actors' motivations. The open nature of the living labs allowed for diverse projects to take root in the community, weave in the experiences, skills and motivations of previous Documenta projects, and expand without an outlook for profit-making, prioritising conviviality instead.

3   Discussion and Conclusions

Erik Olin Wright's (2010) strategic dimensions of transformation are useful in thinking about the prefigurative power of the examples introduced in this chapter. *Symbiotic* transformational strategies aim to work within existing power structures through reformist policy and cooperation with existing institutions. *Interstitial* strategies develop in the cracks of the existing hegemony, operating outside of mainstream actor networks to prefigure desired futures. *Ruptural* strategies transform through confrontation and resistance to first break with existing structures and then create viable alternatives. To a certain degree, each of the examples introduced engages in strategies that align with the three dimensions described by Wright. Schmid (2021) further strengthens Wright's framework through the inclusion of spatial contexts to describe

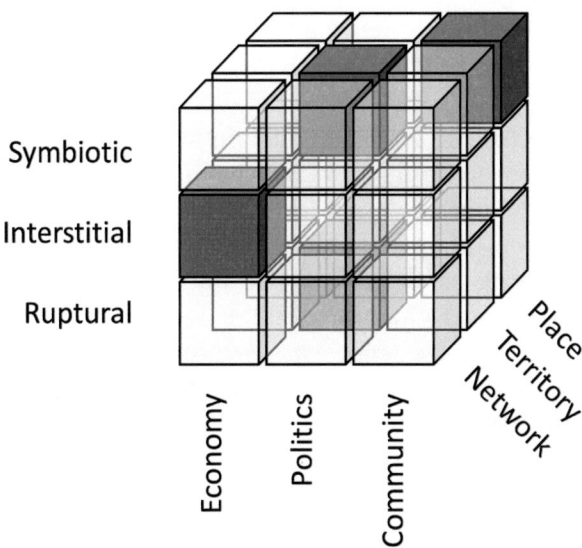

Symbiotic

Interstitial

Ruptural

Economy Politics Community

Place Territory Network

Fig. 2 Socio-spatial post-growth transformation strategies. Dark cubes indicate intersections of three aspects: transformational (left), social (bottom), and spatial (bottom right) (*Source* Schmid [2021, 77])

how post-growth transformations might emerge economically, politically, and socially in a given place (see Fig. 2).

Schmid identifies three spatially connected strategies for post-growth transformations that resonate with the experiences of the three examples above (2021).

3.1 Confronting Territorially Organised Power

Territorial power exercised by municipal governments played an outsized role in negatively influencing the transformative potential in each case. The power of the state to regulate land, whether public or private, through planning stipulations, zoning laws, or access to permission or permits dictated the long-term viability of each project. Since each of these projects is tied to a physical place, the jurisdiction of the state is an inescapable reality that severely limits their prefigurative power. Governments are preoccupied with the interests of capital and furthering the growth paradigm.

FPA used persistent symbiotic and ruptural strategies to approach territorial power. Years of negotiation with the municipality created local momentum, but a solution failed to materialise. To overcome this impasse, the momentum was utilised to create local volunteer groups engaged in sowing sunflowers, walks, and meetings, leading to the registration for tender (Kamsma, 2025). Despite distribution centres looming all around, the tender was granted and the food park can finally begin, albeit on a smaller scale. Supporting the efforts of the activists, Amsterdam's rising land prices also provided financial room for projects like the food park (Kamsma, 2025). R-Urban was forced to relocate after political winds shifted and development plans were put into action. While the project survived through translocation, the example demonstrates the power of the state to repattern interstitial projects according to the logic of capital accumulation and growth. Even though the SDG+ Lab operates as a municipality-sanctioned project, the activities and initiatives generated through its work run into bureaucratic pitfalls, such as adherence to planning and building regulations, dependency on permits and state aversion to take long-term responsibility. Of the three examples, the SDG+ Lab is, perhaps by default, a strong example of a symbiotic approach to transformation that has allowed post-growth values and shifts in everyday practices to emerge under the veneer of mainstream sustainability.

3.2 Post-growth Coalitions: The Place-Relatedness of Symbiotic Strategies

Place-based networks and new communities of practice emerged in each of the cases. As Schmid notes, "direct contacts, trust, personal relations and mutual knowledge are important resources for reflexive and emancipatory cooperation" (2021, 75) in symbiotic approaches to transformation. The networks in each case embraced a post-growth ethic through sharing of land & space, food & recipes, knowledge & skills, or bicycles. However, the quality and embeddedness of the coalitions formed varied and contributes to how effective they prefigure futures.

FPA was born from a strong network of organisations and experts interested in the emancipatory project that FPA represented. FPA also formed ties with political parties, municipal actors, and the local neighbourhood. Working on the land also re-established convivial human-non-human relationships. The persistence and solidarity of FPA networks, operating in policy and societal circles, eventually led to the realisation of the project. R-Urban, originally situated in the heart of the neighbourhood, is composed of strong bottom-up networks that include the neighbourhood and activists. Local, traditional, and context-specific knowledge is not only valorised in the project, but also shared through workshops and collective work such as gardening, cooking, repairing, and building. While these networks were forced to uproot and relocate, they demonstrated the prefigurative power of collective action. The SDG+ Lab enabled the activation of existing or latent neighbourhood networks to realise their projects. The creation of new networks through the experiment planned this year remains to be seen.

3.3   Networked Interstitial Strategies for an Economic Transformation

The creation or reestablishment of local economic circuits and relationships to replace the exploitative, profit-hungry economic systems that dominate today was an overt priority for two of the three cases, and more subtle in the third case. In most instances, these circuits emerge and take root in interstitial spaces to increase autonomy, maintain wealth circulation locally, and limit extraction and unfair practices.

The land where FPA was originally envisioned does generate some food as farms and small holders grow locally distributed produce. If the vision of FPA is realised on the adjacent plot for which the tender has been won, the potential to restructure the local food system in Amsterdam is tremendous. At the same time, winning the tender can spark further initiatives in Amsterdam and beyond showing that alternatives are possible, even within a dominant capitalist paradigm. R-Urban prototyped and demonstrated ways in which multiple provisioning systems might be made more autonomous, local, and sustainable. Alternative ways to use building materials and construction, gardening and livestock farming, sustainable energy production, water management and waste circulation were learned put into practice by participants, inspiring new initiatives elsewhere (e.g. Loughborough Junction gardens/LJWorks, in London). Multiple initiatives realised via the SDG+ Lab also strived to create post-growth economic relationships within their communities: clothing repair, citizen platform on solar energy, tinkering with recycled materials, community cleaning services, and mobile kitchens.

3.4   Reflections on Prefiguring for Post-growth Transformations

Prefiguring interstitial space for post-growth transformations cannot ignore the power of the state in determining how land is used or owned. Alternative forms of communal or multi-owner land ownership and property regimes offer a way to surmount territorial power, but the examples in this chapter also point to potential politically or capital-driven difficulties (Calo et al., 2021). Beyond ties with land, the politics of commoning systems of provisioning is also made difficult by existing governance that tends to support capital over community interests, despite efforts to show the value of commons in generating resilient urban futures (Petrescu et al., 2021).

Politics are also at play in place-based networks and communities of practice, as oftentimes experts or outside groups bringing energy and post-growth ideas fail to navigate the nitty gritty of local, neighbourhood politics. Local champions or "residential researchers" might be able to bridge these two groups, otherwise the process of building trust and a common understanding can take time. We also see the difficulty of "selling" what are seen as radical post-growth visions to local communities, which points to the need for co-developing context specific visions, ultimately increasing local actors' sense of ownership of the process and possible outcomes.

Each case worked with temporary or pop-up spaces, buildings, or ways of engagement to prefigure post-growth futures in practice or as tools for discussion. Collaborative experiments to redesign spaces, such as the one now underway in Kassel, or the pop-up feel of R-Urban allow people an experience of possible futures in a way that abstract visioning cannot (Hopkins, 2020; McGreevy et al., 2022). However, with the reality of limited windows of funding and the difficulty of navigating the regulatory environment, the prefigurative power of these temporary initiatives can be limited. Like interstitial projects operating in the cracks of society, legitimising, maintaining, and making permanent these temporary spaces becomes a critical challenge. Thinking beyond the "local", without falling into growth paradigms of scaling, could be one way to strengthen the prefigurative potential of these initiatives and mobilise them towards wider transformations. For instance, commoning alternative initiatives, particularly the knowledge, imaginaries and practices they encapsulate, as part of wider networks and through communities of practice in open-source ways remains an underexplored possibility for change (Baibarac & Petrescu, 2017). Yet, connecting diverse and locally situated initiatives as part of a coherent system capable of fostering radical transformations would have to consider potential frictions that may occur in the process, mobilising them as levers for change, while preserving and enhancing local ownership and autonomy. At the same time, the temporary nature of these initiatives calls for approaches that can "mediate" their value to act as experiential futures for alternative possibilities in the present and reshape collective imaginaries (Baibarac-Duignan & Medesan 2023).

Competing Interests The author has no conflicts of interest to declare that are relevant to the content of this chapter.

References

Baibarac, C., & Petrescu, D. (2017). Open-source resilience: A connected commons-based proposition for urban transformation. *Procedia Engineering, 198*, 227–239. https://doi.org/10.1016/j.proeng.2017.07.157

Baibarac-Duignan, C., & Medesan, S. (2023). 'Gluing' alternative imaginaries of sustainable urban futures: When commoning and design met in the post-socialist neighbourhood of Mănăştur, Romania. *Futures, 153*, 1–19. https://doi.org/10.1016/j.futures.2023.103233

Calo, A., McKee, A., Perrin, C., Gasselin, P., McGreevy, S., Sippel, S. R., Desmarais, A. A., Shields, K., Baysse-Lainé, A., Magnan, A., Beingessner, N., & Kobayashi, M. (2021). Achieving food system resilience requires challenging dominant land property regimes. *Frontiers in Sustainable Food Systems, 5*, 683544. https://doi.org/10.3389/fsufs.2021.683544

Documenta Fifteen. (2022). Homepage. https://www.documenta.de/de/retros pective/documenta_fifteen. Accessed March 6, 2025.

Finn, E., & Wylie, R. (2021). Collaborative imagination: A methodological approach. *Futures, 132*, 102788. https://doi.org/10.1016/j.futures.2021.102788

Foodpark Amsterdam. (2023). *Harvest for the future—FPA site plan for Lutke-meerpolder.* https://voedselparkamsterdam.nl/wp-content/uploads/2025/01/Bodemplan-Lutkemeer-mei2023.pdf. Accessed March 6, 2025.

Hajer, M., & Pelzer, P. (2018). 2050—An energetic Odyssey: Understanding 'techniques of futuring' in the transition towards renewable energy. *Energy Research & Social Science, 44*, 222–231. https://doi.org/10.1016/j.erss.2018.01.013

Hopkins, R. (2020). *From what is to what if: Unleashing the power of imagination to create the future we want.* Chelsea Green Publishing.

Illich, I. (1973). *Tools for conviviality.* Harper & Row.

Jackson, T. (2011). *Prosperity without growth: Economics for a finite planet.* Edward Elgar.

Kamsma, M. (2025, May 26). *How green wins over boxes* (Hoe het groen wint van de dozen). NRC. https://www.nrc.nl/nieuws/2025/05/26/hoe-het-groen-wint-van-de-dozen-a4894676

McGreevy, S. R., Rupprecht, C. D., Tamura, N., Ota, K., Kobayashi, M., & Spiegelberg, M. (2022). Learning, playing, and experimenting with critical food futures. *Frontiers in Sustainable Food Systems, 6*, 909259. https://doi.org/10.3389/fsufs.2022.909259

Miles, S. (2012). Spaces for consumption: Pleasure and placelessness in the post-industrial city. SAGE Publications Ltd.

Par. (2017, February 20). Agrocité forcibly relocated from Colombes to Gennevilliers (L'Agrocité déménagée de force de Colombes vers Gennevilliers). *Le Parisien.* https://www.leparisien.fr/hauts-de-seine-92/colombes-92700/l-agrocite-demontee-de-force-a-colombes-20-02-2017-6695853.php

Petrescu, D., Petcou, C., & Baibarac, C. (2016). Co-producing commons-based resilience: Lessons from R-Urban. *Building Research & Information, 44*(7), 717–736.

Petrescu, D., Petcou, C., Safari, M., & Gibson, K. (2021). Calculating the value of the commons: Generating resilient urban futures. *Environmental Policy and Governance, 31,* 159–174. https://doi.org/10.1002/eet.1890

Savini, F. (2025). Degrowth as ideology: Making values for the soil of Amsterdam. *Environmental Values.* https://doi.org/10.1177/096327192513 18139

Schmid, B. (2021). Spatial strategies for a post-growth transformation. In B. Lange, M. Hülz, B. Schmid, C. Schulz (Eds.), *Post-growth geographies—Spatial relations of diverse and alternative economies* (pp. 61–84). Transcript.

Steinberger, J., Guerin, G., Hofferberth, E., & Pirgmaier, E. (2024). Democratizing provisioning systems: A prerequisite for living well within limits. *Sustainability: Science, Practice and Policy, 20*(1), 2401186. https://doi.org/10.1080/15487733.2024.2401186

Tribillon, J. (2015, September 11). Why is a Paris suburb scrapping an urban farm to build a car park? *The Guardian.* https://www.theguardian.com/cities/2015/sep/11/paris-un-climate-conference-colombes-r-urban-urban-farm-car-park

University of Kassel. (2022, August). *Knowledge repository—100 ideas for the world of tomorrow* (Wissensspeicher- 100 Ideen für die Welt von morgen). https://www.uni-kassel.de/einrichtung/ukt/wissenschaftsdialog/wissensspeicher. Accessed March 6, 2025.

Wright, E. O. (2010). *Envisioning real utopias.* Verso Books.

Open Access This chapter is licensed under the terms of the Creative Commons Attribution-NonCommercial-NoDerivatives 4.0 International License (http://creativecommons.org/licenses/by-nc-nd/4.0/), which permits any noncommercial use, sharing, distribution and reproduction in any medium or format, as long as you give appropriate credit to the original author(s) and the source, provide a link to the Creative Commons license and indicate if you modified the licensed material. You do not have permission under this license to share adapted material derived from this chapter or parts of it.

The images or other third party material in this chapter are included in the chapter's Creative Commons license, unless indicated otherwise in a credit line to the material. If material is not included in the chapter's Creative Commons license and your intended use is not permitted by statutory regulation or exceeds the permitted use, you will need to obtain permission directly from the copyright holder.

Part II
Spaces of Conviviality and Politics

7

Broadening the Understanding of Deconstruction in Prefigurative Social Spaces

Giuseppe Feola

1 Introduction: Constructive and Deconstructive Processes in Prefiguration

Prefigurative grassroots initiatives are forms of collective action that attempt to create alternative systems of provision and sociomaterial realities in the present while aligning means such as horizontal organisation, deep democracy, social inclusion, and ends such as social justice, ecological sustainability, autonomy, dignity, and sovereignty. Hence, they are often seen as post-capitalist 'nowtopias' and alternatives to capitalism (Monticelli, 2021, 2022; Wright, 2013).

A longer version of this essay appeared in the Journal Sociologica (ISSN 1971–8853) at https://doi.org/10.60923/issn.1971-8853/19919.

G. Feola (✉)
Copernicus Institute of Sustainable Development, Utrecht University, Utrecht, The Netherlands
e-mail: g.feola@uu.nl

© The Author(s) 2026
S. Sareen and S. Juhola (eds.), *Societal Transitions to Sustainability*, Palgrave Studies in Environmental Transformation, Transition and Accountability, https://doi.org/10.1007/978-3-032-07395-2_7

Evidence of the impacts of prefigurative grassroots initiatives abounds and includes the structural reduction of the environmental impact of systems of provisions and consumption practices, the empowerment and socialisation of new subjects to postcapitalist logics, and the creation of political support for deeply ecological ways to address unsustainability (e.g. Frantzeskaki et al., 2016; Henfrey et al., 2023). Most importantly, in the context of sustainability transformation, they also realise social change forms that are 'deep' (e.g. Goncalves, 2025; Henfrey et al., 2023; Schiller-Merkens, 2022) and address the cultural, social and political foundations of capitalist societies and their unsustainable and unjust development model (e.g. Brand & Wissen, 2021; Fraser, 2021).

It is crucial to recognise and explain how grassroots initiatives prefigure alternatives to capitalism. However, it also remains unclear how they concretely undo capitalism, including its structural imperative to grow, and how they unsettle its imaginary (Kallis, 2018; Kallis et al., 2012; Monticelli, 2022).

Two issues remain unresolved and are specifically addressed in this chapter. First, the risk of reproducing an 'innovation bias' exists (Davidson, 2019; Schmid & Taylor Aiken, 2023), as accounts of prefiguration have generally foregrounded constructive, creative, and future-oriented processes (Dinerstein, 2015; Gibson-Graham, 2008; Sales, 2023; Yates, 2015). The innovation bias assumes that the displacement of extant configurations is somehow the automatic consequence of adding socially, technically, or culturally innovative alternatives. Furthermore, this innovation bias implies the intrinsic utility of novel practices, technologies, and experiments while de-emphasising and losing sight of social, socio-ecological and socio-technical relations already in place, as well as past and present forms of solidarity, trust, respect, and cooperation on which transformative change may rest through renewal and re-assemblage (Schmid & Taylor Aiken, 2023; Shove, 2012).

An affirmationist bias is a second risk of accounts of prefiguration. Rather than realising an alternative to capitalism, prefiguration may reproduce capitalism's affirmative logic—and the ensuing imperative of self-valorisation (Dekeyser & Jellis, 2021). From this standpoint, the emancipatory nature of alternatives should reside in disengagement and the withdrawal from capitalism, more than in the affirmation of alternative arrangements (Pellizzoni, 2021). However, it is impossible to think of full withdrawal from capitalism of prefigurative grassroots initiatives that exist 'within' it (Chatterton & Pickerill, 2010; Holloway, 2010). Hence, prefiguration is likely predicated on disabling and weakening rather than severing the cultural and sociomaterial connection to capitalism.

As the discussions around the above-mentioned biases show, deconstructive processes remain debated and undertheorised in the literature on prefiguration (also see Taylor-Aiken, 2025). This is, thus, the question guiding this chapter: how to conceptualise deconstructive processes involved in prefiguration? Taking an interdisciplinary perspective, the objective of this chapter is to advance a refined understanding of deconstructive processes and entanglement with constructive processes in prefigurative grassroots initiatives.

Intervening in the above-mentioned debates on prefiguration, I contend, first, that the genesis of alternatives to capitalism through the social spaces produced by prefigurative grassroots initiatives is inherently entangled with deconstructive processes of cultural differentiation from capitalism. In other words, there is a range of deconstructive processes that disable or weaken relations with capitalist cultural and sociomaterial configurations. My second contention, then, is that to fully comprehend the differentiation from capitalism in prefigurative grassroots initiatives, scholars should build upon, but also move beyond, the concept of refusal as elaborated in the extant prefiguration literature. The conceptual repertoire should encompass unlearning, sacrifice, everyday resistance, decolonisation of the imaginary and defamiliarisation. These are deconstructive processes in the everyday and often covert experience of individuals and collectives prefiguring alternatives to capitalism in concrete places. Third, I contend that such deconstructive processes are generative; they may be preconditions to prefiguration. Underpinning this thinking is a spatial conceptualisation of prefigurative grassroots initiatives that sees them as the production of social space: condensation of sociomaterial relations that occur through processes of cultural differentiation and proximation.

In the remainder of this chapter, I first introduce a spatial under-standing of prefigurative grassroots initiatives as social spaces. In the two subsequent sections, I first discuss the conceptualisation of refusal, which is the predominant understanding of deconstruction in the prefiguration literature, and bring in the unmaking capitalism perspective, allowing me to show how refusal is but one type of process among many ones engaged with by individuals and collectives prefiguring alternatives to capitalism. I then conclude the chapter with some reflections on future research.

2   Prefigurative Grassroots Initiatives as Social Spaces

Prefigurative grassroots initiatives are social formations defined by a multiplicity of social relations bundled so that they are interpreted in everyday life as a unity. They can be insightfully understood as social spaces formed by the assemblage of relations endowed with meaning between individuals and groups, human and non-human, and mate-rial and immaterial cultural elements, including cognitive structures (i.e. symbolic spaces) (Bourdieu, 1985, 1996).

By reading prefigurative grassroots initiatives spatially, we can under-stand them as relatively permanent condensations of relations produced through coexisting processes of (i) differentiation, which involves distancing and dissociation from relations with the material and symbolic elements of the undesired social, cultural, and economic system, and (ii) proximation, which involves the (re)association and tightening of relations with the elements composing the grassroots initiative. In the differentiation lies the potentiality of unity and purposeful mobilisa-tion for common objectives (Bourdieu, 1996), while proximation rests on constructing and activating relations that constitute the social space (Melucci, 1995).

This reading underscores the processual and relational nature of prefig-urative grassroots initiatives as social spaces. Social space is reproduced, constituted, and performed through social action; it does not pre-exist. The identity of such social formation is, following Melucci (1995), not predetermined nor inscribed in some structural (e.g. socio-economic) characteristics of the actors but results from a process of progressive crystallisation since '[c]ollective identity tends to coincide with conscious processes of organisation' (51).

Thinking further with hooks (1989) and Holloway (2010), I understand prefigurative grassroots initiatives as spaces within the capitalist whole but at its margins. The marginal position is chosen by those occupying it rather than imposed by the 'centre' through marginalisation and exclusion, although this may also be true. This 'breaking out' (Chatterton, 2016) allows for more degrees of freedom to create alternative sociomaterial configurations for 'radical openness' (hooks 1989). The marginal position is reflected in an assemblage of relations that weakens—but never fully severs—relations between the grassroots social formation and the broader capitalist system in which it is embedded, even if that initiative exists in some measure 'against' and 'beyond' that system (Chatterton & Pickerill, 2010; Holloway, 2010).

Thus, prefigurative grassroots initiatives can be understood as social spaces extracted from capitalism (Escobar, 2008), which are defined by relatively stable condensations of relations infused with alternative or post-capitalist logics and values. However, they configure no stoppage of or escape from capitalism since they are impossible for social formations that inevitably exist within capitalism. Grassroots initiatives have porous boundaries and face movement across social space (e.g. new members bringing values, artefacts, ideas, and practices from other social spaces), engaging actively in brokerage (e.g. accessing public funding or support), while also defending their space (i.e. guardianship) (Guerrero Lara et al., 2024; Liu, 2021). Prefigurative grassroots initiatives are thus spaces that are produced by differentiation and proximation, but an overlap between such initiatives and spaces of capitalism remains.

3 The Role of Refusal in Prefiguration

Accounts of prefiguration have generally foregrounded constructive, creative, and future-oriented processes. However, various scholars have discussed deconstructive processes, specifically in terms of refusal in the context of prefiguration (Dinerstein, 2012; Graeber, 2004, 2013; Sales,

2023). Refusal is the deliberate rejection of an imposed and taken-for-granted definition of a situation, subjectivity and/or social relation. Refusal is seen as a political act involving concrete everyday distancing practices, such as a withdrawal from work and other capitalist relations. While prefigurative grassroots initiatives may also engage in political strategies of protest and contestation alongside prefiguration (Evans, 2021; Monticelli, 2021; Schiller-Merkens, 2022), here refusal is not understood as a political strategy in addition to prefiguration, but as one of its inherent components, and possibly a precondition.

Much of the prefiguration scholarship supports a generative understanding of refusal. For example, Sales (2023) discussed the distinction between refusal and resistance (or opposition), arguing that the 'productive features of refusal are (a) the recognition of the historically transitory nature of the current state of affairs; (b) the acceptance of the dialectical and unfinished status of the current reality; (c) the invitation to move beyond what is possible in a situated historical moment' (Sales, 2023, 50). Similarly, Dinerstein (2012) highlighted the generative connection between refusal and hope. In her view, refusal is a conscious rejection of identity within capitalism; it establishes the possibility of disagreement and, hence, of politics and political subjects able to create new realities. One of the advantages of Dinerstein's framework is its ability to overcome the risk of understanding refusal as a single denunciatory moment rather than as an ongoing process of differentiation contributing to the production of social space.

Such generative understanding of refusal resonates with Graeber's discussion of refusal concerning autonomous and egalitarian societies. Graeber contended that cultures 'in their origins and to a large extent in their maintenance [can be] self-conscious political projects' (Graeber, 2013, 2). He argued that 'cultures are not just conceptions of what the world is like, not just ways of being and acting in the world, but active political projects which often operate by the explicit rejection of other ones' (Graeber, 2013, 1). Crucially, Graeber posited that differentiation between social formations can be generative of—and intrinsically entangled with—sociomaterial configurations and cultural norms within such formations. The 'conscious rejection of certain forms of overarching political power which also causes people to rethink and reorganize the way they deal with one another on an everyday basis' (Graeber, 2004, 56). Thus, while differentiation is a multidimensional, unfinished, and ambivalent project, this perspective prompts us to appreciate how the internal structuring of new cultural and sociomaterial configurations can find continued motivation in a political project of cultural differentiation (Wengrow & Graeber, 2018).

On the other hand, the empirical evidence suggests that deconstructive processes may not be fully or accurately captured by such notion of refusal. The empirical evidence shows that also lack of action (silence) (Kanngieser & Beuret, 2017), cognitive and material reconfiguration of social practices (Brunori et al., 2012), and embodied experiences of different organisational forms (Ehrnström-Fuentes & Biese, 2022), among other manifestations, may contribute to the diversity of processes at play when prefigurative grassroots initiatives deliberately differentiate themselves from capitalism. For instance, studying solidarity-based purchasing groups in Italy Brunori et al. (2012) revealed how for 'a great part of their daily life, consumers involved [in alternative food networks] live in the same relational context as conventional consumers. The shift from the second category to the first one has, therefore, to be understood as a process of building new networks, *detachment from old networks* and attachment to new ones, and of the creation and *destruction of coherence between sub-spheres of daily life*. [...] In these new networks, changing consumption patterns rest on the change of patterns of relations, the adoption of new rules and *breaking down of old ones*, the use of new artefacts and the *abandonment of old ones*' (4–21, emphasis added).

In sum, considering how refusal has been developed in part of the prefiguration scholarship helps to conceptualise it as a generative process and, to an extent, overcome the risk of understanding refusal as a single act. Nonetheless, the scope remains to articulate deconstructive micro-level processes comprising refusal, but at the same time being equipped to grasp the diversity of processes involved in cultural differentiation as further discussed in the following section.

4   Broadening the Understanding of Deconstructive Processes in Prefiguration: Beyond Refusal

The unmaking perspective provides an analytical basis for understanding the range of interconnected, multi-level, and multidimensional processes that make space for radical alternatives to capitalist modernity (Feola, 2019). These processes include unlearning, sacrifice, everyday resistance, refusal, decolonisation of the imaginary, and defamiliarisation (Feola et al., 2021). Unmaking posits that deconstructive processes may be conditions for, rather than consequences of, sustainability transformation beyond capitalism. In other words, sustainability transformation may not come about through the mere addition of supposed solutions, values, or social imperatives but by deliberately subtracting problematic existing institutions, forms of knowledge, practices, imaginaries, power structures, and human/non-human relations at the outset (Feola, 2019; Feola et al., 2021).

The unmaking perspective hinges on an understanding of deconstruction as historical and situated (Feola et al., 2021; Raj et al., 2024). It is informed by a decolonial and culturally anthropological tradition of thinking about refusal (e.g. McGranahan, 2016; Simpson, 2016), which helps to understand the generative role of deconstructive processes: deliberately refusing capitalist relations reaffirms the primacy of alternative attachments, connections, and shared goals (McGranahan, 2016). Deconstructive processes can be an affirmative investment of hopes and energies in another possibility (Simpson, 2016), thereby enabling the imagination and prefiguration of different futures (Feola, 2019).

This perspective suggests that deconstruction can occur through public actions such as civil disobedience and protests, as well as the development of disruptive public discourse (Feola et al., 2021). However, these processes are often covert because they are bound to the private sphere of everyday life and undermine established and socially accepted order, including institutionalised social expectations, institutions, cultural models, and material infrastructure.

The unmaking perspective also frames social transformations as personal shifts in being, emphasising the importance of approaching deconstruction as a concrete personal experience (Chatterton & Pickerill, 2010). In processes of cultural differentiation from capitalism, cultural models are at stake alongside more mundane, material, and often tentative and inconsistent reconfigurations of everyday life (Raj et al., 2024; van Oers et al., 2023). What is more, deconstructive processes should not be seen as endpoints but as means inscribed in the performance of transformation, which may or may not result in the unmaking of those configurations.

This perspective has informed various studies of prefigurative grassroots initiatives. For example, Feola et al. (2021) examined the processes of construction of a peasant economy based on the principles of solidarity, justice, dignity, a holistic view of human and non-human life, collective participation, autonomy, and sovereignty in Colombia. The study revealed that such an alternative peasant economy was enabled by the rejection of imposed and taken-for-granted peasant identities and imaginary significations, the abstention from using undignifying but routinised and interiorised language, the withdrawal from exploitative market relations, and the expulsion of destructive mining enterprises from the territory.

In a study of work relations in community-supported agriculture in Portugal, Raj et al. (2024) provided evidence of collective and individual processes of refusal, unlearning, sacrifice, everyday resistance, and defamiliarisation enabling the transformation towards non-alienated, non-monetised, and full-of-care work relations. In another study of strategies to de-commodification, de-instrumentalisation and de-monetisation of labour in community-supported agriculture initiatives, Rossi et al. (2024) identified collective attempts at unmaking capitalism in the refusal of market logics and labour exploitation, the challenging of economic valuation of labour and food, and the attempt to place volunteer work outside the constraining scope of the legal system by framing it as a recreational activity. In yet another study, van Oers et al. (2023) detailed the processes of unlearning payment routines, and more importantly, valuation and collective responsibilities, which were activated by the farmers in two Dutch community-supported agriculture farms that implemented solidarity payment schemes.

In sum, the unmaking perspective complements the perspectives on refusal in prefiguration sketched earlier with a proposal that provides a more fine-grained theoretical entry point into (i) the everyday experience of deconstructive at the micro level, (ii) a broader and more diverse range of processes that may be activated beside refusal, and (iii) the often-covert strategies and diverse political grammar of activism involved.

5   Conclusions and Outlook

This chapter renews the case for a perspective that attends to constructive and deconstructive processes in prefigurative grassroots initiatives. It casts new light on the deconstructive processes involved in constructing alternatives to capitalism in and through prefigurative grassroots initiatives. It does so by contending that deconstructive processes, which include but exceed refusal as discussed in the literature, are entangled with constructive ones, as the former may be generative of the latter.

While alternative or post-capitalist logics provide a common grammar and inspiration to prefigurative praxis in grassroots initiatives, the propulsion for sustainability transformation may be found more in a political project of conscious cultural differentiation from capitalism than in the allure of a precisely envisioned post-capitalist future. By calling attention to the political pursuit of cultural differentiation and the construction of alternatives it engenders, we can produce refined empirical analyses and compelling theoretical accounts of the processes through which such initiatives can transform modern capitalist socio-material configurations in concrete places. This perspective avoids the innovation bias and addresses the affirmationist bias by taking a nuanced perspective on the generative functions of deconstruction.

Therefore, future research may apply and test the theoretical perspective proposed in this chapter to build further empirical evidence on a broad range of deconstructive processes including, but not limited to refusal, across different types of prefigurative grassroots initiatives in distinct contexts.

6 Competing Interests

The author has no conflicts of interest to declare that are relevant to the content of this chapter.

Acknowledgements I gratefully acknowledge funding from the European Research Council (Grant 802441) and the Dutch Research Council (Grant 016.Vidi.185.173).

References

Bourdieu, P. (1985). The social space and the genesis of groups. *Theory and Society, 14*, 723–744. https://doi.org/10.1177/053901885024002001

Bourdieu, P. (1996). *Physical space, social space and habitus*. University of Oslo.

Brand, U., & Wissen, M. (2021). *The imperial mode of living: Everyday life and the ecological crisis of capitalism*. Verso.

Brunori, G., Rossi, A., & Guidi, F. (2012). On the new social relations around and beyond food. Analysing consumers' roles and actions. *Sociologia Ruralis, 52*, 1–30. https://doi.org/10.1111/j.1467-9523.2011.00552.x

Chatterton, P. (2016). The Rocky road of post-capitalist grassroots experimentation. In M. Dastbaz & C. Gorse (Eds.), *Sustainable ecological engineering design: Selected Proceedings from the International Conference of Sustainable Ecological Engineering Design for Society (SEEDS)* (pp. 31–44). Springer. https://doi.org/10.1007/978-3-319-32646-7

Chatterton, P., & Pickerill, J. (2010). Everyday activism and transitions towards post-capitalist worlds. *Transactions of the Institute of British Geographers, 35*(4), 475–490. https://doi.org/10.1111/j.1475-5661.2010.00396.x

Davidson, D. J. (2019). Exnovating for a renewable energy transition. *Nature Energy, 4*, 254–256. https://doi.org/10.1038/s41560-019-0369-3

Dekeyser, T., & Jellis, T. (2021). Besides affirmationism? On geography and negativity. *Area, 53*, 318–325. https://doi.org/10.1111/area.12684

Dinerstein, A. C. (2012). Interstitial revolution: On the explosive fusion of negativity and hope. *Capital and Class, 36*, 521–540. https://doi.org/10.1177/0309816812461062

Dinerstein, A. C. (2015). *The politics of autonomy in Latin America: The art of organising hope*. Palgrave Macmillan.

Ehrnström-Fuentes, M., & Biese, I. (2022). The act of (de/re)growing: Prefiguring alternative organisational landscapes of socio-ecological transformations. *Human Relations, 76* (11), 1739–1766. https://doi.org/10.1177/00187267221112241

Escobar, A. (2008). *Territories of difference: Place, movements, life, redes.* Duke University Press.

Evans, H. (2021). Beyond resistance: the role of prefiguration in social movements addressing the climate crisis. *Bath papers in international development and wellbeing.* University of Bath. https://hdl.handle.net/10419/263278

Feola, G. (2019). Degrowth and the unmaking of capitalism: Beyond 'decolonisation of the imaginary'. *ACME: An International Journal for Critical Geographies, 18,* 977–997. https://acme-journal.org/index.php/acme/article/view/1790

Feola, G., Vincent, O., & Moore, D. (2021). (Un)making in sustainability transformation beyond capitalism. *Global Environmental Change, 69,* Article 102290. https://doi.org/10.1016/j.gloenvcha.2021.102290

Frantzeskaki, N., Dumitru, A., Anguelovski, I., Avelino, F., Bach, M., Best, B., Binder, C., Barnes, J., Carrus, G., Egermann, M., Haxeltine, A., Moore, M.-L., Mira, R. G., Loorbach, D., Uzzell, D., Omann, I., Olsson, P., Silvestri, G., Stedman, R., & Rauschmayer, F. (2016). Elucidating the changing roles of civil society in urban sustainability transitions. *Current Opinion in Environmental Sustainability, 22,* 41–50. https://doi.org/10.1016/j.cosust.2017.04.008

Fraser, N. (2021). Climates of capital: For a trans-environmental eco-socialism. *New Left Review, 127,* 94–127. https://newleftreview.org/issues/ii127/articles/nancy-fraser-climates-of-capital.pdf

Gibson-Graham, J. K. (2008). Diverse economies: Performative practices for 'other worlds'. *Progress in Human Geography, 32,* 613–632. https://doi.org/10.1177/0309132508090821

Goncalves, G. E. (2025) Climate change as a crisis of recognition: Prefigurative politics of socio-ecological movements for food sovereignty in Brazil. In S. Sareen & S. Juhola (Eds.) *Societal transitions to sustainability: The prefigurative politics of present transformation.* Palgrave Macmillan.

Graeber, D. (2004). *Fragments of an anarchist anthropology.* Prickly Paradigm Press.

Graeber, D. (2013). Culture as creative refusal. *The Cambridge Journal of Anthropology, 31,* 1–19. https://doi.org/10.3167/ca.2013.310201

Graeber, D., & Wengrow, D. (2018). *How to change the course of human history. Eurozine.* https://davidgraeber.org/wp-content/uploads/2018-Graeber-y-Wengrow-How-to-change-the-course-of-human-history.pdf

Guerrero Lara, L., Feola, G., & Driessen, P. (2024). Drawing boundaries: Negotiating a collective 'we' in community-supported agriculture networks. *Journal of Rural Studies, 106*, Article 103197. https://doi.org/10.1016/j.jrurstud.2024.103197

Henfrey, T., Feola, G., Penha-Lopes, G., Sekulova, F., & Esteves, A. M. (2023). Rethinking the sustainable development goals: Learning with and from community-led initiatives. *Sustainable Development, 31*, 211–222. https://doi.org/10.1002/sd.2384

Holloway, J. (2010). *Crack capitalism.* Pluto Press.

hooks, b. (1989). Choosing the margin as a space for radical openness. *Framework: The Journal of Cinema and Media, 36*, 15–23. https://www.jstor.org/stable/44111660

Kallis, G. (2018). *Degrowth.* Agenda Publishing.

Kallis, G., Kerschner, C., & Martinez-Alier, J. (2012). The economics of degrowth. *Ecological Economics, 84*, 172–180. https://doi.org/10.1016/j.ecolecon.2012.08.017

Kanngieser, A., & Beuret, N. (2017). Refusing the world: Silence, commoning, and the Anthropocene. *South Atlantic Quarterly, 116*, 363–380. https://doi.org/10.1215/00382876-3829456

Liu, S. (2021). Between social spaces. *European Journal of Social Theory, 24*, 123–139. https://doi.org/10.1177/1368431020905258

McGranahan, C. (2016). Theorising refusal: An introduction. *Cultural Anthropology, 31*, 319–325. https://doi.org/10.14506/ca31.3.01

Melucci, A. (1995). *Social movements and culture.* University of Minnesota Press.

Monticelli, L. (2021). On the necessity of prefigurative politics. *Thesis Eleven, 167*(1), 99–118. https://doi.org/10.1177/07255136211056992

Monticelli, L. (Ed.). (2022). *The future is now: An introduction to prefigurative politics.* Policy Press.

Pellizzoni, L. (2021). Prefiguration, subtraction and emancipation. *Social Movement Studies, 20*, 364–379. https://doi.org/10.1080/14742837.2020.1752169

Raj, G., Feola, G., & Runhaar, H. (2024). Work in progress: Power in transformation to post-capitalist work relations in community-supported agriculture. *Agricultural and Human Values, 41*, 1353–1368. https://doi.org/10.1007/s10460-023-10486-8

Rossi, A., Piccoli, A., & Feola, G. (2024). Transforming labour around food? The experience of community supported agriculture in Italy. *Agriculture and Human Values, 41*, 1667–1686. https://doi.org/10.1007/s10460-024-10572-5

Sales, A. L. L. D. F. (2023). *A political psychology approach to militancy and prefigurative activism: The case of Brazil*. Springer International Publishing.

Schiller-Merkens, S. (2022). Social Transformation through Prefiguration? A Multi-Political Approach of Prefiguring Alternative Infrastructures. *Historical Social Research, 47*, 6690. https://doi.org/10.12759/hsr.47.2022.39

Schmid, B., & Taylor Aiken, G. (2023). A critical view on the role of scale and instrumental imaginaries within community sustainability transitions research. *Area, 12884*. https://doi.org/10.1111/area.12884

Shove, E. (2012). The shadowy side of innovation: Unmaking and sustainability. *Technology Analysis and Strategic Management, 24*, 363–375. https://doi.org/10.1080/09537325.2012.663961

Simpson, A. (2016). Consent's revenge. *Cultural Anthropology, 31*, 326–333. https://doi.org/10.14506/ca31.3.02

Taylor-Aiken, G. (2025). Community-based proleptic environmentalism: Food, energy, and Kafka's 'Weg-von-hier'. In S. Sareen & S. Juhola (Eds.), *Societal transitions to sustainability: The prefigurative politics of present transformation*. Palgrave Macmillan.

van Oers, L., Feola, G., Runhaar, H., & Moors, E. (2023). Unlearning in sustainability transitions: Insight from two Dutch community-supported agriculture farms. *Environmental Innovation and Societal Transitions, 46*, Article 100693. https://doi.org/10.1016/j.eist.2023.100693

Wengrow, D., & Graeber, D. (2018). "Many seasons ago": Slavery and its rejection among foragers on the Pacific Coast of North America. *American Anthropologist, 120*, 237–249. https://doi.org/10.1111/aman.12969

Wright, E. O. (2013). Transforming capitalism through real utopias. *American Sociological Review, 78*, 1–25. https://doi.org/10.1177/0003122412468882

Yates, L. (2015). Rethinking prefiguration: Alternatives, micropolitics and goals in social movements. *Social Movement Studies, 14*, 1–21. https://doi.org/10.1080/14742837.2013.870883

Open Access This chapter is licensed under the terms of the Creative Commons Attribution-NonCommercial-NoDerivatives 4.0 International License (http://creativecommons.org/licenses/by-nc-nd/4.0/), which permits any noncommercial use, sharing, distribution and reproduction in any medium or format, as long as you give appropriate credit to the original author(s) and the source, provide a link to the Creative Commons license and indicate if you modified the licensed material. You do not have permission under this license to share adapted material derived from this chapter or parts of it.

The images or other third party material in this chapter are included in the chapter's Creative Commons license, unless indicated otherwise in a credit line to the material. If material is not included in the chapter's Creative Commons license and your intended use is not permitted by statutory regulation or exceeds the permitted use, you will need to obtain permission directly from the copyright holder.

8

Prefigurative Politics in Action: Youth Climate Activism and Arendt's Politics of New Beginnings

Turkan Firinci Orman

1   Introduction

This chapter explores prefigurative politics through the lens of Hannah Arendt's political philosophy, focusing on youth climate activism. Prefiguration—the enactment of desired futures in the present—offers a critical framework for understanding how young activists resist systemic inertia and reimagine socio-political structures. As a key concept in social movement scholarship, prefiguration shapes debates on activism, democratic participation, and alternative political practices (Yates et al., 2024). It highlights how social movements not only demand change but actively embody and experiment with the political and social structures they seek to create (Yates, 2021). Rather than treating youth climate movements as empirical cases, this chapter approaches them as generative sites of political meaning. These movements prefigure just futures, not by prescribing fixed solutions, but by enacting collective agency, embracing plurality,

T. Firinci Orman (✉)
Tampere, Finland
e-mail: turkanfirinci@gmail.com

© The Author(s) 2026   **115**
S. Sareen and S. Juhola (eds.), *Societal Transitions to Sustainability*, Palgrave Studies
in Environmental Transformation, Transition and Accountability,
https://doi.org/10.1007/978-3-032-07395-2_8

and practising alternative ways of living together. The analytical focus is on youth climate activism as a site of prefigurative political practices and meaning-making. Drawing on Arendt's thought, this chapter examines how youth climate activism exemplifies prefigurative politics, enacting visions of a just and sustainable world.

Arendt's political philosophy, among the most influential of the twentieth century, offers a valuable perspective for studying environmental politics (Butler, 2017). While her work helps in understanding democratic movements and grassroots activism (Lederman, 2015), it also resonates with prefigurative practices of resisting systemic oppression and envisioning new political possibilities through utopian thinking (Firinci Orman, 2025). Her critique of totalitarianism underscores how the erosion of public and private boundaries fosters isolation, making individuals more vulnerable to authoritarianism (Wolfe, 2023). This insight is particularly relevant to youth navigating global capitalism while mobilising for climate justice, as they form networks of resistance across political and economic systems (see Holmberg & Alvinius, 2020). Youth climate activism is a dynamic form of environmental engagement that challenges political stagnation and corporate interests, pushing for systemic change amid intergenerational tensions (Thomas et al., 2019). It is inherently intersectional, linking climate justice to broader social inequalities (Bowman & Germaine, 2022; Wright & McLeod, 2023). Through protests, policy advocacy, and sustainable practices, young activists actively shape climate discourse and action.

Scholars have explored how youth climate activism employs utopian imagination, drawing on Ernst Bloch's (1995) concept of concrete utopias. These movements resist static and unreachable ideals, instead embracing transformative possibilities for the future (Friberg, 2022; McKnight, 2024). Through prefigurative politics, these movements resist extractive capitalism and oppression, while prefiguring democratic, intersectional, and eco-logical futures through creative, caring, and radical practices (see McKnight, 2024). For young activists, utopia is not a distant goal, but an evolving process shaped by collective action and shared experiences. This aligns with prefiguration, where visionary ideals manifest in tangible actions (Yates et al., 2024). The act of envisioning new futures is thus inseparable from the everyday practices of youth activists. Though often seen as opposites, reaction and prefiguration often intersect, as youth experiment with horizontal decision-making, regenerative culture, and daily sustainability (for extensive analysis, see Evans, 2021; Lajarthe & Laigle, 2024).

This chapter argues that youth climate activism exemplifies prefigurative politics, wherein young people actively embody and enact the futures they seek to create. Youth climate activism reflects prefigurative politics through "embodied strategies to render desirable futures with immediacy" (Sareen & Juhola, 2025) and enacts just and sustainable worlds through collective action in line with Arendt's politics of new beginnings. Their engagement in protest and political action reflects Arendt's concept of natality and her reimagining of utopia as a participatory, dynamic process (see Firinci Orman, 2025). Prefiguration can thus be understood as world-building, a practice that creates spaces of appearance outside or beyond conventional political structures by enacting alternative futures in the present. Framed through an Arendtian lens, prefiguration emerges as an ontological and existential act: youth movements do not just anticipate change but live new ways of being, appearing, and acting together. In this sense, the means and ends align—the desired world is enacted as the means, not merely pursued as the end.

Following a review of literature on prefiguration in climate activism, the chapter introduces key concepts from Arendt's political philosophy to provide theoretical grounding. It then connects this framework to discussions on youth mobilisation, focusing on movements such as Fridays for Future (FFF) and Extinction Rebellion (XR) as examples of embodied public action. The conclusion turns to Arendt's vision of a people's utopia, highlighting citizen councils and participatory governance. Finally, the chapter reflects on prefigurative politics in action, linking it to Arendt's notion of natality as an expression of new beginnings and possibilities.

2   Prefigurative Politics and Youth Climate Activism

Prefiguration has become a key concept in social movement scholarship, particularly in studies on activism, democracy, and alternative political practices. While prefigurative politics has been widely explored (see Gordon, 2018; Maeckelbergh, 2011; Yates, 2015; Yates et al., 2024; Yates & de Moor, 2022), its role in climate justice activism remains underexamined. Research has analysed its relevance across different temporal and geographical settings (e.g., Evans, 2021; Lajarthe & Laigle, 2024), including studies on young climate activists' experiences (e.g., Arya & Pickard, 2025). Prefigurative politics fosters alternative structures and engagement models that embody activists' values and social relations. Youth-led movements, such as FFF and XR, model new ways of interacting with the world, rejecting top-down approaches in favour of direct, participatory action.

For young climate activists, prefiguration serves as both a strategy and a practice—realising equitable futures through sustainable living, decentralised organising, and direct action (see Firinci Orman, 2025). Their motivations stem from urgency and hope, fostering collective agency through shared goals, democratic participation, and mutual support. Despite structural challenges, young activists invigorate climate discourse by enacting their vision of an environmentally just world in the present, demonstrating that the future they seek is already in motion (see Soler-i-Martí et al., 2022; Nairn et al., 2024).

3   Hannah Arendt's Key Concepts as Analogical Lenses

3.1   Critique of Totalitarianism and Utopias as Blueprints

Arendt critiques both utopias and totalitarianism for their rigid ideologies and fixed societal structures, which suppress political freedom and public engagement. She argues that utopias as blueprints are inherently anti-political, replacing dynamic political life with inflexible systems that curtail civic participation. As Arendt notes, "Although none of the utopias ever came to play any noticeable role in history... they broke down quickly under the weight of reality, not so much the reality of exterior circumstances as of the real human relationships they could not control" (Arendt, 1958a, 227).

Arendt distinguishes totalitarianism from traditional oppression by emphasising its reliance on rigid, pseudo-scientific ideologies that attempt to explain all aspects of human life. These ideologies, maintained by terror, dismantle human plurality, which she views as central to freedom. She writes, "Total domination... is possible only if each and every person can be reduced to a never-changing identity of reactions" (Arendt, 1958b, 438), highlighting the loss of individuality in such regimes. For Arendt, the totalitarian pursuit of absolute power erases human diversity, undermining freedom. As she asserts, "Plurality is the condition of human action," because we are all unique, despite our shared humanity (Arendt, 1958a, 8). This view directly opposes static, top-down utopian visions, aligning instead with her belief in politics as a pluralistic, unpredictable, and collective endeavour.

While Arendt critiques abstract utopias detached from action, she does not entirely reject utopian thinking. She supports a more nuanced, concrete approach to political reconstruction—one that engages with history and reality rather than escaping from them. As Hiruta (2017) notes, Arendt's perspective aligns with Bloch's utopian ideals, where individuals actively engage in protecting their rights and resisting oppressive modern social forces (Firinci Orman, 2025). Bloch's work (1995) emphasises the principle of hope, arguing that utopian thinking is essential for social transformation, as it envisions possibilities beyond the present and guides action toward meaningful change. This aligns with both Bloch's concrete utopias and Arendt's politics of action, offering a prefigurative lens where utopia becomes a lived, strategic practice.

3.2   Political Freedom, Public Action, and Human Plurality

Arendt's political philosophy asserts that genuine freedom is not an internal state but emerges through relational dynamics and communal interactions. She critiques the idea of freedom as individual autonomy, arguing that true freedom manifests through public engagement, where individuals act visibly and collectively (Arendt, 1961). As she states, "We first become aware of freedom or its opposite in our intercourse with others" (Arendt, 1961, 148). For Arendt, political freedom is the ability to act publicly, initiating change through discourse and collective action. Unlike labour, which focuses on repetitive tasks for survival, and work, which creates lasting objects or structures, political action is an end in itself, emphasising engagement and participation over any specific outcome.

This perspective challenges traditional liberal notions of freedom by positioning it within collective, transformative public action. Arendt underscores that freedom unfolds in the public sphere, where dialogue, creativity, and citizenship can be expressed, especially during social and political change (Agnihotri, 2024; Calhoun, 2022). Youth climate movements exemplify this dynamic, using the public sphere to mobilise, collaborate, and advocate for systemic transformation (Marquardt, 2020; O'Brien et al., 2018). These spaces foster meaningful discussions on social justice, inequality, and climate change, reinforcing the idea that freedom is enacted, not merely possessed.

For Arendt, political action is the essence of politics, initiating new beginnings and deriving its value from engagement, not outcomes (Arendt, 1961, 146; Arendt, 1958a, 188). Human plurality—the uniqueness of individual perspectives—plays a crucial role in shaping a shared world through interaction. Youth climate movements embody this plurality, as diverse voices engage in transformative dialogue. These interactions generate new narratives, elicit responses, and sustain a dynamic, participatory public sphere. Arendt (1958a) defines power as the force upholding the public realm, emerging not from economic or military strength but from the collective efforts of citizens acting together (Gregg, 2019).

Arendt further connects the public realm to council democracy, where citizenship is expressed through radical participation, particularly in revolutionary moments (Agnihotri, 2024). A formal public sphere, characterised by face-to-face interaction, is essential for meaningful acts of citizenship, enabling individuals to shape collective politics (Smith, 2019). Youth activists bring this participatory model to life, fostering inclusive spaces for dialogue and mobilisation.

3.3   Natality

Arendt's reflections on brokenness in *The Origins of Totalitarianism* (1958b) resonate with contemporary concerns about the climate crisis as a total collapse, echoed by Bargues et al (2024) in their comparison between the Holocaust and the Anthropocene, both representing the end of the world. This comparison aligns with Arendt's notion of natality, the human capacity for new beginnings, which she sees as central to human life and action (Arendt, 1965). She emphasises that natality underpins all forms of vita activa—labour, work, and especially action (Arendt, 1958a). She distinguishes birth as the initial emergence into the world from action as a second birth, where individuals affirm their existence in the public sphere, realising freedom and initiating new beginnings (Arendt, 1958a). Natality fosters human plurality through dialogue, enabling individuals to act and transcend the cycle of life and death. In this context, Rafuse (1997) argues that revolutions fail when they neglect conditions for new beginnings, such as public space, freedom, and plurality.

In youth activism, natality captures the generative force of young people entering the political sphere. Youth climate movements manifest natality, as new generations act publicly to create transformative possibilities for the future. By virtue of their societal position, youth challenge established norms, critique current environmental policies, and devise novel approaches to climate justice. Arendt's concept of natality, as the capacity to introduce something new, is key to prefigurative politics. Youth climate activists embody the futures they wish and hope to see, experimenting with sustainable living, community organising, and responsible consumption, rather than waiting for institutional change. This aligns with Arendt's view (1958a) that politics should focus on action, not bureaucracy. Prefiguration, in this sense, can be seen as an act of beginning, where young people bring their envisioned futures into the present.

Arendt's political philosophy sharpens this view by framing politics as an end in itself—valued not for its outcomes, but for its enactment of freedom and plurality. From this standpoint, prefigurative politics is not validated by what it achieves in the long term, but by its enactment of political life in the present. Youth movements thus prefigure by doing: transforming spaces, building solidarities, and acting in concert in the here and now.

4   Youth Climate Activism: A Political Force for Transformation

4.1   Public Advocacy and Action

Youth-led climate movements like FFF and XR are reshaping public discourse and policy through advocacy and direct action. These grass-roots movements challenge political structures and demand urgent climate policies. Inspired by Greta Thunberg's 2018 school strike, FFF quickly mobilised millions globally, marking a shift in climate activism and drawing attention to youth-driven demands for immediate action (Thomas et al., 2019). Similarly, XR uses civil disobedience to disrupt political systems and highlight the urgency of the climate crisis (Berglund, 2023), reflecting Arendt's concept of political freedom. These movements advocate not just for environmental policies but for a broader climate justice framework addressing both environmental and social injustices (Bowman & Germaine, 2022; Wright & McLeod 2023). Youth activists, as critics of political inaction, are agents of collective change, reshaping the public sphere through solidarity and shared judgment, challenging the powers that delay climate action (Nairn et al., 2024).

Greta Thunberg's "How dare you?" speech at the UN Climate Action Summit underscores the moral urgency of the climate crisis and youth activism's role in holding leaders accountable (Thunberg, 2019). Similarly, XR's nonviolent actions, such as road blockages and sit-ins, disrupt daily life to force governments and corporations to confront the ecological crisis (Richardson, 2020). These disruptive actions create space for new political possibilities and push for systemic change (see Friberg, 2022). Youth-led activism has shifted public consciousness, especially through events like the Global Climate Strikes and XR's disruptive actions, though the movement's momentum has not yet led to lasting transformative change.

4.2 Embodied Activism

Youth climate activism embodies Arendt's view that politics is about transformation, not control. Activists act as if a better world is already possible, enacting their ideals in the present. Protests, strikes, and civil disobedience transform public spaces into democratic arenas (Axon, 2019; Marquardt, 2020). Juris (2015) highlights how these performances shape oppositional discourse, building solidarity and forming new activist identities. Public spaces become platforms where youth assert their right to shape the future, making climate justice visible.

Arya and Pickard (2025) analyse UK youth activism as prefigurative politics, focusing on FFF and XR's intersectional approach to environmental, racial, economic, and gender injustices. Pickard (2022) shows how these movements challenge institutional inaction through direct action and DIO politics, advocating urgent policies and emphasising climate change's disproportionate impact on marginalised communities, linking ecological and social justice. As such, youth environmental citizenship is seen as performative (e.g., Firinci Orman, 2022), with young people expressing agency through both protest and subtler forms of activism—not as exclusive acts, but as complementary expressions of agency—embodying freedom in line with Arendt's concept of transformative action. Ginsburg (2024) shows how Global Climate Strikes serve as calls for action and demonstrations of political freedom. Youth activists centre intersectionality, framing their struggles within broader inequalities and their activism as systemic change. They embody prefigurative politics through spatial and performative activism, manifesting change via space, actions, and experimentation.

For example, Friberg (2022) shows how these movements create micro (utopian) communities through decentralised hubs, pop-up assemblies, protests, and youth-led campaigns. These spaces embody a radically democratic, ecologically conscious world, where care and solidarity replace apathy and market logic. FFF's school strikes collapse the gap between present and future, demanding immediate climate action. XR activists in London have transformed public spaces into political arenas, with actions like "Tree Planting" in Parliament Square symbolising ecological values. Thus, when youth climate movements transform streets into democratic spaces or model alternative social arrangements in their organising practices, they do not merely anticipate a just future, they act it out now.

5   Prefigurative Politics in Action

Arendt critiques mass democracy when it becomes bureaucratic and detached from participatory engagement, reducing citizens to passive spectators within representative institutions (Arendt, 1958a, 220–223; 1965, 269–275). In contrast, her concept of councils—rooted in her study of revolutions—highlights grassroots political spaces as the foundation of participatory democracy. Built on horizontal governance, these councils provide direct avenues for political engagement. In this context, youth climate activism embodies Arendt's "people's utopia" (see Lederman, 2019), where power is decentralised, and collective decision-making prevails. By rejecting traditional politics in favour of self-organised assemblies, consensus-based decision-making, and direct action, youth activists' grassroot organising aligns with Arendt's vision of council democracy. Their activism extends beyond policy advocacy to reimagining the political process itself, shifting power from institutions to the collective. Grumiau and Hughes (2025) apply Arendt's concept of natality to the Chilean citizen assemblies, illustrating how collective action, driven by a "starting by doing" ethos, fosters open-ended prefigurative politics that resonate with the transformative potential of youth climate activism.

A notable example is Extinction Rebellion UK (n.d.), which advocates for citizens' assemblies on climate and ecological justice to enhance direct public involvement in decision-making. They cite past examples such as electoral reform (Canada 2006), abortion rights (Ireland 2016), climate action (France 2020), and air quality regulation (UK 2019). Additionally, they promote informal community assemblies at the local level, engaging people in grassroots decision-making.

In Arendt's framework, these councils serve as transformative models of active citizenship, countering the elitism of representative democracy (Lederman, 2015). They create collective spaces where individuals engage directly with political realities, exercising both freedom and agency. This emphasis on participation and prefiguration reflects Arendt's belief in political action as a force for shaping the future (see Isaac, 1994). Accordingly, youth-led climate movements prefigure a radically democratic future rooted in justice, equality, and environmental sustainability (Firinci Orman, 2025).

While these movements can be seen as echoing the revolutionary spirit like the 1871 Paris Commune and the 1905 Russian Revolution (Arendt, 1965, 264–265), Arendt cautions that grassroots initiatives often struggle to endure. Radical democratic transformation frequently encounters resistance from institutional forces, raising a critical question for youth climate activism: Can prefigurative practices of young people evolve beyond protest to establish lasting governance models? As Arendt warns, councils, though powerful in moments of revolution, often succumb to institutional pressures (Arendt, 1965, 247–252). Thus, this key question persists for future endeavours.

5.1   Concluding Words

The concept of natality—the potential for new beginnings and the capacity for human action to initiate change—deeply resonates with youth activism as both a prefigurative strategy and practice. Through collective action and imagination, youth climate movements embody the idea of starting anew, creating spaces where alternative futures can be envisioned and enacted (Bowman & Germaine, 2022). Their activism in the public sphere is not merely a reaction to present crises but an expression of freedom and plurality, illustrating the dynamic interplay between means and ends. Each protest, strike, and collective act signals a new beginning—an empowered, hopeful future aligned with Arendt's vision of participatory democracy (Isaac, 1994).

The Arendtian lens resituates prefigurative politics within a philosophical tradition that is both humanist and radically emancipatory. It invites us to see youth activism as more than strategic resistance—as an ontological force that asks what it means to act, to appear, and to begin anew together (Firinci Orman, 2025), illustrating the possibility of radical democratic change (see Lederman, 2019). For Arendt, for action to be truly political, it must create public spaces that cultivate plurality, political discourse, and diversity, rather than isolation. Prefiguration, in this sense, is the lived expression of concrete utopias (see Friberg, 2022) that embody natality, driven by collective agency and public engagement—not distant ideals, but experimental community-driven practices unfolding in the present.

Competing Interests   The author has no conflicts of interest to declare that are relevant to the content of this chapter.

References

Agnihotri, S. (2024). *Arendtian constitutional theory: An examination of active citizenship in democratic constitutional orders* (Doctoral dissertation). London School of Economics and Political Science.

Arendt, H. (1958a). *The human condition* (2nd ed.). University of Chicago Press.

Arendt, H. (1958b). *The origins of totalitarianism* (2nd ed.). Meridian Books.

Arendt, H. (1961). *Between past and future: Six exercises in political thought.* Viking Press.

Arendt, H. (1965). *On revolution.* Penguin Books.

Arya, D., & Pickard, S. (2025). Young people's climate and environmental activism in Britain: Cooperation, confrontation, and fragmentation in intersectional prefigurative politics. In E. Avril, L. Cossu-Beaumont, D. Fée, & F. Mourlon (Eds.), *Fragmented powers* (pp. 251–267). Emerald Publishing Limited. https://doi.org/10.1108/978-1-83608-412-920251018

Axon, S. (2019). Warning: Extinction ahead! Theorizing the spatial disruption and place contestation of climate justice activism. *Environment, Space, Place, 11*(2), 1–26. https://doi.org/10.5749/envispacplac.11.2.0001

Bargues, P., Chandler, D., Schindler, S., & Waldow, V. (2024). Hope after "the end of the world": Rethinking critique in the Anthropocene. *Contemporary Political Theory, 23*(2), 187–204. https://doi.org/10.1057/s41296-023-00649-x

Berglund, O. (2023). Disruptive protest, civil disobedience & direct action. *Politics.* Advance Online Publication. https://doi.org/10.1177/02633957231176999

Bloch, E. (1995). *The principle of hope.* MIT Press.

Bowman, B., & Germaine, C. (2022). Sustaining the old world, or imagining a new one? The transformative literacies of the climate strikes. *Australian Journal of Environmental Education, 38*(1), 70–84. https://doi.org/10.1017/aee.2022.3

Butler, R. E. (2017). *Between nature and artifice: Hannah Arendt and environmental politics* (Master's thesis). University of Victoria.

Calhoun, C. (2022). 2 facets of the public sphere: Dewey, Arendt, Habermas: Views on the Nordic model. In F. Engelstad, H. Larsen, J. Rogstad, & K. Steen-Johnsen (Eds.), *Institutional change in the public sphere* (pp. 23–45). De Gruyter Open Poland. https://doi.org/10.1515/9783110546330-003

Evans, H. (2021). Beyond resistance: The role of prefiguration in social movements addressing the climate crisis. *Bath papers in international development and wellbeing* (no. 66, pp. 1–20).

Extinction Rebellion UK. (n.d.). *Citizens' assembly.* https://extinctionrebellion.uk/decide-together/citizens-assembly/

Firinci Orman, T. (2022). Youth's everyday environmental citizenship: An analytical framework for studying interpretive agency. *Childhood, 29*(4), 495–511. https://doi.org/10.1177/09075682221107750

Firinci Orman, T. (2025). Rethinking utopias through Hannah Arendt: Youth climate activism and the politics of possibility. *Philosophy & Social Criticism.* Advance Online Publication. https://doi.org/10.1177/01914537251317117

Friberg, A. (2022). Disrupting the present and opening the future: Extinction rebellion, fridays for future, and the disruptive utopian method. *Utopian Studies, 33*(1), 1–17. https://doi.org/10.5325/utopianstudies.33.1.0001

Ginsburg, J. (2024). *"To create communities where care is enjoyed and valued in a very high regard": Motivations, approaches, and impacts of youth-led climate collectives* (Doctoral dissertation). Concordia University.

Gordon, U. (2018). Prefigurative politics between ethical practice and absent promise. *Political Studies, 66*(2), 521–537. https://doi.org/10.1177/003232 1717722363

Gregg, J. (2019). Motivating climate activism through framing: Hope, fear, injustice, and sacrifice. *Scholarly Horizons: University of Minnesota, Morris Undergraduate Journal, 6*(2), Article 2. https://doi.org/10.61366/2576-2176.1077

Grumiau, B., & Hughes, I. (2025). Starting by doing: Ethical orientations for open-ended prefigurative action. In S. Sareen & S. Juhola (Eds.), *Societal transitions to sustainability: The prefigurative politics of present transformation.* Palgrave Macmillan.

Hiruta, K. (2017). An 'anti-Utopian age?': Isaiah Berlin's England, hannah arendt's America, and Utopian thinking in dark times. *Journal of Political Ideologies, 22*(1), 12–29. https://doi.org/10.1080/13569317.2016.1258773

Holmberg, A., & Alvinius, A. (2020). Children's protest in relation to the climate emergency: A qualitative study on a new form of resistance promoting political and social change. *Childhood, 27*(1), 78–92. https://doi.org/10.1177/0907568219879970

Isaac, J. C. (1994). Oases in the desert: Hannah Arendt on democratic politics. *American Political Science Review, 88*(1), 156–168. https://doi.org/10.2307/2944888

Juris, J. S. (2015). Embodying protest: Culture and performance within social movements. In A. Flynn & J. Tinius (Eds.), *Anthropology, theatre, and development* (pp. 43–58). Palgrave Macmillan. https://doi.org/10.1057/978113 7350602_4

Lajarthe, F., & Laigle, L. (2024). Bringing the future back to the present: The role of prefiguration in European climate justice activism. *Futures, 160*, Article 103384. https://doi.org/10.1016/j.futures.2024.103384

Lederman, S. (2015). Councils and revolution: Participatory democracy in anarchist thought and the new social movements. *Science & Society, 79*(2), 243–263. http://www.jstor.org/stable/24583897

Lederman, S. (2019). *Hannah Arendt and participatory democracy: A people's utopia*. Palgrave Macmillan. https://doi.org/10.1007/978-3-030-11692-7

Maeckelbergh, M. (2011). Doing is believing: Prefiguration as strategic practice in the alterglobalization movement. *Social Movement Studies, 10*(1), 1–20. https://doi.org/10.1080/14742837.2011.545223

Marquardt, J. (2020). Fridays for Future's disruptive potential: An inconvenient youth between moderate and radical ideas. *Frontiers in Communication, 5*, Article 48. https://doi.org/10.3389/fcomm.2020.00048

McKnight, H. (2024). World-building enactments of the school strike movements during the pandemic: Reading youth climate crisis movements through a micro- and nano-utopian lens. In H. Alberro, E. Atasoy, N. Castle, R. Firth, & C. Scott (Eds.), *Utopian and dystopian explorations of pandemics and ecological breakdown: Entangled futurities* (pp. 1–14). Routledge.

Nairn, K., Showden, C. R., Matthews, K. R., Kidman, J., & Sligo, J. (2024). Scaffolding collective hope and agency in youth activist groups: "I get hope through action." *The Sociological Review*. Advance Online Publication. https://doi.org/10.1177/00380261241245546

O'Brien, K., Selboe, E., & Hayward, B. M. (2018). Exploring youth activism on climate change: Dutiful, disruptive, and dangerous dissent. *Ecology & Society, 23*(3), Article 42. https://doi.org/10.5751/ES-10287-230342

Pickard, S. (2022). Young environmental activists and Do-It-Ourselves (DIO) politics: Collective engagement, generational agency, efficacy, belonging, and hope. *Journal of Youth Studies, 25*(6), 730–750. https://doi.org/10.1080/13676261.2022.2046258

Rafuse, J. A. (1997). *Natality as foundation for the capacity to initiate in the thought of Hannah Arendt* (Master's thesis). Université de Montréal.

Richardson, B. J. (2020). Editorial: Climate strikes to extinction rebellion: Environmental activism shaping our future. *Journal of Human Rights and the Environment, 11*(3), 1–9. https://doi.org/10.4337/jhre.2020.03.00

Sareen, S., & Juhola, S. (2025). The prefigurative politics of present transformation. In S. Sareen & S. Juhola (Eds.), *Societal transitions to sustainability: The prefigurative politics of present transformation*. Palgrave Macmillan.

Smith, B. (2019). Citizenship without states: Rehabilitating citizenship discourse among the anarchist left. *Citizenship Studies, 23*(5), 424–441. https://doi.org/10.1080/13621025.2019.1620688

Soler-i-Martí, R., Fernández-Planells, A., & Pérez-Altable, L. (2022). Bringing the future into the present: The notion of emergency in the youth climate movement. *Social Movement Studies, 23*(4), 517–536. https://doi.org/10.1080/14742837.2022.2123312

Thomas, A., Cretney, R., & Hayward, B. (2019). Student strike 4 climate: Justice, emergency, and citizenship. *New Zealand Geographer, 75*(2), 96–100. https://doi.org/10.1111/nzg.12229

Thunberg, G. (2019, September 23). *The UN 2019 climate summit, New York City* [Video]. YouTube. https://youtu.be/KAJsdgTPJpU

Wolfe, E. (2023). *Unearthing a voice: Arendt on action, public space, and identity, a feminist analysis* (Master's thesis). University of New Brunswick.

Wright, K., & McLeod, J. (Eds.). (2023). *Childhood, youth and activism: Demands for rights and justice from young people and their advocates.* Emerald Publishing Limited.

Yates, L. (2015). Rethinking prefiguration: Alternatives, micropolitics and goals in social movements. *Social Movement Studies, 14*(1), 1–21. https://doi.org/10.1080/14742837.2013.870883

Yates, L. (2021). Prefigurative politics and social movement strategy: The roles of prefiguration in the reproduction, mobilization, and coordination of movements. *Political Studies, 69*(4), 1033–1052. https://doi.org/10.1177/0032321720936046

Yates, L., & de Moor, J. (2022). The concept of prefigurative politics in studies of social movements: Progress and caveats. In L. Monticelli (Ed.), *The future is now: An introduction to prefigurative politics* (pp. 179–190). Bristol University Press.

Yates, L., Daniel, A., Gerharz, E., & Feldman, S. (2024). Introduction to the special issue: Foregrounding social movement futures: Collective action, imagination, and methodology. *Social Movement Studies, 23*(4), 429–445. https://doi.org/10.1080/14742837.2024.2343683

Open Access This chapter is licensed under the terms of the Creative Commons Attribution-NonCommercial-NoDerivatives 4.0 International License (http://creativecommons.org/licenses/by-nc-nd/4.0/), which permits any noncommercial use, sharing, distribution and reproduction in any medium or format, as long as you give appropriate credit to the original author(s) and the source, provide a link to the Creative Commons license and indicate if you modified the licensed material. You do not have permission under this license to share adapted material derived from this chapter or parts of it.

The images or other third party material in this chapter are included in the chapter's Creative Commons license, unless indicated otherwise in a credit line to the material. If material is not included in the chapter's Creative Commons license and your intended use is not permitted by statutory regulation or exceeds the permitted use, you will need to obtain permission directly from the copyright holder.

9

A Multi-Political Reflection on the Transformative Potential of Citizen Panels: Insights from Five Case Studies

Maria Luisa Lode, Samyajit Basu, and Cathy Macharis

1   Prefigurative and Citizen Panels

Prefigurative politics are a specific type of politics based on the assumption that politics are a form of collective action aimed at bringing about the desired future social change in the present (Yates, 2015). In contrast to social activism and protest, prefigurative politics are composed of actions and actors that already experiment and live the envisioned alternative to current (political) practices (Monticelli, 2021). Prefigurative politics are grounded in core elements of prefiguration.

Prefiguration entails the experimentation and providing a lived example of functioning alternatives to our current economy-dominated organisation of our society (Monticelli, 2022). Building on this, societal

M. L. Lode (✉) · C. Macharis
House of Sustainable Transitions, Vrije Universiteit Brussel, Brussels, Belgium
e-mail: Maria.Luisa.Lode@vub.be

S. Basu
Mobilise, Vrije Universiteit Brussel, Brussels, Belgium

© The Author(s) 2026   **133**
S. Sareen and S. Juhola (eds.), *Societal Transitions to Sustainability*, Palgrave Studies
in Environmental Transformation, Transition and Accountability,
https://doi.org/10.1007/978-3-032-07395-2_9

change is closely connected with local actions, where alternative practices are implemented by movements, communities, and collectives. Such practices can expand and upscale (Adloff & Neckel, 2021) ultimately leading to the erosion of current unsustainable practices, rather slowly over time. The transformative potential of prefiguration lies at the core of practising the envisioned change in the future already in the present.

Furthermore, prefigurative politics also build on the notion of direct and representative democracy where potential solutions are deliberated upon and without power or repressive dynamics within these processes (Fians 2022). For prefigurative politics, decision-making is, ideally, composed of "participant[s] to act and express oneself not as a representative of given political agendas, social status, or cultural backgrounds, but as individuals who autonomously question, for instance, oppressive, sexist, and colonialist regimes of truth" (Fians 2022, 8). Relating to this, prefiguration and the living of alternative practices also build on the assumption of liberation and justice from current oppressions and inequalities. Since prefigurative politics do not (only) build on the critique of current practices through protest and civic action, but also engage in envisioning and living alternative practices in decision-making, citizen assemblies are studied as one such alternative decision-making form (Fominaya 2014). Such citizen assemblies take place as open spaces for deliberation and sharing different points of views. In contrast to institutional politics, there is usually no strategic agenda for decision-making, no representativeness (e.g., being led by activists/individuals), and an "open-endedness" of decision-making and solutions. At the same time, such an approach has been criticised for its time-consuming nature and lack of actionable implementation results.

Relating to the concept of prefigurative citizen assemblies, citizen panels have emerged as democratic tools to facilitate direct participation and integrate local perspectives into political discussions (Crosby et al., 2008). These panels engage diverse and representative groups, including marginalised communities, through deliberation. While citizen panels can incorporate elements of prefigurative politics by implementing future visioning and enabling decentralised and non-representative decision-making, they are also often linked with institutional politics. This is especially the case if they are institutionalised bodies that serve as an input for political decision-making processes, which is already in practice in various countries such as Australia, Belgium, Colombia, and France (OECD, 2021).

Non-institutionalised citizen panels are not as closely linked to institutional decision-making bodies. While building on the notion of deliberation and inclusive participation in decision-making and practice on alternative ways of democracy, they usually still follow a more strategic, thematic approach for certain practices within the society (e.g., on dominant mobility and energy systems) (Fians, 2022). Therefore, it can be argued that certain types of citizen panels can be considered as open spaces for prefigurative politics by practising an alternative way of decision-makingfor an alternative design of societal practices. This is specifically applicable to citizen panels that are actively engaging with prefigurative practices (such as envisioning, radical inclusiveness, and participation).

Most citizen panels are embedded or connected with current decision-making bodies or, even if independent, are influenced and affected by existing structures opposing the prefigured alternative practice, as in this case, decision-making (Swain, 2019). While most prefigurative practices and politics build on the notion of practising the desired change, the experimentation or activation of alternative practices is heavily influenced by present power dynamics, existing infrastructure, and economy (Schiller-Merkens, 2024). In contrast to the conceptualisation of prefigurative politics, looking into existing examples of prefigurative politics shows that alternative practices rather exist and operate at the intersection with contentious and institutional politics (Schiller-Merkens, 2022). Non-institutionalised citizen panels are also confronted by such dynamics, e.g., by being directly impacted by counter-decisions on a higher regional, and politically legitimate decision-making body and struggling with alternative values that affect the potential for transformative change at least in the short-term but also long-term (Abels, 2007).

The political dynamics surrounding citizen panels and how they relate to prefigurative, institutional (counter), and contentious politics (such as protest and civic unrest) have not been studied. Yet, researching these dynamics might help to understand how prefigurative practices, especially in the political realm, can facilitate transformative change.

This study is two-fold; first, we elaborate how prefigurative practices were implemented in five case study citizen panels, then we reflect on the interplay contentious, institutional, and prefigurative practices in the case studies using the multi-political approach to prefiguration by Schiller-Merkens (2022). The five case studies are all characterised by challenges surrounding just low-carbon transition policies and have all implemented the same approach to deliberation and participatory democracy.

In the following, we briefly describe the approach to the citizen panels and provide a description of the case studies, then explain how the multi-political approach to prefiguration was used to analyse the citizen panels, before presenting the results. We conclude with challenges encountered in the implementation of the citizen panels, which potentially affected their transformative potential.

2   Methodology

2.1   Integrated Panel Approach

For the exploration of the transformative potential of non-institutionalised citizen panels, we are building on the insights of citizen panels which were held in five different European countries supported by the Horizon Europe project TANDEM (TANDEM, 2023). The panels addressed key low-carbon transition policies in the respective regions where the citizen panels were held. The case studies addressed the energy and heating transition in Innsbruck, Austria, the gradual phase-out of fossil fuel vehicles in Brussels Capital Region, Belgium, the phase-out of peat as an energy carrier in Ostrobothnia, Finland, the phase-out of coal and wood as a means for heating in Krzywcza, Poland and the expansion of renewable energy generation in Pujalt, Spain. The case studies were selected because they were all experiencing the contestation of low-carbon transition policies and contestation at the local level, especially highlighting the unequal impacts of the policies across space and society.

To systematically address inequalities in low-carbon transitions, the case studies applied a transdisciplinary research approach integrating various methodologies into a sequence of three citizen panels across all five case studies with a minimum of 6–26 people participating in each of the total 15 citizen panels (in total, more than 150 people were involved across the case studies). The citizen panels took place over the course of 1.5 years starting in October 2023 and ending in February 2025. Fig. 1 summarises the overall approach of the citizen panels and the content of each panel.

Before the first panel took place, each case study conducted a stakeholder mapping and power analysis of the identified actors. Furthermore, characteristics of vulnerability were defined in each case study to randomly select citizens who fulfil certain pre-defined characteristics in each case study. In each case study, 20–30 randomly chosen citizens were invited to the panel recruited based on socio-demographic characteristics.

In the first panel, participatory system mapping was conducted (Klingler et al., 2024). The primary goal was to define one or several problem statements relevant in the context of the transition policies and then derive important injustices perceived by the citizens. Finally, different leverage points were identified in the system maps where interventions into the behaviour of the system are most relevant and impactful. In the second panel, different interventions were combined in intervention bundles. The bundles were, for example, clustered according to official interventions stated in the transition policies and citizen-developed interventions. Using a stakeholder-based impact scoring methodology (te Boveldt, 2019), the impacts of the different intervention bundles were assessed by gathering the impacts of the citizens and a high-level evaluation of the different bundles on the citizens and their impact factors, as well as additional entities that were added in the evaluation as personas (e.g., nature, future generations, a specific citizen). The impacts were discussed among the panellists but also with invited authorities and/or experts on the topic or certain impacts.

In the last panel, a visioning and backcasting exercise was conducted to see if the current transition policies align with a preferred future vision in

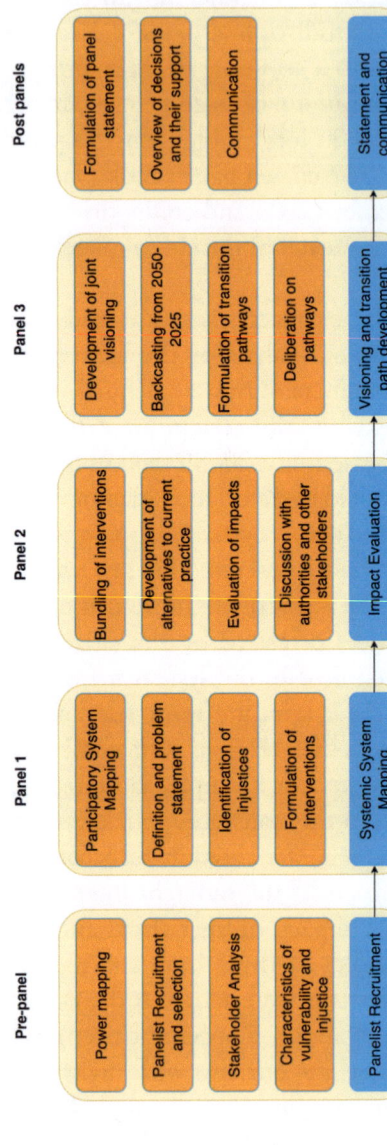

Fig. 1 Panel approach. *Source* Authors. Alt text: Panel approach visualised in five columns

2050 (Robinson, 2003). Bridging between present and future, transition pathway(s) were developed with milestones and activities needed to reach that vision. Overall, the citizen panels followed concepts of deliberation, an informed open discussion on policies and their impacts.

After the panels, a joint panel statement was phrased and confirmed by the panellists, which could then be shared to policymakers and practitioners. Next, the case studies are described in more detail.

3   Case Studies

Innsbruck, Austria: Austria aims for climate neutrality by 2040, and also the city of Innsbruck seeks energy autonomy by 2050. To achieve this goal, it would require large-scale building renovations, passive house standards, and district heating expansion. At the same time, Innsbruck faces a housing crisis due to rising real estate investment, housing demand, and limited supply, disproportionately affecting low-income households. Energy poverty already affects around 10,000 people in Tyrol, which might become an even more challenge over time as policies like the Renewable Heat Act increase pressure on housing affordability. This case focused on the intersection of housing and energy policies with socio-economic inequalities.

Brussels Capital Region, Belgium: Brussels, home to 1.2 million inhabitants, faces high congestion and poor air quality. Ranked among the 10 most congested cities in 2024, it struggles with pollutants like NO_x, particulate matter, and black carbon. The Low Emission Zone (LEZ), introduced in 2018, has improved air quality but remains contested due to its perceived socio-economic impact on lower-income residents and businesses. Though studies confirm its effectiveness, political debates persist, leading to the postponement of the next phase from 2025 to 2027. This case focused on the social justice implications of mobility transitions.

South Ostrobothnia, Finland: Finland plans to halve peat use by 2030 to reach climate neutrality by 2035, but declining peat viability is forcing a faster transition than planned, causing economic and social challenges. Peat production, crucial in North and South

Ostrobothnia, is tied to regional identity, employment, and culture. The region faces workforce decline, outmigration, and job losses, despite an entrepreneurial economy. This case explores how national climate targets and the EU Emission Trading System affect peat-dependent communities.

Krzywcza, Poland: Poland's 2040 transition plan phases out coal and firewood as sources for heating, yet rural areas like Krzywcza struggle with lack of energy alternatives and infrastructure. The commune, reliant on agriculture, has minimal gas grid connections and limited renewable energy investment. Its ageing population and declining workforce further complicate the transition. This case examines how national decarbonisation goals affect remote, underserved areas.

Pujalt, Spain: Pujalt, Catalonia, faces land use conflicts over wind and solar expansion. With 105 wind turbines and 17 more under construction, concerns about landscape impacts, biodiversity, and lack of community involvement have fuelled NIMBY opposition. The region, affected by ageing demographics and limited social services, highlights tensions between national energy policies and local resistance. This case explores the social and environmental justice dimensions of renewable energy transitions.

4   Multi-Political Framework for Prefiguration

All five case studies followed the described sequence of citizen panels. To go beyond the findings of the citizen panels itself, we use the multi-political approach of prefiguration by Schiller-Merkens (2022) as a framework to analyse the dynamics of non-institutionalised citizen panels in relation to institutional and contentious politics. The framework builds on the assumption that prefigurative politics does not happen in a silo, but rather is affected by and affects different forms of politics (such as contentious and institutional politics) that co-exist simultaneously. The three different types of politics that are differentiated in the framework include (i) contentious politics, which brings about ruptural transformation through activism, protest and contestation; (ii) institutional politics, which in contrast, relate to existing political and institutional practices and institutions, as well as decision-making elites or bodies, creating symbiotic transformations; and (iii) prefigurative politics, in between the other two, which embody practices that can lead to interstitial transformations. The framework shows that different activities and actors in the different political spheres affect each other and might be mutually reliant to actualise transformative potential for change over time.

By examining the interplay of contentious, institutional, and prefigurative politics, we explore how these dynamics enable or constrain transformative change in the context of the citizen panels and just low-carbon transitions. We want to place special focus on the aspect of deliberation, shared visions, representation, mutual understanding, lived actions, and navigating institutional power dynamics.

5   Results

5.1   Participation, Representation, Deliberation

As explained under methodology, the citizen panels followed and integrated prefigurative practices in several steps of the entire process. Participating panellists were intended to be selected through civic lottery (such as panelot.org), which followed more traditional concepts of representative democracy. However, due to practical limitations or geographic conditions (some regions had less inhabitants with less socio-demographic diversity), the concept of representative panels was only fulfilled in the Finnish case study. In contrast to that, the participation to the citizen panels in Austria, Belgium, and Spain was made entirely open to people who were interested in participating. While the participants to the citizen panels were therefore not representative on certain socio-demographics (age, gender, income), the participants all represented characteristics of vulnerability to low-carbon transitions that were pre-defined before the panel recruitment. The panel participation was guided and structured by academia, think tanks and/or non-government organisations and participants also received €60 in the form of vouchers or local currencies. The citizen panels built on participatory democracy and deliberation and adapted several key values at the beginning of the panels (mutual respect, inclusiveness, humour). While first intended, not all participants took part in all three citizen panels, again leading to a more flexible form of participation.

The citizen panels were based on decision-making enabled through open deliberation and provision of data to make informed decisions. Such information was provided by the researchers involved in the project, or organisations as well as authorities active in the context. Authorities and decision-making bodies were, however, not present in all phases of open deliberation and only provided their perspective or data when invited for discussion (two times in each case study across all three panels). In the second panel, these actors accorded more importance to obtain insights into the expected impacts of the transition policies and the developed alternatives to the impact factors stated by the participants. In the last panel, the main output of the panel was a joint panel statement and an overview of which alternatives to the analysed transition policies were preferred by the participants.

5.2   Shared Visions, Mutual Understanding

In the third panel, a joint vision for the context of the transition policies was defined for 2050 in each case study. Only aspects of that vision accepted by all participants were included in the overall vision statement. The panel applied various approaches to support inclusive participation and translation of scientific knowledge, e.g., by using personas that represent entities that cannot be presented through a citizen (e.g., using a tree in the room to think about non-human actors and nature). Participants also used pre-developed persona drawings/cards to discuss jointly how they think the persona (both human and non-human entities) are affected by the transition policy. While there were disagreements across the panellists, the personas enabled discussions beyond their own perspectives. After the deliberation process of the second panel, the discussion of the impacts on different societal entities was repeated, with differences compared to the pre-deliberation assessment. For example, more participants weighted the importance of open decision-making, awareness, and education across all case studies higher after than before the deliberation.

5.3 Transformative Change and Actions

Two moments within the three citizen panels in the case studies were dedicated to defining interventions to the current system, which were then summarised to intervention bundles, so a bundle of alternative practices that should enable a more just and fairer low-carbon transition (end of panel 1). Then, in panel 3, through backcasting, transition pathways were developed to see if following the developed alternatives can reach the desired future in 2050. The different alternatives across the case studies only pertain to limited actions and interventions that address transformative system change. Most alternatives 1) increased exceptions to the current transition policies and 2) limited adaptations to the existing system (e.g., focusing on the roll-out of electric vehicles in contrast to reducing overall car-ownership, (Belgian Citizen Panel). Compared to this, the transition pathways developed after the future visioning included much more radical suggestions in fewer time intervals (e.g., starting to implement 20-minute city design and strategic plans in 2025, to achieve an accessible city to everyone by 2050).

After the sequence of the three panels, the two main outputs of the panels were (i) a joint statement of the citizen panel summarising all findings and decisions, and (ii) citizens who were more informed and engaged on the topic of the transition policy at stake.

6 Discussion

6.1 Multi-Political Approach to Prefiguration

Figure. 2 summarises the identified dynamics in the case studies across different political practices, building on the multi-political framework (Schiller-Merkens, 2022).

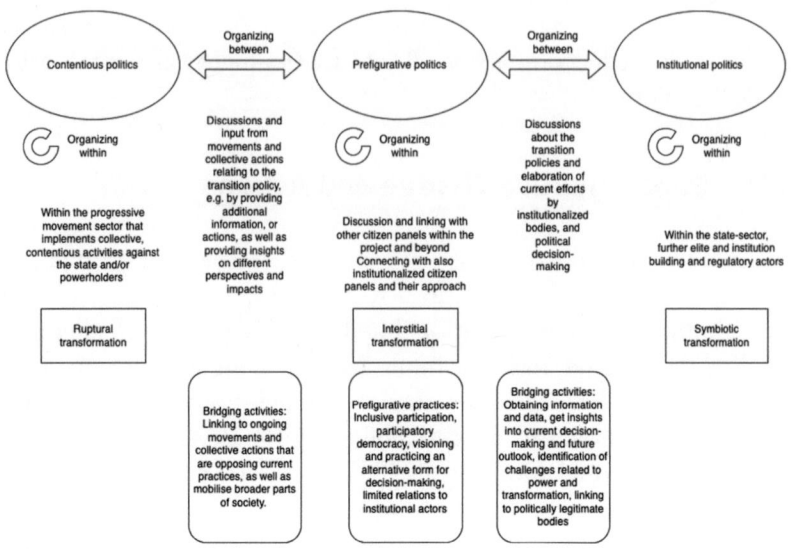

Fig. 2 Multi-political dynamics across the case studies. *Source* Authors' work building on Schiller-Merkens (2022). Alt text: Figure depicting linkages between contentious, prefigurative and institutional politics, and ruptural, interstitial and symbiotic transformation.>

Not all case studies showcased the same extent of these dynamics, but they were observed across all citizen panels. While all panels followed prefigurative practices (e.g., inclusive participation, participatory democracy, deliberation) and were supported through future visioning and backcasting exercises, the panels cannot be seen entirely separated from institutional and contentious political entities. First, the fact that all case studies address highly politicised transition policies, often accompanied by social protest and unrest, places the panels in the context of contentious politics with various individual and collective actions that oppose the decision-making from an institutionalised politics perspective. Thus, the failure or perception of injustice from an institutional perspective enabled momentum for contentious politics. While protests to the institutional decision-making bodies encouraged contentious activities, there might be the need to bridge between contentious and institutional activities. Here, the prefigurative practices can both be seen as a bridging practice between these dimensions, as well as a bridging of practices within prefiguration. As mentioned with regard to participation, representation, and deliberation, the panels ultimately did not follow the concept of representative participation because of limited capacity to mobilise a representative set of participants. The lack of representative participation reduces the political potential of the citizen panels from an institutional politics perspective, especially considering that the analysed case studies are non-institutionalised. The citizen panels also relied on input from actors across all political levels (e.g., authorities for data and obtaining information on political decisions surrounding the policy, as well as being aware of existing blockages towards alternative implementation and collective actions that brought the challenges and injustices of the transition policy to the core of public debate).

In summary, the citizen panels were affected by institutionally decided transition policies (both on local, national and international levels), notably outrage at the local level inciting contestation and protesting the decisions taken. The citizen panels unfolded with the input of institutional actors through data and information and were supported by civic collectives that provided examples for alternative practices. The citizen panels built on prefigurative actions specifically about decision-making, but were limited in both their transformative potential in terms of bringing forward transformative alternatives to the transition policies, as well as from a political perspective, since the panels were limited in their time of operation (1.5 years) and were not intended to be prolonged by the participants in early 2025.

In the case studies, the overall political context of contestation of institutionalised decisions taken on low-carbon transitions seemed to benefit the emergence or need of prefigurative politics through citizen panels. Furthermore, contentious politics created more urgency and need for alternative processes to the institutionalised processes. In the same context, institutionalised actors also seemed to be more reliant on the input of affected citizens regarding the impacts of and alternative approaches to policies. So, actors from both political realms seemed supportive of the implementation of citizen panels and their bridging prefigurative elements (participation, visioning, and alternative democratic practice). What became more difficult was maintaining the prefigurative realm from within; several participants started to doubt the "political potential" of the citizen panels as alternative approaches to current decision-making, specifically referring to the output of a political statement of the panel. By design, the citizen panels aimed to inform institutional politics, but in doing so, they might have reduced their credibility as independent and more transformative entities. The most agreed upon benefit of the process was the creation of shared understanding of the panellists, communication with various actors, increased knowledge and awareness of transition policies, and the experience of participating in alternative forms of democracy.

7   Conclusion

This research introduced non-institutionalised citizen panels as an alternative means to current representative democracy on just low-carbon transitions. The study builds on the insights of three citizen panels in each of five case studies across Europe, which have all followed a transdisciplinary approach to citizen panels, encompassing methodologies such as participatory system mapping, impact assessment, and deliberation supported by arts-based methods such as personas and the involvement of non-human actors. The citizen panels were analysed using the multi-political approach to prefiguration.

It was shown that citizen panels cannot only be defined through prefigurative practices, since they followed ideas that are specifically valued by institutional politics (e.g., representative democracy and political legitimacy). At the same time, we consider the output of the citizen panels less reliable from an institutional perspective due to their non-institutionalised character and their practical challenge to implement representative participation and political legitimacy. We found that the citizen panels and prefigurative practices thrived in an environment of institutional contestation through contentious practices, which increased awareness of existing systemic failure. The learning process of the panellists about transition policies and potential alternatives, the creation of a more common understanding founded on mutual respect and deliberation, as well as the practice of alternative means of citizenship through citizen panels, emerge as the biggest benefits of these prefigurative practices.

Competing Interests The authors have no conflicts of interest to declare that are relevant to the content of this chapter.

References

Abels, G. (2007). Citizen Involvement in Public Policy-making: Does it Improve Democratic Legitimacy and Accountability? The Case of pTA. *Interdisciplinary Information Sciences, 13*(1), 103–116. doi:https://doi.org/10.4036/iis.2007.103.

Adloff, F., & Neckel, S. (2021). Futures of sustainability: Trajectories and conflicts. *Social Science Information, 60*(2), 159–167. https://doi.org/10.1177/0539018421996266

Crosby, N., Kelly, J. M., & Schaefer, P. (2000). Citizens Panels: A New Approach to Citizen Participation. In *The Age of Direct Citizen Participation*. Routledge.

Fians, G. (2022). Prefigurative politics. *Cambridge Encyclopedia of Anthropology*. https://doi.org/10.29164/22prefigpolitics

Fominaya, C. F. (2014). *Social Movements and Globalization: How Protests, Occupations and Uprisings are Changing the World*. Palgrave Macmillan. https://www.academia.edu/6789346/Social_Movements_and_Globaliza tion_How_Protests_Occupations_and_Uprisings_are_Changing_the_World

Klingler, M., De Fontana, F., Gerdes, D., Plöchl, J., Scherhaufer, P., & Spittler, N. (2024). Systemic impacts of low-carbon transition policies for housing in Innsbruck. *Mapping the intersections of vulnerability and social justice with affected citizens and stakeholders*. https://doi.org/10.31223/X5QH7R

Monticelli, L. (2021). On the necessity of prefigurative politics. *Thesis Eleven, 167*(1), 99–118. https://doi.org/10.1177/07255136211056992

Monticelli, L. (2022). Prefigurative politics within, despite and beyond contemporary capitalism. In *The future is now: An introduction to Prefigurative politics* (pp. 15–31). Bristol University Press. https://doi.org/10.51952/9781529215687.ch001

OECD. (2021). Eight ways to institutionalise deliberative democracy. https://www.oecd.org/content/dam/oecd/en/publications/reports/2021/12/eight-ways-to-institutionalise-deliberative-democracy_e1f898a0/4fcf1da5-en.pdf

Robinson, J. (2003). Future subjunctive: Backcasting as social learning. *Futures, 35*(8), 839–856. https://doi.org/10.1016/S0016-3287(03)00039-9

Schiller-Merkens, S. (2022). Social transformation through prefiguration? A multi-political approach of prefiguring alternative infrastructures. *Historical Social Research, 47*(4), 66–90. https://doi.org/10.12759/hsr.47.2022.39

Schiller-Merkens, S. (2024). Prefiguring an alternative economy: Understanding prefigurative organizing and its struggles. *Organization, 31*(3), 458–476. https://doi.org/10.1177/13505084221124189

Swain, D. (2019). Not not but not yet: Present and future in Prefigurative politics. *Political Studies, 67*(1), 47–62. https://doi.org/10.1177/0032321717741233

TANDEM. (2023). *Tandem\Tandem horizon europe Just Transition Policies Co-Created with Communities.* https://tandem-heu.eu/

te Boveldt, G. (2019). *All aboard?* Vrije Universiteit Brussel.

Yates, L. (2015). Rethinking prefiguration: Alternatives, micropolitics and goals in social movements. *Social Movement Studies, 14*(1), 1–21. https://doi.org/10.1080/14742837.2013.870883

Open Access This chapter is licensed under the terms of the Creative Commons Attribution-NonCommercial-NoDerivatives 4.0 International License (http://creativecommons.org/licenses/by-nc-nd/4.0/), which permits any noncommercial use, sharing, distribution and reproduction in any medium or format, as long as you give appropriate credit to the original author(s) and the source, provide a link to the Creative Commons license and indicate if you modified the licensed material. You do not have permission under this license to share adapted material derived from this chapter or parts of it.

The images or other third party material in this chapter are included in the chapter's Creative Commons license, unless indicated otherwise in a credit line to the material. If material is not included in the chapter's Creative Commons license and your intended use is not permitted by statutory regulation or exceeds the permitted use, you will need to obtain permission directly from the copyright holder.

10

Laughter, Role-Play, and Immersion: Rediscovering the Audience in Prefigurative Politics through Carnival

Frank Eckerle

"Celebration in adverse times feels counter-intuitive but it remains key to discovering what we're capable of." Lucy Neal (2014, 279)

1 Introduction

The purpose of this chapter is to reconsider the role that prefiguration has on its (intended or unintended) audience, which I believe has not yet received enough attention. Prefigurative politics was originally coined to describe movements which exemplify a democratic socialist utopia within their internal structure, where the effects on outsiders were of second importance (Boggs, 1977). In contrast, the working definition of

F. Eckerle (✉)
Department of Psychology, University of Klagenfurt, Klagenfurt am Wörthersee, Austria
e-mail: Frank.Eckerle@aau.at

© The Author(s) 2026
S. Sareen and S. Juhola (eds.), *Societal Transitions to Sustainability*, Palgrave Studies in Environmental Transformation, Transition and Accountability,
https://doi.org/10.1007/978-3-032-07395-2_10

prefiguration offered in the introduction to this book, "embodied strategies to render desirable futures with immediacy" (Sareen and Juhola, this
book), implies the potential of prefiguration to *communicate* desirable
futures through their embodiment. While prefiguration is routinely theorised to also affect societal norms, institutionalised politics and policy,
and to contribute to radical change (Dinerstein, 2017; Törnberg, 2021;
Yates, 2015), research on the specific psychological processes that are triggered when people participate in to experience prefiguration is still in its
infancy (Clarke & Drury, 2025). Notably missing from the literature are
accounts on the psychological effects that witnessing prefiguration has
on unassuming bystanders (but see Eckerle et al. 2025). This gap seems
even wider, when we compare it to the breadth of literature available on
people's reaction to other forms of collective action (for a recent review,
see Shuman et al., 2024).

I focus here on prefigurative disruption of everyday life because these
tactics most likely already include considerations of public reactions.
These are, for example, Critical Mass events (i.e., cyclists come together
to ride their bike and blockade streets usually full of cars) and Parking
Days (i.e., installing urban living spaces in car parks). Although prefigurative experiments are often small in scale, the shared identity and
collective empowerment people experience within them provide the
perfect seeds for movement growth and therefore carry huge potential
to garnering people's solidarity (Burrows et al., 2023). To draw out this
potential even more, it might make sense to embrace the close relationship between prefigurative politics and the political nature of the
carnivalesque (Erickson, 2021). That is, prefiguration, which laughs in
the face of the status quo and invites people to join-in on the fun and to
play with new identities—such prefiguration might be able to legitimise
its own disruption and, most importantly, spread hope.

2 Overcoming the Present by Becoming the Future

By encountering prefiguration, people realise that a certain utopian ideal is not just a futile daydream but a concrete thing that can be experienced and interacted with; utopia becomes an "objectively-real possibility" (Bloch 1986; see also Haugland & von Wirth, this book). Such encounters create tension between an unsatisfying reality and the utopian future that reality implies (i.e., the parts that reality is missing to become utopia), which is where Bloch located *educated hope* (Bloch 1986; see also Dinerstein 2022). He termed this hope *educated*, because to him, hope includes awareness of the steps required to make the implicit future become explicit reality and because it recognises the possibility of failure, thus requiring action in the here and now. The psychological approach taken in hope theory defines hope in rather similar terms: as "the perceived capability to derive pathways to desired goals, and motivate oneself via agency thinking to use those pathways" (Snyder 2002, 249). Curiously, the original publication of hope theory fails to reference Bloch while building almost word for word on his philosophy, which is a good example of how psychology can sometimes be ignorant of social theory. That aside, in both accounts, hope is differentiated from optimism because optimism does not imply that action is needed to bring the desired future about, one only has to wait.

The task for prefigurative activists who consider their audience is then to reflect how their prefigurative tactics can help translating between implicit future, explicit reality, and required action (Dinerstein, 2017). Or, in more psychological terms, they might reflect on how their prefigurative tactic can help linking perceived discontents to desired goals and thus empower moving towards them (Clarke and Drury 2025). Bloch provided an illustrative example for this task by asking, "How can this concrete utopia be a light before my feet, so that I can walk and not only dream, so that I act and not only contemplate" (Bloch & Ueding, 1975, 43).

This key distinction between moving and acting versus contemplating and dreaming points towards a central difference between *imagining abstract* utopias and *experiencing concrete* utopias: when imagining an abstract utopia, people tend to envision a fully realised society that has successfully overcome the most pressing issues of the present (Fernando et al. 2023). While this image might fill oneself with optimism and inspiration answering what-if's, it provides no answers to the more action-relevant what-now's. Therefore, optimism without a pathway for action can be dangerous because it may lead to despair (Seligman, 1972). In contrast, concrete utopias are usually flawed to some degree, constantly contested by the presence of outside and inside forces of the status quo. In other words, they are in constant state of becoming, and this becoming can be experienced physically (Swain, 2019). It is through this tangible incompleteness that concrete utopias help translating between the what-if and the what-now (Hopkins, 2019): the utopian future they present is firmly rooted in the discontents of the here and now. Recent research corroborated this, showing that participation in prefiguration often motivates people to double down on their activist identity (Bleh et al., 2024; Clarke & Drury, 2025).

3   Harnessing Carnival to Increase the Reach of Prefiguration

As indicated earlier, our insight is limited in whether experiencing prefiguration has similarly mobilising effects on bystanders. As a starting point, I was inspired by a recent publication, "What if there were no cars"? in which Cox (2024) discussed the possibility that spatial (Civlovía) or temporal (COVID-19 lockdowns) prefigurative experiences of car-free streets may have so-called *carnivalesque* effects (for a similar idea, see also Haugland & von Wirth, this book). The carnivalesque is closely related to prefiguration. Both concepts imply that the world is turned on its head for a moment in time, which provides space for a concrete utopia to flourish. In fact, carnival has some unique qualities which might greatly increase the infectiousness of prefiguration and thus its potential to contribute to excess change. That is, change that lasts beyond the lifespan of the prefigurative experiment itself (Dinerstein and Alfaro 2024).

The origins of carnival are somewhat debated. Some locate it in pre-Christian Mediterranean societies (Godet, 2020), others argue that this ignores indigenous festivals outside Europe that may just as well have affected European carnival (Liverpool 1998). Regardless of origins and history, the fact remains that nowadays Carnival is celebrated in many societies in the form of an upside-down world which mocks the powerful, empowers the regulars, and liberates the bodies (Marshall et al., 2018). While institutionalised Carnival in Europe has lost most of its political connotation since the late eighteenth century, the liberating spirit of carnivalesque festivals is alive and well in many postcolonial societies (Godet 2020), but also within some activist tactics. For instance, the Occupy movement (Tancons 2014) used carnivalesque tactics when employing the power of laughter, mockery, and role reversal to achieve liberation—giving origin to the term "protestival" (Hammond, 2020; St John, 2008).

Perhaps most famously, Bakhtin (1984, 34) commented on the transformative function of carnival in terms of:

"[...] to consecrate inventive freedom, to permit the combination of a variety of different elements and their rapprochement, to liberate from the prevailing point of view of the world, from conventions and established truths, from cliches, from all that is humdrum and universally accepted. This carnival spirit offers the chance to have a new outlook on the world, to realize the relative nature of all that exists, and to enter a completely new order of things."

In similar vein, Erickson (2021, 516) stated that "[t]he utopian nature of Carnival is demonstrated not only through abundance and temporary freedom from day-to-day responsibilities but also freedom from the need to treat higher ups with deference". Just as in prefiguration, then, carnival allows people to creatively experiment with new political meanings and aspirations, planting seeds for change by collectively imagining what it would be like if things were different.

Infusing prefiguration with the carnivalesque is hardly a new idea (e.g., Graeber 2004). We can find this in Critical Mass events in San Francisco (Blue 2012), the Gezi Park Protests in Istanbul (Salman 2022), the Tooting Tooting Trashcasters Carnival and other projects of the Transition Movement (Neal 2015), and radical clowning within the global justice movement (Bogad 2010). What all these events have in common is that they rock the assumption that the status quo is without alternative, sparking psychological processes concerned with the what-if: "The improbable and exceptional are gateways to the imagination" (Keltner & Stamkou 2024, 336).

I see three qualities of carnival that can be harnessed to increase the reach and social change potential of prefiguration: the liberating effect of laughter, the playing with and reversals of roles, and the breaking down of the dichotomy between actors and spectators.

3.1   The Liberating Effect of Carnivalesque Laughter

Monsters in Medieval carnival were grotesque metaphors of oppression which were to be ritually overcome: "The people play with terror and laugh at it" (Bakhtin 1984, 91). This laughter, much like educated hope, does not stem from a reaction to the present but from an anticipation of the future. This intriguing parallel between anticipated hope and anticipated joy can also be found in anarchist theory (see also bergman & Montgomery, 2017), for instance when Emma Goldman (1969, 26) argued that anarchism "aims to make work an instrument of joy, of strength, of color, of real harmony, so that the poorest sort of a man [sic!] should find in work both recreation and hope". Critical mass protests, for example, provide spaces for hope by joyfully reclaiming the streets (Blue 2012). It is easy to imagine how such protests could further emphasise their carnivalesque character, and it becomes necessary when we consider Deleuze's observation that "[t]he tyrant needs sad spirits in order to succeed, just as sad spirits need a tyrant in order to be content and to multiply" (Deleuze 1988, 25).

3.2   The Liberating Effect of Carnivalesque Roleplay

Play is at the same time a fun and deeply serious activity (Turner, 2008). During play, actors and potential audiences are both themselves and not themselves, creating a liminal (in-between) spacetime (Stenner, 2017). This liminality can increase the receptiveness to alternative ideas and practices, because in this space, people are not fully themselves. Few people, for instance, would exchange kisses with a stranger in public, but if a zone of play was to be created around them, they might be much more ready to do so if the rules of the game required it.

Prefiguration can be understood as such a zone of play. By embodying roles within the prefigurative experiment, activists are also negating "the singularity of the individual person in the world" (Grübel 2022, 891), thus fundamentally opposing the dominant neoliberal ideology and its disregard for the collective (Eckerle et al. 2024). As noted by Salazar-Sutil (2008, 8) while watching the carnival in Bielsa: "Everybody can see, as I stand in the balcony facing the laughing crowds, that my academic self is a questionable if not ridiculous performance in the context of carnival".

3.3   The Liberating Effect of Carnivalesque Immersion

Carnivalesque prefiguration brings out an important function of concrete utopias: breaking down the dichotomy between spectator and actor (Shahar, 2024). This is because spectators are easily incorporated through laughter (Bakhtin 1984) or identity reflection (Salazar-Sutil, 2008).

The power of immersion is illustrated by the Serbian student protests that are ongoing for over a year as of writing this chapter. Triggered by a tragic accident in which 15 people lost their lives, these student protests employ direct democracy in the form of decentralised assemblies, have no designated representative, and organise joyful, nonviolent events and protests. Most strikingly, in their *student marches*, they have been walking huge distances from one place in Serbia to another, participating in activities organised in their support by locals in what can be described as a highly festive atmosphere (called *celebration of the liberators*). The performative nature of their protests often motivates locals to interact with the activists, expressing their support materially without directly joining the movement. By immersing the public, the student protests have gained a massive solidarity among the Serbian population, which might be why the challenged regime has been hesitant about using overt aggression and repression (although there are repressive actions of the regime directed at civil society organisations and non-student activism (De La Feld, 2025), and some violent means are used to discredit students (Miholjcic Ivkovic, 2025)).

4   How Prefiguration Affects Observers: An Experimental Proof of Concept

Colleagues and I have conducted an initial test of the claim that prefiguration is not only transformative for observers, that is, people who are not part of the movement or project (see Eckerle et al., 2025 for details). Specifically, we conducted two online experiments ($N=645$) where we randomly presented to participants either a prefigurative protest (e.g., activists drew a bicycle lane overnight) or a similar non-prefigurative (indirect) protest (e.g., activists sprayed paint on the street to call on the local government to install a bicycle lane). This experimental setup allowed us to perform pairwise comparisons between the prefigurative and non-prefigurative versions of three different pairs of prefigurative versus non-prefigurative protests.

In both experiments, the prefigurative versions were perceived to be more joyful. Most of them were also rated more legitimate and increased the ability to imagine a traffic-reduced future. Interestingly, and in line with Bloch's concrete utopias, only protests, which increased the ability to imagine a traffic-reduced future also increased hope. In conclusion, we found some promising initial evidence for the notion that prefigurative traffic disruption can have transformative effects on observers by eliciting positive emotions. However, the experiment was not able to ascertain whether joy was indeed an important mechanism behind the other positive effects. While ethnographic accounts of protestivals (Acar & Uluğ, 2023; Bogad, 2010; Neal, 2015; Salman, 2022) and our research strongly suggest that the unique properties of carnivalesque prefiguration render such protests more convincing, the question of causality remains open for future research.

5 Mocking the Present from the Future

Humans live in the future: our present actions are constantly tied to our imagination of the future (Kashima et al., 2025), and other worlds (Keltner & Stamkou, 2024). This implies that one of humanity's core qualities is the ability to collectively anticipate and creatively adapt to external changes. One might call this delusional considering the lack of serious transformation efforts in the face of looming climate and ecological catastrophes and rising authoritarianism. However, that we are stuck in a dysfunctional system is not a failure inbuilt to humanity but evidence for its shackles (Fox, 1985). These shackles are tied to the stark global inequalities and power differentials carbon-capitalism has produced (Piketty 2020) as well as capitalism's negation of any future beyond itself (Holloway, 2010; Shahar, 2024). I have argued here that an underappreciated path towards liberation lies in the mocking of the present *from* the future. This strategy requires only two resources, which human collectives usually have in abundance: the ability to imagine and the joy in playing pretend.

The hopeful message is that this path is already being traversed (bergman & Montgomery, 2017). A myriad of prefigurative projects or "beautiful solutions" (Williams et al., 2024) already exist, which focus on care, conviviality, and the subversive joy of resistance (see also Batterbury et al., this book). They range from local spaces for community healing (Malherbe, 2023) to prefiguring a new global economic system that is based on solidarity (Dinerstein, 2017). They come in the form of resistance and defence movements against authoritarian leadership and corruption (such as in Serbia), participatory art (Burke et al., 2018), and protestivals (Neal, 2015) and many more that are covered in this book.

My contribution to this is a call to embrace the playfulness and absurdity of this endeavour in order to invite bystanders to join in on the fun of transforming. The more prefiguration manages to spread the joy of already being free, the more it invites playing with identities, and the more it dilutes the dichotomy between activists and spectators, the brighter will be its light before our feet.

Acknowledgements The author wishes to thank Tijana Karić who contributed the example of Serbian student protestors, and Francesca Aarons who helped with clarity and style. Both provided very helpful comments to an early version of this chapter.

Competing Interests The author has no conflicts of interest to declare that are relevant to the content of this chapter.

References

Acar, Y. G., & Uluğ, Ö. M. (2023). Ten years after the Gezi Park protests: Looking back on their legacy and impact. *Social Movement Studies, 1–6.* https://doi.org/10.1080/14742837.2023.2267996

Bakhtin, M. (1984). *Rabelais and his world.* Indiana University Press.

Batterbury, S., Chakori, S., & Uxo, C. (2025). Prefiguring consumption reduction in shadow networks: Recirculating resources, knowledge and skills in urban tool libraries and community bike workshops. In (eds. S. Sareen. & S. Juhola) *Societal transitions to sustainability: The prefigurative politics of present transformation (this book).* Palgrave Macmillan.

bergman, c., & Montgomery, N. (2017). *Joyful militancy: Building thriving resistance in toxic times.* AK Press.

Bleh, J., Masson, T., Köhler, S., & Fritsche, I. (2024). From imagination to activism: Cognitive alternatives motivate commitment to activism through identification with social movements and collective efficacy. *British Journal of Social Psychology.* https://doi.org/10.1111/bjso.12811

Bloch, E. (1986). *The principle of hope* (Vol. 1). MIT Press.

Bloch, E., & Ueding, G. (1975). Utopien des kleinen Mannes und andere Tagträume. Ein Gespräch mit Gerd Ueding. In (eds. R. Traub and H. Wiesner) *Gespräche mit Ernst Bloch* (pp. 41–57). Suhrkamp.

Blue, J. A. (2012). *Anarchy, Play, and Carnival in the Neoliberal City: Critical Mass as Insurgent Public Space Activism.* https://repository.usfca.edu/thes/1121

Bogad, L. M. (2010). Carnivals against capital: Radical clowning and the global justice movement. *Social Identities, 16*(4), 537–557. https://doi.org/10.1080/13504630.2010.498242

Boggs, C. (1977). Marxism, prefigurative communism, and the problem of workers control. *Radical America, Winter, 6*, 99–122.

Burke, M., Ockwell, D., & Whitmarsh, L. (2018). Participatory arts and affective engagement with climate change: The missing link in achieving climate compatible behaviour change? *Global Environmental Change, 49*, 95–105. https://doi.org/10.1016/j.gloenvcha.2018.02.007

Burrows, B., Uluğ, Ö. M., Khudoyan, K., & Leidner, B. (2023). Introducing the collective action recursive empowerment (CARE) model: How small-scale protests led to large-scale collective action in Armenia's velvet revolution. *Political Psychology, 44*(2), 397–418. https://doi.org/10.1111/pops.12858

Clarke, D., & Drury, J. (2025). Emergent prefigurative politics and social psychological processes: A systematic review and research agenda. *Journal of Community & Applied Social Psychology, 35*(1), 1–17. https://doi.org/10.32388/MAMYEK.3

Cox, P. (2024). What if there were no cars? Prefigurative cycling activism for a degrowth world. *Ephemeral Journal, 24*(1), 143–168.

De La Feld, S. (2025, February 26). *Serbia, Vučić increasingly repressive: Police raid four NGO's offices to please Trump*. Eunews. https://www.eunews.it/en/2025/02/26/serbia-vucic-increasingly-repressive-police-raid-four-ngos-offices-to-please-trump/

Deleuze, G. (1988). *Spinoza, practical philosophy*. City Lights Books.

Dinerstein, A. C. (2017). Co-construction or prefiguration? The problem of the 'translation' of social and solidarity economy practices into policy. In (eds. P. North & M. S. Cato), *Towards Just and Sustainable Economies* (pp. 57–72). Policy Press.

Dinerstein, A. C. (2022). Decolonizing prefiguration: Ernst Blochs philosophy of hope and the multiversum. In L. Monticelli (Ed.), *The future is now: An introduction to prefigurative politics* (pp. 47–64). Bristol University Press.

Dinerstein, A. C., & Alfaro, M. J. V. (2024). Intricacies of prefiguration: A conversation about hope, resistance, critical theory, open Marxism, and weaving. *Ephemeral Journal, 24*(1), 215–244.

Eckerle, F., Clarke, E. J. R., & Lutz, A. E. (2024). Wider die Reproduktion neoliberaler Ideologien in der Umwelt-und Sozialpsychologie: Ein Aufruf zur kritischen Reflexion und Präfiguration. In (eds. Arbeitskreis Kritische Umweltpsychologie der Initiative Psychologie im Umweltschutz e.V. and Psychologists/Psychotherapists for Future e.V.), *Kritische Umweltpsychologie: Krisen verstehen, Handlungsfähigkeiten entwickeln* (pp. 61–86). Psychosozial-Verlag.

Eckerle, F., Clarke, E. J. R., & Landmann, H. (2025). When the means are no dead end: Effects of wittnessing direct collective action for traffic transformation. Manuscript under review. https://osf.io/preprints/psyarxiv/xsbm5

Erickson, B. (2021). Grotesque logic: Catalan carnival utopias and the politics of laughter. *Visual Studies, 36*(4–5), 507–523. https://doi.org/10.1080/1472586X.2020.1798810

Fernando, J. W., Burden, N., Judge, M., O'Brien, L. V., Ashman, H., Paladino, A., & Kashima, Y. (2023). Profiles of an ideal society: The utopian visions of ordinary people. *Journal of Cross-Cultural Psychology, 54*(1), 43–60. https://doi.org/10.1177/00220221221126419

Fox, D. R. (1985). Psychology, ideology, utopia, and the commons. *American Psychologist, 40*(1), 48–58. https://doi.org/10.1037/0003-066X.40.1.48

Godet, A. (2020). Behind the masks: The politics of carnival. *Journal of Festive Studies, 2*(1), 1–31. https://doi.org/10.33823/jfs.2020.2.1.89

Goldman, E. (1969). *Anarchism and other essays*. Dover Publications.

Graeber, D. (2004). *Fragments of an anarchist anthropology*. Prickly paradigm press.

Grübel, R. (2022). Carnival, carnivalism and Bakhtin's culture of laughter. In M. Mrugalski, S. Schahadat, & I. Wutsdorff (Eds.), *Central and eastern European literary theory and the west* (pp. 887–898). De Gruyter. https://doi.org/10.1515/9783110400304-047

Hammond, J. L. (2020). Carnival against the capital of capital: Carnivalesque protest in Occupy Wall street. *Journal of Festive Studies, 2*(1), 265–288. https://doi.org/10.33823/jfs.2020.2.1.47

Haugland & von Wirth. (2025). Pursuing and producing politics through prefiguration: the case of urban experimentation. In (eds. S. Sareen, & S. Juhola) *Societal transitions to sustainability: The prefigurative politics of present transformation* (this book). Palgrave Macmillan.

Holloway, J. (2010). *Crack capitalism*. Pluto.

Hopkins, R. (2019). *From what is to what if: Unleashing the power of imagination to create the future we want*. Chelsea Green Publishing.

Kashima, Y., Bain, P. G., Fernando, J. W., & Truong, A. (2025). Collective future thinking in cultural dynamics. *European Review of Social Psychology, 1–45*. https://doi.org/10.1080/10463283.2025.2458961

Keltner, D., & Stamkou, E. (2025). Possible worlds theory: How the imagination transcends and recreates reality. *Annual Review of Psychology, 76*, 329–358. https://doi.org/10.1146/annurev-psych-080123102254

Liverpool, H. U. (1998). Origins of rituals and customs in the Trinidad carnival: African or European? *The Drama Review, 42*(3), 24–37.

Malherbe, N. (2023). Returning community psychology to the insights of anarchism: Fragments and prefiguration. *Journal of Social and Political Psychology, 11*(1), 212–228. https://doi.org/10.5964/jspp.9385

Marshall, E. Z., Farrar, M., & Farrar, G. (2018). Popular political cultures and the Caribbean carnival. *Soundings: A Journal of Politics and Culture, 67*(67), 34–49.

Miholjcic Ivkovic, N. (2025, January 22). *Changing the Narrative: How the Serbian State Works to Demonise Protesters*. Balkan Insight. https://balkanins ight.com/2025/01/22/changing-the-narrative-how-the-serbian-state-works-to-demonise-protesters/

Neal, L. (2015). *Playing for time: Making art as if the world mattered*. Oberon Books.

Piketty, T. (2020). *Kapital und Ideologie*. C.H.Beck.

Salazar-Sutil, N. (2008). Carnival post-phenomenology: Mind the hump. *Anthropology Matters, 10*(2), 1–11. https://doi.org/10.22582/am.v10i2.34

Salman, S. (2022). *The Gezi protests: Between carnivalesque and prefiguration* [Master Thesis, Middle East Technical University]. https://open.metu.edu.tr/handle/11511/99425

Sareen, S., Juhola, S. (2025). The prefigurative politics of present transformation. In (eds. S. Sareen & S. Juhola) *Societal transitions to sustainability: The prefigurative politics of present transformation* (this book). Palgrave Macmillan.

Seligman, M. E. P. (1972). Learned helplessness. *Annual Review of Medicine, 23*(1), 407–412.

Shahar, K. (2024). The no-where and the now-here: Ernst Bloch's concrete utopia. *Critical Horizons, 25*(4), 315–328. https://doi.org/10.1080/144 09917.2024.2431789

Shuman, E., Goldenberg, A., Saguy, T., Halperin, E., & van Zomeren, M. (2024). When are social protests effective? *Trends in Cognitive Sciences, 28*(3), 252–263. https://doi.org/10.1016/j.tics.2023.10.003

Snyder, C. R. (2002). Hope theory: Rainbows in the mind. *Psychological Inquiry, 13*(4), 249–275. https://doi.org/10.1207/S15327965PLI1304_01

St John, G. (2008). Protestival: Global days of action and carnivalized politics in the present. *Social Movement Studies, 7*(2), 167–190. https://doi.org/10.1080/14742830802283550

Stenner, P. (2017). *Liminality and experience: A transdisciplinary approach to the psychosocial*. Palgrave Macmillan.

Swain, D. (2019). Not not but not yet: Present and future in prefigurative politics. *Political Studies, 67*(1), 47–62. https://doi.org/10.1177/003232171 7741233

Tancons, C. (2014). Occupy Wall street: Carnival against capital? Carnivalesque as protest sensibility. In I. P. Werbner, M. Webb, & K. Spellman-Poots (Eds.), *The political aesthetics of global protest* (pp. 291–319). Edinburgh University Press. https://doi.org/10.1515/9780748693504-016

Törnberg, A. (2021). Prefigurative politics and social change: A typology drawing on transition studies. *Distinktion: Journal of Social Theory, 22*(1), 83–107. doi:https://doi.org/10.1080/1600910X.2020.1856161.

Turner, V. (2008). *From ritual to theatre: The human seriousness of play* (6. print). PAJ Publ.

Williams, E., Plattus, R., Feghali, E., & Schneider, N. (2024). *Beautiful solutions: A toolbox for liberation.* OR Books.

Yates, L. (2015). Rethinking prefiguration: Alternatives, micropolitics and goals in social movements. *Social Movement Studies, 14*(1), 1–21. https://doi.org/10.1080/14742837.2013.870883

Open Access This chapter is licensed under the terms of the Creative Commons Attribution-NonCommercial-NoDerivatives 4.0 International License (http://creativecommons.org/licenses/by-nc-nd/4.0/), which permits any noncommercial use, sharing, distribution and reproduction in any medium or format, as long as you give appropriate credit to the original author(s) and the source, provide a link to the Creative Commons license and indicate if you modified the licensed material. You do not have permission under this license to share adapted material derived from this chapter or parts of it.

The images or other third party material in this chapter are included in the chapter's Creative Commons license, unless indicated otherwise in a credit line to the material. If material is not included in the chapter's Creative Commons license and your intended use is not permitted by statutory regulation or exceeds the permitted use, you will need to obtain permission directly from the copyright holder.

11

Community-Based Proleptic Environmentalism: Zero-Waste and Kafka's 'Weg-von-hier'

Gerald Taylor Aiken

1   Introduction

Ich befahl mein Pferd aus dem Stall zu holen. Der Diener verstand mich nicht. Ich ging selbst in den Stall, sattelte mein Pferd und bestieg es. In der Ferne hörte ich eine Trompete blasen, ich fragte ihn, was das bedeutete. Er wusste nichts und hatte nichts gehört. Beim Tore hielt er mich auf und fragte: "Wohin reitet der Herr?" "Ich weiß es nicht", sagte ich, "nur weg von hier, nur weg von hier. Immerfort weg von hier, nur so kann ich mein Ziel erreichen." "Du kennst also dein Ziel", fragte er. "Ja", antwortete ich, "ich sagte es doch: ‚Weg-von-hier '—das ist mein Ziel." "Du hast keinen Eßvorrat mit", sagte er. "Ich brauche keinen", sagte ich, "die Reise ist so lang, daß ich verhungern muß, wenn ich auf dem Weg nichts bekomme. Kein Eßvorrat kann mich retten. Es ist ja zum Glück eine wahrhaft ungeheure Reise."

G. T. Aiken (✉)
Research Institute for Sustainability (RIFS), Potsdam, Germany
e-mail: gerald.aiken@rifs-potsdam.de

© The Author(s) 2026   **165**
S. Sareen and S. Juhola (eds.), *Societal Transitions to Sustainability*, Palgrave Studies
in Environmental Transformation, Transition and Accountability,
https://doi.org/10.1007/978-3-032-07395-2_11

I asked for my horse to be brought from the stable. The servant didn't understand me. So, I went into the stable, saddled my horse and got on it. Far away I heard a bugle sound out. I asked my servant what it meant but he didn't know and hadn't heard. By the gate he stopped me and asked: 'Where are you riding to?' I answered, 'away from here, away from here, always away from here. Only by going that way can I reach my goal.' 'Then you know your target?', he asked. 'Yes', I said, 'I have already said so, "Away-From-Here", that is my destination.' 'But, you have no provisions with you,' he said. 'I don't need any,' I said. 'The journey is so long that I will die of hunger if I do not get something along the way. It is, fortunately, a truly immense journey.'
(Kafka, der Aufbruch [author's translation]).

There has been a surfeit of writings on prefigurative politics in recent years, including this very book(!). Less attention though has focused on the way in which prefigurative action can be a *reaction to an unwanted future event*. That is, social movements, activists, and community initiatives foresee an undesirable situation coming into view, predicting an oncoming situation heading down the line that they then react to and *away from*. What they are moving away from has not yet come to pass, and so in a way they are reacting to something which has not yet happened. This is difficult to grasp as they first need to imaginatively bring this 'not yet' into the now, and then react away from it. This particular temporal interrelationality is a crucial and underacknowledged part of prefigurative politics, which I want to dive into in this chapter.

Prefigurative politics are often described as requiring a shared future vision to be enacted (Jeffrey & Dyson, 2020). In the case of *Ouni*, we see a particular expression of a social movement focused not on reacting to an event, or seeking to enact or bring into being a new arrangement. Rather, *Ouni* are primarily focused on moving away, of deconstructing and dismantling (similar to Feola, this book). Moving away is a key and neglected part of prefigurative politics. This adds an important layer to understandings of refusal and deconstruction in practices of prefiguration, which then has implications for temporality (see also Feola, 2019, 2025).

It helps to see that many community-based initiatives, in this specific and programmatic sense, have anticipation at the heart of their orientation. In how these community initiatives relate to the future, they are prefigurative, in that they lead by example—both in terms of making do with less, but also how they participate and organise themselves. More than this, though, they are also proleptic. Community initiatives are often claimed to be 'prefigurative', in the sense that they embody and enact the future state of affairs they wish to bring into being (Pickerill & Chatterton, 2006). However, for community initiatives such as *Ouni*, they are as much about anticipation as they are about enacting. Community initiatives, such as *Ouni*, are not only prefigurative, but proleptic.

Prolepsis is the anticipation of a new thing and is also associated with answering a possible object before it is raised. To prefigure by contrast is to imagine beforehand, to be an early indication of something. Literally and etymologically, to be prefigurative is to 'represent beforehand'; whereas prolepsis is to 'anticipate'. Acting proleptically, is acting as if the future is already present, even when it may not be. Prefiguration is bringing the future into being (by enacting), prolepsis is engaging with and responding to the future context ahead of time (through staging), before it is properly here. Preppers, or living without oil, as many eco-community movements in the West do, in the expectation that peak oil will inevitably hit, is strictly speaking not prefigurative, but proleptic. Likewise moving away from a particular stretch of coastline under the assumption that is will (soon) be inundated, is not a prefigurative act, but a proleptic one. Here, shifting eating habits in order to respond to anticipated future shocks to the food system is also proleptic. It helps to understand this proleptic move in the manner of Kafka's rider—the desire is to anticipate and move away from here.

This chapter will then, first, introduce *Ouni*, who they are, and what they focused on. Swiftly following, the chapter then dives into how this research was carried out, including contextualising information about the context, and myself as researcher. There then follows an approach to how the temporal politics of prefigurative action played out in *Ouni*, in the light of this short story from Kafka. Centrally, this is through an appreciation of *Ouni* as a proleptic social movement.

2   Ouni

Ouni—Luxembourgish for 'without'—started in 2016 in Luxembourg City. It began as a vanguard sustainability shop that offered zero-waste and eco-friendly groceries. By emphasising its plastic-free character, it was the first shop in Luxembourg that required visitors to bring their own bags and containers to buy their products. These were mostly foodstuffs such as grains, nuts, herbs and spices, oils, and a range of cleaning and personal hygiene products. The shop also had a small café and bakery space, which served as a social hub. *Ouni* closed in 2021, the pandemic having taken its toll on the social enterprise.

Part of the motivation for starting *Ouni* was the desire to (re)connect consumers with the agricultural production of food. A key awareness mentioned in interviews with those who initiated and those who then staffed and volunteered in *Ouni* was the idea that we 'have only 60 harvests left'.[1] This claim from the UNs International Year of Soils in 2015 has been much debated. However, those who started this community enterprise though took it very much as a warning and at face value. Participants' motivations included the desire to participate with like minds in a collective action project, were attracted by a new economic model introducing a profit-free way to carry out grocery shopping, and for some, *Ouni* was a way to promote veganism and make easier and more visible legume-based food cultures. Their worry about this imagined future of a world without productive soil for harvests was starkly realised and then fed into their imagining and building of *Ouni*.

Ouni was a proleptic example of responding to this challenge. They anticipated declining soil productivity and sought out a response. Their response was akin to Kafka's Away-From-Here. *Ouni* were not *stricto sensu* trying to represent something new, or bring into being an already worked out settled state of affairs. Rather, akin to a via *negativa*, they sought to reject and forestall the future. *Ouni* did seek to 'represent beforehand', and bring into being a positive vision of zero-waste, pesticide-free, community-owned grocery shopping. Yet, the way *Ouni* was designed and conceived was as much in moving away from a negative

---

[1] https://www.fao.org/soils-2015/events/detail/en/c/338738/

state of affairs as towards a positive alternative vision. In this way, their moving away from an imagined negative is better described as proleptic than prefigurative.

When environmental movements have been described as prefigurative, it is often because they are trying to build something new. Alternatively, seeing them as deconstructively proleptic allows an alternative perspective. Jeffrey and Dyson, in a key paper, define prefigurative politics as: 'the self-conscious channeling of energy into modelling the forms of action that are sought to be generalised in the future in circumstances characterised by power, hierarchy, and conflict' (2020). The prefigurative politics of *Ouni* however are not about developing generalisable models, but about deconstructing existing models, resisting patterns of behaviour or ways of living they identify as the root cause of harmful living. Exploring *Ouni* in the light of this short story from Kafka sheds light on how we can understand prefigurative politics today.

3   Researching Ouni

This research is part of a wider set of explorations addressing the community-based economies of Luxembourg. In terms of positionality, I am broadly sympathetic with the aims and objectives of these groups and want to see them succeed. It was a moment of deep sadness in our family's life when we realised *Ouni* would shut during the pandemic. I was never a formal member of *Ouni*, but was a regular customer and near neighbour, involved only through 'the strength of weak ties', before carrying out the more formal part of the research here. This more formal part consisted of semi-structured interviews with those involved in *Ouni* across all levels of involvement: from key players to 'regular' volunteers and members. Observational notes fleshed this out in a research diary drawing on my wide variety of experiences of this shop. Much of this chapter and the research underpinning it came from the time I spent with this Kafka short story as a German learner, and reflected on the connections and comparisons between this work of fiction and the practical example of *Ouni*.

I must also add here that due to being a neighbour, my regularly shopping and passing by the shop front with my two young children led to connections and conversations with those involved in *Ouni*, unlike any of those I have experienced in other research encounters where my family and personal life are less present. I feel certain that this led to barriers lowering and connections being made. Nevertheless, it feels an important part of gaining access and trust to participants here. Certainly, in most of the research encounters here I was primarily seen as a local resident and father, rather than an 'expert' or researcher. In Luxembourg too, given its multilingual character and the politics of language choice and fluency, my status as a primarily anglophone 'outsider' also helped to navigate some of the challenges between the more locally sedimented roles and expectations (Luxembourgish vs French vs Portuguese).

Partly, my 'outsider' status also emerged from the typical character of those involved in *Ouni*. Without any statistical overview, many of those involved spoke Luxembourgish and were well set up for life in what can be a very expensive place to live. By this I mean it was not untypical that either they or their partner would have a privileged state civil service job, or be a benefactor of inherited wealth, and especially property, enabling their daily life and subsistence, or ability to volunteer. But this was by no means universal, and it was also not uncommon to have those working full time but also volunteering 2 hours a month, for example. Or volunteers who had recently moved to Luxembourg, including those who only spoke one language.

4   Prefiguration, Prolepsis, and Responding to Events

Even when social movements are claimed to emerge 'spontaneously', there is often a trigger event, such as the self-immolation of Tunisian stallholder Mohamed Bouazizi, in sparking the 'Arab Spring'. Environmental social action groups such as *Ouni* have an interesting relationship to the idea of the 'event', distinct from the 'event' of Mohammad Bouazizi. *Ouni* is proleptic in the sense that they anticipate future events, such as climate change and future states of environmental degradation, acting as if they are already on their way. Not only do they anticipate, but these initiatives also 'act ahead of time' as one staff member put it. They behave in a way not literally necessary or applicable to the present (such as life without oil, or consuming food without degrading soils), in order to proleptically inaugurate such a future. The 'event' these groups are proleptically responding to is an as-yet-to-come, imagined apocalyptic vision of a world without soil, battered by climate change. In this, these groups perform a curious loop of imagining a future towards which society, or the world, is heading. Instead of seeing this future as destination though, eco-social initiatives like *Ouni* re-imagine this future into the present. Like the rider in Kafka's short story above, their direction of travel, is 'Away-from-here' (*Wegvonhier*). That is their goal (*das Ziel*). It is this fidelity to a proleptically inaugurated future event, which is crucial to grasping their direction of travel and relationship between time and community.

The Transition movement, which forms the backdrop to the groups studied here in general, was founded as 'a *response* to the twin threats of climate change and peak oil' (Hopkins, 2008), this is what motivates and drives them. This raises the question, what is it exactly they are responding to? Climate change, peak oil, or a world without productive soil are not 'events' understood in conventional terms. That is, they are 'events' that are yet to be fully unveiled, events only in so much as they are imagined, projected, and planned for. These events are then responded to, only in so much as the response is to an imagined, inaugurated future, a future realised in the present.

Social movements generally travel towards a specific desired future goal, or away from a definite event in the past. For instance, the Communist Party travels towards the common ownership of the means of production. Although away from private ownership, it is their intended future goal, which is clear, motivating, and inspiring. Other movements see some dissatisfaction in the present or near past, and intent to move away from it. Many environmental justice campaigns have operated in this way. The 'events' symbolised by Mohamed Bouazizi above fit this pattern too. The transition that these environmental groups envision doesn't, however, strictly fit into either of these categories. These events, such as a future world without soil, are imagined into the present, or near past, before reacting to them. In this way *Ouni*—akin to the peace movement and nuclear war—are moving away from a future event, a subtle but important distinction.

Ouni and Kafka offer a way into understanding prefigurative political action operating in deconstructive manner and differently to how prefiguration is generally understood. Jeffrey and Dyson (2020) outline prefiguration as how: 'people enact a vision of change—through organisation, design, architecture, practices, bodies, or something as simple as a gesture or demeanour—and promote this as indicative of an imminent or more distant "future"'. But, what *Ouni* is doing is not so much enacting a vision of change as attempting to forestall it.

Environmental initiatives such as *Ouni* do intend to travel, but they are not travelling directly towards something. Like Kafka's rider, at first, they appear to be travelling somewhere—a world without packaging, with healthy soils, low-carbon society perhaps. But on closer inspection, it appears that somewhere they are travelling to, is an 'Away-From-Here'. *Ouni* travels away from a world of easy consumption, damaging environmental externalities, a fast, stressful living, without time to slowly cook and eat together. This was mostly seen in the literal and metaphorical figure of the pulses they sold in their shop. Often those living in the city imagined they had not enough time to soak these, and plan ahead to cook the nutrient-rich legume recipes. These legumes were tied up with a vegan, or low-meat diet, and multiple aspects of these ingredients—the cooking time, the integrated planning of cooking and eating, the lack of instant, and a convenience culture perhaps summed up through the online food delivery meals—in a symbolic fashion.

Ouni's sense of purpose if in its own way utopian. Their goal, like the rider's is to be away from here. To be away-from-here though, is not simply to be in another place, an 'over there'. It is, as Butler's interpretation of this short story indicates, to be '*free of the spatiotemporal conditions of "here". We would not only have to be elsewhere, but that very elsewhere would have to transcend the spatiotemporal conditions of any existing place*' (Butler, 2011). In this way, utopia—literally *u topos* (no place)—according to Thomas More, is this very *Wegvonhier*, which the rider attempts to reach. *Ouni* seek to reach and attain a utopian away-from-here. A place of food justice where everyday calorie consumption is achieved in right relationship to a whole ecology, economics and value chain.

5   Prolepsis and Building Community

Why does this matter? *Ouni,* as a community-owned and run enterprise, relied on and purposively built solid social bonds and connections between those who worked there. As outlined above, social movement theory recognises that an initial event, or moment of clarity or crisis is often very helpful as a coalescent point to build a moment on or to gel social relations. While journeying towards a zero-waste world, as quickly as possible, but without a definitive point to achieve this by, can be motivating and for some can serve this function. The urgency of moving away from a real or imagined immediate event, such as running out of harvests, can motivate, encourage, and provide a focus for action in a more purposive sense. For *Ouni,* the specific event mattered, but also the will to continually deconstruct, journey onwards, and keep *moving away* from present and anticipated future injustices.

Butler calls Kafka's 'away from here' a gesture, in the tradition of critical theory. A gesture because: 'a gesture opens up a horizon as a goal, but there is no actual departure and there is surely no actual arrival' (Butler, 2011). Prefiguration concerns acting to bring about a desired circumstance or set of affairs, prolepsis involves acting *as if* these circumstances or sets of affairs are already there. Acting proleptically fits nicely with the general tone of the Kafka story then: restless, continual pursuit of a more desirable future. The desired end point, the eventual goal, is always shifting, elusive but still within sight. This means that whatever state of affairs one has brought into being, can again be critiqued, improved upon and responded to.

Ouni never rested in their aims and ambitions. After becoming plastic free, they then turned their attention to selling as many local, organic and fairly traded items as possible. After selling as wide a range of foodstuffs as they possibly could, they then moved towards selling cleaning products and also cosmetics. After making progress with their environmental and economic goods they sold, they then returned to their social model, to the ways in which they included volunteer labour and their pricing structure for members and non-members. In each of these tweaks and improvements, it's more accurate to say their imagined destination was away-from-here. That is, they did not set up an ideal or preferred set of outcomes, or set about bringing into being a set state of affairs, but raised a critique against contemporary city living environmentally, economically, and socially. They then always sought to move away from this object of critique. The destination (if there was one) was not a named place, or circumstance, but away from a present undesirable reality.

It is drawing attention to the proleptic aspects of *Ouni's* present transformation that one can fully appreciate their restless continual pursuit of food justice, rather than only setting up a better or more preferable institution. Kafka's story helps us to pay attention to the usefulness of having a goal to motivate for urgent action. But also that it is important that this goal is not a specific once-and-for-all ideal destination. But rather can continually shift and evolve if one is to stay true to the founding principles and desires of the initiative.

6   Conclusion

I want to conclude this chapter with a short reflection on the last part of this enigmatic short story of Kafka, particularly apt to *Ouni*: *'The journey is so long that I will die of hunger if I do not get something along the way. It is, fortunately, a truly immense journey'*. *Ouni's* journey is so long, and they are also focused on finding something to eat along the route. It is often recognised in writing on the role of food and food culture in prefigurative politics that food has a particular immediacy in binding and gelling participants together in a way other sectors, such as energy, do not. We each have a direct and collective relationship to the food we eat; we (often) need to eat every day; we put it inside ourselves; our habits and practices of eating involve a whole tradition, culture, and way of being that connects us.

In the spirit of Kafka's story, there is a parallel with the Jewish story of the manna from heaven that provided and sustained the Israelites in the desert. This sustenance was provided each morning, and there those who needed had their fill. Importantly, this could not be stored up or preserved, not only because the wanderers needed to travel light but because the manna would spoil and go off. Collecting and eating the manna thus became a daily activity that enabled the travellers to go lightly but also not go without. For Kafka's rider, who no doubt also comes from a similar storytelling tradition, the ability to wayfare, to navigate through one's journey and find provisions *en route*, was crucial to travelling light, but also sets a parameter for developing a food culture of daily gathering and consuming. *Ouni*—the Luxembourgish word for without—had touched on this by purposively developing a travel light ethos, not gathering too much in the awareness that it could spoil, but finding and consuming regularly the food they needed for daily life. By regularly returning to the *Ouni* shop to top-up the shoppers' kitchens, face-to-face encounters took place, and a collective daily life, and involvement of others within their food culture. In short, the surrounding food culture coalescing around *Ouni* produced its own community. A community based on the shared interest of daily sustenance.

Travelling somewhere new, and seeking sustenance on the way, are both parts of this 'truly immense journey' of community-based prefigurative and proleptic politics. Neither though requires a specific desired future, or a fully planned or known in advance provisioning. All that was required is that they knew their destination was Away-From-Here, and they would need to continually restore themselves on the way. In this way not only *Ouni*, but also Kafka, have some valuable lessons for community-based environmentalists today

Competing Interests The author has no conflicts of interest to declare that are relevant to the content of this chapter.

References

Butler, J. (2011). Who owns Kafka? *London Review of Books, 33*(5). 3rd March.

Feola, G. (2019). Degrowth and the unmaking of capitalism: Beyond 'decolonisation of the imaginary'. *ACME: An International Journal for Critical Geographies, 18,* 977–997. https://acme-journal.org/index.php/acme/art icle/view/1790

Feola, G. (2025). Broadening the understanding of deconstruction in prefigurative social spaces. In S. Sareen & S. Juhola (Eds.), *Societal transitions to sustainability: The prefigurative politics of present transformation* (this book). Palgrave Macmillan.

Hopkins, R. (2008). *The transition handbook: From oil dependency to local resilience.* Green Books.

Jeffrey, C., & Dyson, J. (2020). Geographies of the future: Prefigurative politics. *Progress in Human Geography, 45*(4). https://doi.org/10.1177/030913 2520926

Kafka, F. (1936). Der Aufbruch. In: *Beschreibung eines Kampfes: Novellen, Skizzen, Aphorismen aus dem Nachlaß* Heinr. Mercy Sohn.

Pickerill, J., & Chatterton, P. (2006). Notes towards autonomous geographies: Creation, resistance and self-management as survival tactics. *Progress in Human Geography, 30*(6), 730–746. https://doi.org/10.1177/030913250 6071516 (Original work published 2006).

Open Access This chapter is licensed under the terms of the Creative Commons Attribution-NonCommercial-NoDerivatives 4.0 International License (http://creativecommons.org/licenses/by-nc-nd/4.0/), which permits any noncommercial use, sharing, distribution and reproduction in any medium or format, as long as you give appropriate credit to the original author(s) and the source, provide a link to the Creative Commons license and indicate if you modified the licensed material. You do not have permission under this license to share adapted material derived from this chapter or parts of it.

The images or other third party material in this chapter are included in the chapter's Creative Commons license, unless indicated otherwise in a credit line to the material. If material is not included in the chapter's Creative Commons license and your intended use is not permitted by statutory regulation or exceeds the permitted use, you will need to obtain permission directly from the copyright holder.

12

Imagining Futures in the Present: Conceptualising Prefigurative Impulses Through the Politics of Organising

Kavitha Ravikumar ⓘ

1   Introduction

Concepts around prefiguration are discussed in many forms (Wilson, 2024; Yates et al., 2024), and their underlying ethos all prioritise working towards change in the present over waiting for systems change to occur (Wilson, 2024). However, scholars have cautioned against instrumentalising actions and experiments with formal mechanisms, highlighting the ability of prefiguration to be spontaneous and include alternative and flexible organisational practices (Kruglanski, 2024). They also emphasise that the prefigurative approach can nurture community-based social relations by transcending institutional norms, reimagining struggles for a better future through imaginative practices and fostering trust in change through improvisation and collective action (Brissette, 2016; Moskovitz & Garcia-Lorenzo, 2016). At the same time, discourses

K. Ravikumar (✉)
Institute for International Management and Entrepreneurship & Institute for Creative Futures at Loughborough University, London, UK
e-mail: K.Ravikumar@lboro.ac.uk

© The Author(s) 2026
S. Sareen and S. Juhola (eds.), *Societal Transitions to Sustainability*, Palgrave Studies in Environmental Transformation, Transition and Accountability,
https://doi.org/10.1007/978-3-032-07395-2_12

around sustainability recognise the power of envisioning the future as a stimulus for acting for change (Wiek & Iwaniec, 2014) as deliberate anticipation can help individuals and communities to envision long-term changes. This in turn can influence their present actions and decisions (Mische, 2014; Pereira, 2021).

Hence, this chapter starts with the premise that directed action requires a thought or an initial conception of that idea as 'ideas do play an important part in the determination of action' (Parsons 1938, 652) and proposes that these ideas or imaginative thoughts when occurring within the spaces of prefigurative politics can also be called *prefigurative impulses*. The Oxford English Dictionary includes 'a thrust, a push' (Dictionary n.d.) as definitions for the term impulse in its use as a noun. Leach (2013) calls 'the prefigurative impulse' as one that occurs historically at the beginning stages of rebellions, and Törnberg (2021) attributes their emergence to political failures and current issues that create pressurised conditions and produce vulnerabilities. While the term has been used loosely in prefiguration studies (Brissette, 2016; Maeckelbergh, 2016), the concept of prefigurative impulses remains underexplored.

To illustrate the broader reach of the concept, this chapter introduces the term 'extra-work' events, for a variety of interstitial spaces where organising and participation in collective interactions are acts of spatial prefigurative politics. It then traces how the imaginings that emerge can be called prefigurative impulses in organisational contexts. This chapter seeks to contribute to theoretical perspectives in both the literatures around prefiguration and organising by further developing the concept of 'prefigurative impulses' as emerging due to the spatial performance of prefigurative politics through acts of organising.

2   The Spaces of Collective Prefigurative Politics

Organisational actors often experience a convergence between their sustainability-oriented work and a personal commitment to social change (Augustine, 2021; Welbourne et al., 2017). In supportive workplaces, they form internal teams around social or sustainability interests

(Welbourne et al., 2017), but in siloed or unsupportive environments, they lack internal networks and support (Augustine, 2021). This drives them to seek connections beyond work hours and across formal organisational boundaries, fostering dialogue, empathy, and identity within a broader movement (Tonkinwise, 2024; Welbourne et al., 2017). Emerging in response to the complexity of global challenges, these spaces integrate individual and cultural visions of the future with organisational processes, offering a holistic approach to change (Conway, 2012).

The recognition of space in this sense goes beyond it being just a physical backdrop to social action; space here relationally both produces and is a product of interactive and interconnecting social practices (McGregor, 2003). Organising as a concept can involve processes for reducing ambiguity or vagueness amongst actors (Tsoukas & Chia, 2002), and the resultant spaces reflexively account for each other (Cnossen & Bencherki, 2018). Using prefigurative politics in a collective and spatial sense has been recognised both an organisational tool and as a simultaneous act of refusal and creation in organisation studies (Schiller-Merkens, 2022, 463). Collectives and collective interactions can be both a product and a creator of space. Collectives not only consist of different stakeholders, but they also incorporate individuals with more than one skill-set (Vangkilde & Rod, 2020) where the attitudes and expertise of individuals enter settings where collective imagining emerges as a process of negotiation (Holland & Roudavski, 2016). Collectives can be loose communal undertakings connected to projects or teams or even coordinated actions of unidentifiable or ad-hoc groups (Botero, 2013, 22; Escobar, 2017, 211).

A way that such practices manifest includes the performative and spatial genres of prefigurative politics where people enact a vision of change and 'through organisation, design, architecture, practices, bodies, or something as simple as a gesture or demeanour' and promote this as suggestive of future states (Jeffrey & Dyson, 2020, 643). Contemporary literature on prefigurative politics tends to follow practical studies of prefigurative practices, concrete manifestations of alternative organisations or utopias (Schiller-Merkens, 2022) or identification of social movements (Kruglanski, 2024; Swain, 2017) often in connection with anarchism or other political ends (Habersang, 2024; Jeffrey & Dyson,

2020; Van de Sande, 2013). However, prefigurative politics is not just about expressing political ideas through actions; it can also be about creating experimental or 'alternative' ways of organising that anticipate certain aspirations (Yates, 2015) and enable individuals to support one another in radical ways, fostering life in the present (Engler & Engler, 2014) while embracing improvisation (Jeffrey & Dyson, 2020).

In mainstream economics and management literature, strategic futures are prioritised and the interventions that are studied are often proposed by small groups of external or senior people within the system (Moskovitz & Garcia-Lorenzo, 2016, 199). Prefigurative politics differentiates itself from strategic politics by downplaying instrumental goals of future improvements or change and becomes more attuned to the social and emotional needs of participants by providing spaces of mutual support through processes and networks that sustain people in the present (Engler & Engler, 2014). Social movements alter both politics and culture (Moskovitz & Garcia-Lorenzo, 2016), and when actors recognise the future orientation of their present activities, these can be described as prefiguring even if they do so without being explicit (Swain, 2017). For instance, horizontal movements rejecting hierarchies or specific demands, as in the case of Occupy London, can be viewed as a form of prefigurative organising (Reinecke, 2018). Prefigurative politics can proceed through multiple forms of improvisation, and successful prefigurative politics can become internal to organisations and movements to help reform from the inside (Jeffrey & Dyson, 2020).

The understanding of prefigurative politics as a human experience-oriented, socio-cultural practice that sits within ongoing political structures (Schiller-Merkens, 2022) recognises prefigurative politics as pursuing transformation interstitially 'circumventing, rather than challenging, existing relations of domination' (Pellizzoni, 2021, 365). This understanding also offers a process-based view on how practices develop, moving the focus from a linear definition of success to a more value-based notion of significance and intent (Van de Sande, 2013), where prefigurative politics can be an example of interstitially bypassing existing power structures rather than confronting them directly (Pellizzoni, 2021).

Prefigurative politics can hence be mobilised across diverse contexts in fluid and pluriversal ways, where moments of contestation, compromise, and possibility may flicker and fade or take root and grow (Sareen and Juhola 2025). By making the present an object of intense reflection, prefigurative politics can allow for different perspectives to be generated (Jeffrey & Dyson, 2020) and foster a politics of affinity by 'nurturing empathy, through individuals connecting with others in community, giving them opportunities to see where they may fit in the struggle, and together, create the future they would like to see in the present' (Croswell, 2024, 203).

Research has shown that imagining utopia and prefigurative practices as well as politics are tied together (Habersang, 2024; Van de Sande, 2013), leading to an understanding that the seeds of prefiguration in rebellions and social movements might indeed lie in the imagining of radical possibilities as a starting point, that can then be termed as the prefigurative impulse (Brissette, 2016; Leach, 2013; Maeckelbergh, 2016).

3 Data, Methods, and Fieldwork

Field-sites consisting of interstitial spaces that allowed for informal, disparate, and limited time interactions around common activities (Furnari, 2014) were confirmed through exploratory work and informal conversations with professionals in sustainability allied roles. Extra-work events were identified as short-time bound events occurring outside traditional working hours and formal organisational structures, where individuals engaging in sustainability-related work find camaraderie, inspiration, and validation outside the structured mundane of day-to-day work. They tend to be organised around a common theme or activity, reflect collective interests and include contexts like organised social discussions, talks, informal get-togethers or social events around a sustainability theme with low entry barriers.

The research design used a para-ethnographic approach alongside a complementary use of multi-sited ethnography. Multi-sited ethnography involves the following of an inquiry across physical and virtual sites, embracing plural realities and trans local contexts (Ciuk et al., 2018; Holmes & Marcus, 2006; Islam, 2015) while the para-ethnographic approach acknowledges the fluid, insider expertise of professionals as relevant to scholarly work in a reflexive and collaborative engagement between researchers and practitioners (Ciuk et al., 2018; Falzon, 2016; Islam, 2015; Marcus, 1995, 2012).

The fieldwork was conducted over 18 months and involved participant observation at fifteen different events held both online and in physical spaces in the United Kingdom and Finland. Fieldwork at extra-work events enabled the collection of several articulations of desirable future imaginings that were the result of collective interactions and satisfied the conditional nature of being consciously oriented towards the future. Field notes combined handwritten text with ethnographic maps, sketches, and photos to capture and convey sensory and structural data in compact form. Whenever possible, these imaginings were recorded verbatim.

4   Imagining and Prefigurative Impulses

Prefigurative impulses arise due to the spatial dimension of prefigurative politics that allows for spaces where the underlying frustrations that drive them are collectively explored and result in imaginings that are impulses towards change. Historically, the beginning stages of rebellions have been characterised by impulses that Leach (2013) calls 'the prefigurative impulse'. Prefigurative impulses for change emerge when current issues, often political, have failed to find solutions, and the resulting pressurised conditions throw open vulnerabilities in current situations (Törnberg, 2021), as in the case of the sustainability crisis.

Extra-work events around sustainability topics attract a particular subset of professionals interested in broadly moving towards sustainable or desirable futures, and though routes might be different, the ultimate wish for change is often similar and reflects a space of prefigurative politics. Participants and organisers clearly voiced this as being crucial to the success of these spaces with observations like 'feeling like you are part of a bigger community of people who are also working on similar kinds of things is important'. This reflects the emotional investment of sustainability professionals, which, when blocked or sidelined, often leads to frustration and stress. As one participant noted, 'I think that when you think about this sustainability subject, it is also a matter of having these kinds of safe spaces to load your batteries (...) it is hard to push the boundaries outside of these kinds of events. That is because these are issues that get politicised and that there are a lot of situations where you have to kind of argue your own status and the discussion never progresses'. This is again echoed in this sentiment that 'everybody shares the same idea that this is a top priority (...) you don't have to kind of like start from the beginning and argue everything'.

The future imaginings are conscious descriptions of future aspirations, arising from collective interaction due to the intense contact and reflection in the present enabled by prefigurative politics (Jeffrey & Dyson, 2020) through the spatiality of organising. They have a focus on the future while looking for ways to subvert or change the frustrating circumstances of the present, allowing them to be conceptualised as prefigurative impulses. Table 1 summarises a selection of the imaginings recorded verbatim from various multi-site studies during data collection and provides a context of the frustration and dissatisfaction with present-day situations that provides a contextual push to their conceptualisation as prefigurative impulses.

The overarching themes of the impulses reflect a mix of hopeful visions for the future with a distinct frustration with current systems and practices from across various social spheres that lead to alternative organising and networks as a form of prefigurative politics. They reflect the current state in modernity of technological integration in daily life. The frustration includes the energy crisis, the failure of economic and social policies, lack of sustainability-related leadership, separation from nature, lack of systemic integration in materials and production, and concerns around work/life balance and unsustainable working. Since they arise from collective dialogue, these ideas have already undergone a degree of interaction and debate, adding a layer of collective social rigour in terms of their desirability.

Table 1 Selection of impulse themes, imaginings, and frustrations. *Source* Author

Future action or impulse themes	Verbatim imagining articulation emerges after collective interactions	Emotions around unsustainable present conditions as a contextual push
Science based: Biological tools	The Future will see biological rather than chemical war on pests. For example, Nematodea (Slender worm) kills pests such as slugs, vine weevil and ants	Frustration with artificial/ unsustainable agricultural practices
Social: Human, non-human symbiosis	Could nature and humans live side by side in the same building? What happens when we make space for the non-human within human settlements?	Frustration around separation from nature in urban spaces
Science based: Data centralisation with technology	To be a walking Non fungible token (NFT). All data would be stored on the body and not in the cloud/on paper. Data verification would be easy when entering a building	Frustration with process-based structures of constant data filling and sharing
Science based: Energy futures	If all motion in the world could be converted and captured as energy to be stored, the reliance on fossil fuels would decrease drastically	The energy crisis and dependence on fossil fuels. Unaffordability of energy prices due to war situations
Social: Ecological knowledge is considered a life skill and becomes formalised	Ecology could be taught early in schools like English or Maths, leading to a more feminine way of nurturing with care	Frustration with education systems that seem outdated in dealing with the sustainability crisis

(continued)

Table 1 (continued)

Future action or impulse themes	Verbatim imagining articulation emerges after collective interactions	Emotions around unsustainable present conditions as a contextual push
Social: Educational institutions personalise education and prioritise leadership	Coaching and creating space to support visionaries is vital; leadership training and how this is taught should be facilitated by educational institutions	Frustration with current systems of education that focus on conformity and rigid learning
Social: Work is aligned with natural rhythms and cycles	One day, the more feminine way of nurturing, involving contracting for some time before expanding, hence mimicking a growing cycle in work, might happen	Tiredness and frustration with the current pace and intensity of working life
Social: Free vaccinations	Would like to think of a future when vaccines would be free for all humankind	Frustration with unequal access to health initiatives especially vaccines during the coronavirus pandemic

The frustrations discussed are part of larger societal frustrations and reflect current reality, but the collectiveness of the space takes the negative source of tension and allows for impulses with undertones of action to emerge. Rather than going completely with current ideology, or in opposition, and making demands of it, these prefigurative impulses reshape the struggle by simply organising, acting or articulating differently (Brissette, 2016, 115). However, they are not isolated from other forms of politics and are part of the prefigurative impulse of broader movements (Maeckelbergh, 2016, 129). Research shows that positive emotions like urgency, hope and solidarity can motivate people to collective action (Ganz, 2010; Moskovitz & Garcia-Lorenzo, 2016). The imagining of alternative possibilities as the initial stages of possible action further deepens the links between imagining utopia and prefigurative practices or politics (Habersang, 2024; Van de Sande, 2013).

While imagining actively might seem tame compared to political and social revolutions, the organising of extra-work events can create and expand cracks in the surface of capitalist thinking and represent diverse struggles (Van de Sande, 2013), blurring means and ends as the organising by itself is an action of politics, enabling interaction and conversations around frustrations, and pushing the emergence of prefigurative impulses in real time.

When viewed as multi-sensory, emotionally safe spaces, the organising of extra-work events as alternative interstitial spaces is a form of spatial prefigurative politics that enables collective interaction among peers in more informal and less hierarchical situations. These spaces generate a set of imaginings about desired futures that can be conceptualised as prefigurative impulses as they are radical and arise from contemporary frustrations and unsustainability in organisational life.

5   Conclusion and Possibilities for Further Research

This chapter situates the acts of organising and participating in extra-work events as performances of prefigurative politics that provide spaces of support, camaraderie and collective interactions to deal with frustrating circumstances. This spatiality of prefigurative politics expressed through organising enables the emergence of imaginings that are conceptualised as prefigurative impulses. This conceptualisation expands on the concept of prefigurative impulses by connecting their emergence to the spatiality and performance of prefigurative politics brought in by organising, and hence contributes to the study of prefiguration in current organisational contexts.

Further exploration of these impulses and alternative origins could reveal catalysts for inspiring action within professional or formal organisational settings. Additionally, examining the emotional and affective elements of collectives and organising could enhance our understanding of emotions as not just reactionary, but as generative in shaping future possibilities. The researcher also hopes that this view of prefigurative politics and understanding of emergent impulses will also have a broader impact on praxis as they highlight how organising and collective imagining can support the emergence of transformative impulses and push for organisational and societal change.

Acknowledgements This is a condensed and textually distinct version of a longer, empirically detailed dissertation chapter, and it engages more directly with the premise of the book. The author would like to thank the editors for their invaluable support, the chapter discussants for their constructive feedback, as well as M. Koria, J. Rintamäki, and N. Moushouttas for their comments during the research process.

Competing Interests
The author has no conflicts of interest to declare that are relevant to the content of this chapter.

References

Augustine, G. (2021). We're not like those crazy hippies: The dynamics of jurisdictional drift in externally mandated occupational groups. *Organization Science, 32*(4), 1056–1078. https://doi.org/10.1287/orsc.2020.1423

Botero, A. (2013). Expanding design space(s): Design in communal endeavours. *School of Arts, Design and Architecture.* https://urn.fi/URN:ISBN:978-952-60-5174-1

Brissette, E. (2016). The prefigurative is political: On politics beyond 'the state.' In A. Dinerstein (Ed.), *Social sciences for an other politics: Women theorizing without parachutes* (pp. 109–119). Palgrave Macmillan. https://doi.org/10.1007/978-3-319-47776-3_8

Ciuk, S., Koning, J., & Kostera, M. (2018). Organizational ethnographies. In C. Cassell, A. L. Cunliffe, & G. Grandy (Eds.), *The SAGE handbook of qualitative business and management research methods* (p. 270). SAGE. https://doi.org/10.4135/9781526430212.n17

Cnossen, B., & Bencherki, N. (2018). The role of space in the emergence and endurance of organizing: How independent workers and material assemblages constitute organizations. *Human Relations, 72*(6), 1057–1080. https://doi.org/10.1177/0018726718794265

Conway, M. (2012). Sustainable futures: What higher education has to offer. *Social Alternatives, 31*(4), 35. https://search.informit.org/doi/10.3316/inf ormit.34283785478548

Croswell, K. (2024). Putting the word out there: The sub. Media collective, Indigenous resurgence, and prefigurative politics in the parallel present. *Ephemera: Theory & Politics in Organization, 24*(1), 197–211. https://www.proquest.com/scholarly-journals/putting-word-out-there-sub-media-collective/docview/3111283660/se-2

Engler, M., & Engler, P. (2014). Should we fight the system or be the change? *New Internationalist*. https://newint.org/features/web-exclusive/2014/06/04/fight-system-or-be-the-change

Escobar, A. (2017). Stirring the anthropological imagination: ontological design in spaces of transition. In A. J. Clarke (Ed.), *Design anthropology: Object cultures in transition* (pp. 201–216). Bloomsbury Academic. https://doi.org/10.5040/9781474259071.ch-014

Falzon, M.-A. (2016). Multi-sited ethnography: Theory, praxis and locality in contemporary research. In M.-A. Falzon (Ed.), *Multi-sited ethnography* (pp. 1–23). Routledge. https://doi.org/10.4324/9781315596389

Furnari, S. (2014). Interstitial spaces: Microinteraction settings and the genesis of new practices between institutional fields. *Academy of Management Review, 39*(4), 439–462. https://www.jstor.org/stable/43699259

Ganz, M. (2010). Leading change: Leadership, organization, and social movements. In N. Nohria, R. Khurana, N. Wasserman, B. Anand, W. A. Friedman, R. Ely, J. W. Lorsch, M. E. Porter, R. M. Kanter, L. A. Hill, E. Stecker, & S. A. Snook (Eds.), *Handbook of leadership theory and practice: A Harvard Business School centennial colloquium* (pp. 527–568). Harvard Business Review Press.

Habersang, A. (2024). Utopia, future imaginations and prefigurative politics in the indigenous women's movement in Argentina. *Social Movement Studies, 23*(4), 479–494. https://doi.org/10.1080/14742837.2022.2047639

Holland, A., & Roudavski, S. (2016). Design tools and complexity: mobile games and collective imagination. In *Proceedings of the 34th International Conference on Research in Computer Aided Architectural Design in Europe (ECAADe): Complexity and Simplicity* (pp. 555–564).

Holmes, D. R., & Marcus, G. E. (2006). Para-ethnography and the rise of the symbolic analyst. In M. S. Fisher & G. Downey (Eds.), *Frontiers of capital: ethnographic reflections on the new economy* (pp. 33–57). Duke University Press. https://doi.org/10.1515/9780822388234-003

Islam, G. (2015). Practitioners as theorists: Para-ethnography and the collaborative study of contemporary organizations. *Organizational Research Methods, 18*(2), 231–251. https://doi.org/10.1177/1094428114555992

Ivemark, B., & Ambrose, A. (2021). Habitus adaptation and first-generation university students' adjustment to higher education: A life course perspective. *Sociology of Education, 94*(3), 191–207. https://doi.org/10.1177/003 80407211017060

Jeffrey, C., & Dyson, J. (2020). Geographies of the future: Prefigurative politics. *Progress in Human Geography, 45*(4), 641–658. https://doi.org/10. 1177/0309132520926569

Kruglanski, A. (2024). Spontaneity and the prefigurative feel. *Ephemera, 24*(1), 71–106. https://ephemerajournal.org/sites/default/files/2024-09/24. 1.Kruglanski.pdf

Leach, D. K. (2013). *Prefigurative politics. The Wiley-Blackwell encyclopedia of social and political movements* (pp. 1–3). Wiley. https://doi.org/10.1002/978 0470674871.wbespm167

Maeckelbergh, M. (2016). The prefigurative turn: The time and place of social movement practice. In Dinerstein, A. (Eds.), *Social sciences for an other politics* (pp. 121–134). Palgrave Macmillan. https://doi.org/10.1007/978-3-319-47776-3_9

Marcus, G. E. (1995). Ethnography in/of the world system: The emergence of multi-sited ethnography. *Annual Review of Anthropology, 24*(1), 95–117. https://doi.org/10.1146/annurev.an.24.100195.000523

Marcus, G. E. (2012). Multi-sited ethnography: Five or six things i know about it now 1. In *Multi-sited ethnography* (pp. 16–32). Routledge. https://doi.org/ 10.4324/9780203810156

Mcgregor, J. (2003). Making spaces: Teacher workplace topologies. *Pedagogy, Culture & Society, 11*(3), 353–377. https://doi.org/10.1080/146813603002 00179

Mische, A. (2014). Measuring futures in action: Projective grammars in the Rio+ 20 debates. *Theory and Society, 43*(3–4), 437–464. http://www.jstor. org/stable/43694727

Moskovitz, L., & Garcia-Lorenzo, L. (2016). Changing the NHS a day at a time: The role of enactment in the mobilisation and prefiguration of change. *Journal of Social and Political Psychology, 4*(1), 196–219. https://doi.org/10. 5964/jspp.v4i1.532

Oxford University Press. (n.d.). Impulse, n. In *Oxford English dictionary*. Retrieved May 16, 2025, from https://doi.org/10.1093/OED/4869546940

Pellizzoni, L. (2021). Prefiguration, subtraction and emancipation. *Social Movement Studies, 20*(3), 364–379. https://doi.org/10.1080/14742837.2020.175 2169

Pereira, L. (2021). Imagining better futures using the seeds approach. *Social Innovations Journal, 5*. https://socialinnovationsjournal.com/index.php/sij/article/view/694

Reinecke, J. (2018). Social movements and prefigurative organizing: confronting entrenched inequalities in occupy London. *Organization Studies, 39*(9), 1299–1321. https://doi.org/10.1177/0170840618759815

Sareen, S. & Juhola S. (2025) The prefigurative politics of present transformation. In S. Sareen & S. Juhola (Eds.), *Societal transitions to sustainability: The prefigurative politics of present transformation (this book)*. Palgrave Macmillan.

Schiller-Merkens, S. (2022). Prefiguring an alternative economy: Understanding prefigurative organizing and its struggles. *Organization, 31*(3), 458–476. https://doi.org/10.1177/13505084221124189

Swain, D. (2017). Not Not but Not yet: Present and future in prefigurative politics. *Political Studies, 67*(1), 47–62. https://doi.org/10.1177/003232171 7741233

Tonkinwise, C. (2024). Finding Nature, HQ for sustainability professionals to connect and be nourished in meaningful ways. In *LinkedIn*. LinkedIn.

Törnberg, A. (2021). Prefigurative politics and social change: a typology drawing on transition studies. *Distinktion: Journal of Social Theory, 22*(1), 83–107. https://doi.org/10.1080/1600910X.2020.1856161

Tsoukas, H., & Chia, R. (2002). On organizational becoming: Rethinking organizational change. *Organization Science, 13*(5), 567–582. https://doi.org/10.1287/orsc.13.5.567.7810

Van de Sande, M. (2013). The prefigurative politics of Tahrir Square–An alternative perspective on the 2011 revolutions. *Res Publica, 19*(3), 223–239. https://doi.org/10.1007/s11158-013-9215-9

Vangkilde, K. T., & Rod, M. H. (2020). Para-ethnography 2.0: An experiment in design anthropological collaboration. In *Design anthropological futures* (pp. 155–168). Routledge.

Welbourne, T. M., Rolf, S., & Schlachter, S. (2017). The case for employee resource groups. *Personnel Review, 46*(8), 1816–1834. https://doi.org/10.1108/PR-01-2016-0004

Wiek, A., & Iwaniec, D. (2014). Quality criteria for visions and visioning in sustainability science. *Sustainability Science, 9*(4), 497–512. https://doi.org/10.1007/s11625-013-0208-6

Wilson, M. (2024). Reorganising the alternatives: What lies ahead for prefiguration? *Ephemera, 24*(1), 1–17. https://www.proquest.com/scholarly-journals/reorganising-alternatives-what-lies-ahead/docview/3111283645/se-2

Yates, L. (2015). Rethinking prefiguration: Alternatives, micropolitics and goals in social movements. *Social Movement Studies, 14*(1), 1–21. https://doi.org/10.1080/14742837.2013.870883

Yates, L., Daniel, A., Gerharz, E., & Feldman, S. (2024). Introduction to the special issue: Foregrounding social movement futures: collective action, imagination, and methodology. *Social Movement Studies, 23*(4), 429–445. https://doi.org/10.1080/14742837.2024.2343683

Open Access This chapter is licensed under the terms of the Creative Commons Attribution-NonCommercial-NoDerivatives 4.0 International License (http://creativecommons.org/licenses/by-nc-nd/4.0/), which permits any noncommercial use, sharing, distribution and reproduction in any medium or format, as long as you give appropriate credit to the original author(s) and the source, provide a link to the Creative Commons license and indicate if you modified the licensed material. You do not have permission under this license to share adapted material derived from this chapter or parts of it.

The images or other third party material in this chapter are included in the chapter's Creative Commons license, unless indicated otherwise in a credit line to the material. If material is not included in the chapter's Creative Commons license and your intended use is not permitted by statutory regulation or exceeds the permitted use, you will need to obtain permission directly from the copyright holder.

13

Inhabitation as Prefigurative Politics and Source of Political Transformations

Jens Brandt and Jan Lilliendahl Larsen

1 Introduction: Christiania, Inhabitation, and the Bureaucratic Turn

In the field of alternative urban experiments, Freetown Christiania in Copenhagen stands as a unique example of political prefiguration enacted through inhabitation. Established in 1971 through the occupation of abandoned military barracks, Christiania has long challenged dominant models of urban development and governance. For decades, its inhabitants practised a form of politics rooted in the direct, collective shaping of their living environment—embodying the principle that political transformation must begin in the present, through lived, spatial practices. This chapter examines how Christiania's tradition of self-managed inhabitation functioned as political prefiguration, and how the

J. Brandt (✉)
School of Architecture, Tampere University, Tampere, Finland
e-mail: Jens.Brandt@vub.be

J. L. Larsen
Praxis, Copenhagen, Denmark

© The Author(s) 2026

195

S. Sareen and S. Juhola (eds.), *Societal Transitions to Sustainability*, Palgrave Studies in Environmental Transformation, Transition and Accountability,
https://doi.org/10.1007/978-3-032-07395-2_13

imposition of bureaucratic planning procedures—particularly building permit requirements introduced after the 2012 legalisation agreement—has substantially reduced this prefigurative potential.

Importantly, Christiania has addressed the threat of market-driven dominance through a unique model of collective land ownership. In 2012, as part of its negotiated legalisation, Christiania established a community land trust structure—formally acquiring most of the land through the Christiania Foundation. This move effectively removed the land from the speculative real estate market, safeguarding it as a common and ensuring that property could not be individually owned or sold for profit. In doing so, Christiania responded to what Lefebvre (1991) describes as **capitalist domination** of space: the transformation of land into a commodity governed by exchange value. By securing collective tenure, the community resisted one axis of domination—market-based dispossession—through institutional innovation rooted in common ownership.

However, while market forces have been countered through this collective land trust, **bureaucratic domination** has grown more entrenched. Applying Boggs' (1977) foundational critique, this chapter shows how prefigurative spatial practices become constrained when absorbed into formal institutional frameworks. As Boggs warned, revolutionary movements risk reproducing the very hierarchies and administrative structures they seek to overcome. In Christiania, the transition from informal, consensus-based building practices to externally regulated planning procedures exemplifies this shift. What was once a living, iterative process of building through collective negotiation has been replaced by codes, permits, and compliance mechanisms administered by professionals and municipal authorities.

This analysis also responds to Yates and De Moor's (2022) observation that the concept of prefigurative politics has been applied in divergent and sometimes imprecise ways across contexts, making it difficult to assess its transformative potential. By examining the concrete mechanisms—legal, spatial, and procedural—through which bureaucratic rationality has reshaped inhabitation in Christiania, this chapter offers a focused case study that contributes conceptual clarity. It shows how prefiguration can endure through specific institutional forms (such as the community land trust), while also being constrained or reoriented by others (such as formalised regulatory frameworks).

Inhabitation—understood as the situated, material practice through which people dwell in and co-produce space—was central to Christiania's founding ethos. Building homes and common facilities was not simply a matter of shelter; it was an embodied expression of autonomy, creativity, and mutual responsibility. As a resident, recalls: "Yes yes, so you just built and maybe also changed it a bit along the way." This adaptive, participatory approach made inhabitation an active and political mode of commoning—a counterpoint to the rigid, top-down planning models prevalent in the surrounding city (Traganou, 2022).

This chapter argues that while Christiania has protected itself from commodification through collective land ownership, it has become increasingly vulnerable to bureaucratic incorporation. The shift toward formalised planning procedures introduces layers of abstraction that distance residents from the act of shaping their environment—turning what was once a direct path from dream to house into a circuit mediated by forms, permits, and external approvals. As one resident put it: "Now it has to go through paper, where someone with a stamp has to say good or bad for it. So, it becomes abstract." This process illustrates what both Lefebvre and Boggs identified: the substitution of lived spatial engagement with abstract control, where decisions are made at a distance from those who inhabit and transform space.

2 Theoretical Framework: Inhabitation, Space, and Bureaucracy

Henri Lefebvre's distinction between "abstract space" and "differential space" provides a crucial theoretical framework that complements Boggs' analysis of bureaucratic domination. Abstract space, produced by state and capital, seeks to homogenise, quantify, and control territory through planning regulations and standardised designs. It prioritises exchange value and administrative legibility over lived experiences. In contrast, differential space emerges from the bottom-up, through everyday practices and creative appropriations of inhabitants. It is heterogeneous, qualitative, and resistant to complete codification, embodying use value and the "right to the city"—the right of inhabitants to shape their own environment.

Christiania, in its early decades, represented a powerful enactment of differential space. The self-built houses, communal workshops, and collectively managed areas were not merely physical structures but material manifestations of alternative social relations and political values. Building was a concrete, aesthetic, and democratic participation in public space, or political prefiguration. This direct production of space allowed inhabitation itself to function as political prefiguration, where the desired future of autonomy and collective self-management was practised in the present.

The concept of prefigurative politics, originating in Boggs' (1977) critique of traditional revolutionary strategies, emphasises creating social relations and institutions in the present that mirror the desired future society. In Christiania, this was deeply intertwined with spatial practice. The physical act of building without permits, using recycled materials, adapting structures organically over time, and negotiating designs collectively was about prefiguring a society based on direct action, mutual aid, and participatory decision-making. This approach exemplifies what Yates and De Moor (2022) describe as "open-ended prefigurative politics" that remains responsive to changing conditions and collective desires rather than adhering to rigid blueprints.

However, the integration of Christiania into formal administrative frameworks, particularly following the 2012 agreement, represents the encroachment of abstract space upon differential space and illustrates Boggs' warning about how prefigurative projects often fall prey to bureaucratisation when they interact with state institutions. The introduction of building permit requirements, planning regulations, and heritage protection guidelines introduces layers of bureaucratic mediation that fundamentally alter the process of spatial production. As a civil servant details, the authorities initially viewed Christiania through the lens of abstract space: "So there were some shocked case workers who came back and said 'well, it was completely, completely terrible.' It was almost like a destroyed heritage protected rampart because visually it was very different from what we were used to from other heritage protected monuments."

The requirement to translate lived needs and creative visions into formal drawings for approval introduces an abstract detour. The process shifts from direct, embodied engagement with materials and space to a mediated interaction through plans and permits. The resident's experience designing her new house illustrates this shift: the initial playful idea of a "Smiley house" becomes constrained by the perceived complexities of building round windows within the new system and the reliance on standardised dimensions: "What can I say? Then there's someone who draws it, and actually, I don't think much about a 50 × 50 window. It's actually quite small." This transformation exemplifies Boggs' concern about how bureaucratic procedures impose standardised forms that suppress spontaneity and creativity.

3 Inhabitation as Political Prefiguration in Christiania

To understand the significance of Christiania's inhabitation practices as political prefiguration, we must reconsider what prefiguration means beyond static, programmatic ideals. As Yates (2015) argues, prefigurative politics is often misunderstood as merely creating small-scale models of an ideal society. Instead, he proposes understanding prefiguration as a dynamic, generative process that combines "collective experimentation, the imagining, production and circulation of political meanings, the creation of new and future-oriented social norms or 'conduct', their consolidation in movement infrastructure, and the diffusion and contamination of ideas, tactics and practices" (Yates, 2015: 4).

This understanding of prefiguration as generative praxis rather than static model aligns with what Christiania represented in its early decades. The act of building was not merely about creating physical structures according to a predetermined plan but about engaging in an ongoing, collective process of experimentation and meaning-making. This process-oriented approach allowed inhabitation to function as what Yates and De Moor (2022) call an "open-ended prefigurative politics" that remains responsive to changing conditions and collective desires rather than adhering to rigid blueprints.

In early Christiania, the act of building was inseparable from broader political expressions of autonomy, creativity, and collective self-determination. Without formal building codes or permit requirements, residents could directly translate their needs, desires, and political values into physical structures. This direct relationship between political imagination and material creation allowed building to function as a form of what Lefebvre calls "spatial autogestion" or self-management of space (Lefebvre, 1991; Purcell, 2014).

A long-time resident describes this approach: "In the past, it was such that you could just get started. And then you could build as you wanted. And then you could change it along the way if you found out that it wasn't so smart." This adaptive, iterative approach embodied what Lefebvre calls "lived space"—space as directly experienced and shaped by its inhabitants rather than as conceived by planners and administrators. It also exemplifies the kind of direct participation and spontaneous creativity that Boggs saw as essential to prefigurative politics as a counter to bureaucratic domination.

4　Bureaucratic Domination and the Reduction of Political Transformation

The 2012 legalisation agreement marked a significant turning point in Christiania's development, introducing formal building permit requirements and heritage protection regulations that fundamentally altered the relationship between inhabitation and political prefiguration. This agreement, reached after decades of contested legal status and periodic conflicts with authorities, represented a compromise that granted Christiania's residents collective ownership rights through a foundation structure in exchange for compliance with Danish building regulations, heritage protection laws, and other legal frameworks.

The agreement required Christiania to establish a foundation ("Fonden Fristaden Christiania") that would purchase most of the land from the state at below-market rates, while some areas with historical ramparts remained state property with usage rights granted to the foundation. Residents would pay into the foundation, which would manage the property and ensure compliance with regulations. Crucially, the agreement stipulated that new construction and significant renovations would require building permits from municipal authorities and, for areas affecting the historic ramparts, approval from heritage conservation officials.

This bureaucratic incorporation exemplifies what Boggs (1977) identified as a key mechanism through which prefigurative alternatives are neutralised: their absorption into dominant institutional frameworks that impose bureaucratic procedures and hierarchies. The resident describes this shift: "Yes, it's a huge difference. I mean, in the past, it was such that you could just build. And then there were of course some rules for how to build, but they were such internal rules. But now it's such that you need to have a building permit. And it's also a huge difference that you need to have a building permit from the municipality plus you often also need to have a building permit from the conservation authorities."

This bureaucratic turn has reduced the political transformative potential of inhabitation in several key ways that directly illustrate Boggs' theory of bureaucratic domination:

First, it has introduced layers of abstraction and mediation that distance inhabitants from direct engagement with their built environment. A resident articulates this loss: "That's what's so unfortunate. That you can no longer have that direct sense of what works. Now everything has to go through paper and approvals, and then it's only afterwards that you find out if it actually works in reality." He describes this as "a form of alienation from the building process itself. You no longer build with your hands and your body, but with paper and ink cartridges." This alienation exemplifies Boggs' concern about how bureaucratic procedures separate people from direct control over their conditions of life.

Second, while it was once possible for technical experts—both from Christiania and, for example, the Copenhagen municipality—to understand each other and create a kind of "nerd-to-nerd" understanding that enabled creative solutions to regulatory challenges, this dynamic has become increasingly rare. It was as if they shared a common language that could, at times, cut through bureaucratic thinking, as seen in the case of the root-based wastewater treatment system, which received approval despite conflicting with existing regulations. However, recent changes in planning frameworks have shifted decision-making power away from collective, participatory processes and toward external authorities and technical professionals. Another resident of Christiania describes how this shift has transformed the process into a closed loop of technical communication, excluding the very people who inhabit the spaces in question: "It becomes this thing where the architects talk to the heritage people, who talk to the building authorities, and it's all in this technical language that most of us don't understand. That leaves out the people who actually have to live in these spaces."

Third, the shift has reduced the democratic character of spatial production. Prior to the 2012 legalisation agreement with the state, Christiania maintained its own internal process for approving construction. This included discussions within the local area, publication in *Ugespejlet* (Christiania's newspaper), and approval at the community's financial meeting, which also provided support for the construction. Once consensus was reached, a building site was designated, and construction could begin. This participatory process embodied what Lefebvre describes as *autogestion*, or self-management—the direct, collective control of space by its inhabitants. Its replacement by external bureaucratic procedures exemplifies Boggs' critique of bureaucratic domination, in which direct participation is supplanted by distant forms of representation.

5   The Shift from Presentation to Representation

Christiania's spatial practices have undergone a fundamental transformation—from *presentation* to *representation*. Where construction once served as an aesthetic, personal, and political expression of inhabitation, shaped through lived, participatory processes, it is now increasingly subjected to the logic of formal planning permissions and regulatory compliance. This shift is not merely procedural; it signals a deeper reordering of spatial politics. *Presentation* involves the direct embodiment of values, needs, and identities in the built environment—it is situated, experimental, and open-ended. *Representation*, by contrast, abstracts these practices into codified plans and technical documents, distancing spatial production from those who dwell and build. As a result, the capacity for collective self-determination is diminished. Bureaucratic procedures now mediate what was once negotiated face-to-face, replacing commoning with consultation, and self-building with standardisation. This transition reflects a broader concern raised by Lefebvre and Boggs: that bureaucratic domination displaces democratic inhabitation by substituting lived experience with administrative representation.

In its early decades, building in Christiania functioned as direct presentation—an immediate, unmediated expression of inhabitants' needs, desires, and political values. As the reference notes, "The concrete expression in the form of self-builds is a personal expression and a unique collectively developed culture around construction." This culture of building as direct presentation allowed inhabitation to serve as a form of political prefiguration, where alternative social relations could be directly enacted through spatial practices.

The bureaucratic turn following the 2012 agreement has transformed this direct presentation into mediated representation. As the reference explains, "The introduction of a new planning paradigm in connection with the agreement on the legalisation of Christiania from 2012 meant that new construction requires a building permit as in the rest of Denmark." This requirement for formal representation through plans and permits fundamentally alters the relationship between inhabitants and their built environment, introducing layers of abstraction and mediation that distance people from direct engagement with space.

This shift from presentation to representation exemplifies what Yates and De Moor (2022) identify as a key problem in the application of prefigurative politics: the reduction of prefiguration from a dynamic, generative process to a static, formalised model. The bureaucratic requirements imposed by the 2012 agreement have transformed Christiania's open-ended prefigurative politics into a more constrained form that must operate within the parameters of state regulation. This demonstrates how the diversity in applications of prefiguration that Yates and De Moor critique can be understood through the concrete mechanisms of bureaucratic domination that Boggs identified.

6   Lefebvre and the Domination of Space Through Representation

This shift from presentation to representation can be understood through Lefebvre's critique of representations of space and their role in maintaining power relations. For Lefebvre (1991), representations of space—the conceptualised space of planners, bureaucrats, and technical experts—impose an abstract, homogenising logic that serves dominant interests while suppressing the lived, differential space of everyday life. Building permits, technical drawings, and heritage regulations are not neutral tools but instruments of what Lefebvre calls *abstract space*: space that is fragmented, homogenised, and hierarchised according to the imperatives of state power and capital accumulation.

The *domination* exercised by these representations lies in their capacity to reduce the rich, multidimensional reality of lived space to abstract, quantifiable parameters that can be administered from a distance. As the resident's experience with her window placement illustrates, the translation of embodied needs and desires into standardised measurements fundamentally alters the relationship between inhabitants and their environment. The consequences of these abstract decisions only become apparent once construction is complete, at which point changes become difficult or impossible: "Yes, I wouldn't be able to look out of it from inside."

The reference explicitly identifies this problem, stating that the project investigates whether "the abstract representations in the planning process—such as formal plans for use in building permits—have overlooked consequences for the built environment in the form of a reduction in the social and political complexity." This reduction is precisely what Lefebvre critiques as the *colonisation* of lived space by abstract space: a process through which bureaucratic and technocratic logics displace the plural, contested, and embodied practices that constitute spatial life. In this sense, the domination of representation over presentation is not merely technical—it is political, entailing the substitution of formalised abstractions for concrete, lived relations, as also diagnosed by Boggs in his critique of bureaucratic power.

7   The Mechanisms of Bureaucratic Constraint

The reduction of political transformation through bureaucratic domination operates through multiple mechanisms, each of which can be understood in terms of Lefebvre's analysis of the production of space and Boggs' critique of bureaucratic domination:

First, bureaucratic procedures replace what Lefebvre calls "spatial practice"—the direct, embodied engagement with space—with abstract representations that privilege technical expertise over lived experience. As a resident describes the pre-2012 approach to window placement: "So I think like this: We stood there physically and felt it out. And also very much like—does it make sense—can I look out the window." This embodied spatial practice is replaced by abstract calculations and standardised parameters that distance inhabitants from direct engagement with their environment. This exemplifies Boggs' concern about how bureaucratic domination separates people from direct control over their conditions of life.

Second, bureaucratic requirements enact what Lefebvre critiques as the dominance of *conceived space* over *lived space*—a process in which spatial experience is reduced to abstract representations that serve administrative control. The mandate for formal drawings and technical plans privileges the visual and measurable dimensions of space, sidelining the embodied, tactile, and relational qualities that once defined building practices in Christiania. In doing so, it narrows spatial engagement to what can be seen, mapped, and regulated—replacing sensory richness and collective experimentation with legibility and compliance. This reflects Boggs' concern that bureaucratic procedures impose standardised forms that stifle spontaneity, suppress creativity, and ultimately disconnect people from the spaces they inhabit.

Third, bureaucratic temporality—with its linear sequence of planning, approval, and implementation—replaces what Lefebvre calls the "rhythms" of everyday life, with their cyclical patterns and organic development. The spontaneous, adaptive approach to building that a resident describes—"so you just built and maybe also just changed it a bit along the way"—becomes impossible when plans must be approved in advance and changes require formal authorisation. This transformation exemplifies Boggs' concern about how bureaucratic domination imposes rigid procedures that constrain spontaneous action and adaptation.

Fourth, bureaucratic procedures fragment space into discrete, specialised domains managed by different authorities and experts. As the 2012 agreement illustrates, obtaining approval now involves multiple agencies and authorities, from Christiania's internal procedures to the State and Cultural Agency, the Copenhagen Municipality, and various other administrative bodies. This fragmentation of space into separate domains contradicts what Lefebvre calls the "unitary theory of space"— the understanding of space as a coherent, interconnected whole produced through social relations. It also exemplifies Boggs' critique of how bureaucratic domination fragments social life into specialised domains managed by experts.

8   Conclusion: The Reduction of Political Transformation Through Bureaucratic Domination

The case of Christiania reveals the profound impact of bureaucratic domination on the potential for inhabitation to generate political transformation. The requirement for building permits and compliance with heritage regulations has significantly constrained the capacity for inhabitation practices to prefigure alternative social arrangements, introducing layers of abstraction and mediation that distance inhabitants from direct engagement with their built environment and from the collective processes through which alternative social relations might be enacted.

This constraint operates through the mechanisms that Lefebvre identifies in his critique of abstract space and that Boggs analyses in his theory of bureaucratic domination: the privileging of representations over practice, the fragmentation and homogenisation of space, the imposition of linear temporality over cyclical rhythms, and the concealment of power relations behind technical procedures. Together, these mechanisms significantly reduce the political transformative potential of inhabitation, integrating practices that once challenged dominant spatial arrangements into the very logic they sought to contest.

The shift from direct presentation to bureaucratic representation captures the essence of this transformation. What was once an immediate, unmediated expression of inhabitants' needs, desires, and political values has become a mediated, abstract process governed by external standards and authorities. This shift fundamentally alters the relationship between inhabitation and political transformation, as the direct, embodied engagement with space that characterised early Christiania gives way to the abstract, technical procedures of formal planning.

In the end, the case of Christiania demonstrates that bureaucratic domination, particularly through building permit requirements, does not merely regulate or formalise inhabitation practices but fundamentally alters their political character and transformative potential. By subjecting these practices to the abstract logic of administrative procedures, technical expertise, and standardised parameters, bureaucratic domination significantly reduces their capacity to challenge existing power relations and prefigure alternative social arrangements. The result is a profound reduction in the potential for inhabitation to serve as a source of political transformation—a reduction that reveals the deeply political nature of seemingly technical requirements and the hidden power dynamics embedded in spatial production.

This analysis contributes to Yates and De Moor's (2022) critique of the diversity in applications of prefigurative politics by providing a concrete case study of how prefiguration can be transformed and constrained through bureaucratic incorporation. By applying Boggs' (1977) concept of bureaucratic domination to the specific context of Christiania's spatial practices, it demonstrates how prefigurative politics can be reduced from a dynamic, generative process to a more limited form when absorbed into dominant institutional frameworks. This sharpens our understanding of prefiguration not as a static model but as a contested practice that must constantly negotiate the tension between autonomy and incorporation, between direct action and bureaucratic mediation.

Competing Interests The authors have no conflicts of interest to declare that are relevant to the content of this chapter.

References

Boggs, C. (1977). Marxism, prefigurative communism, and the problem of workers' control. *Radical America, 11*(6), 99–122.

Lefebvre, H. (1991). *The production of space* (D. Nicholson-Smith, Trans.). Blackwell Publishers.

Purcell, M. (2014). Possible worlds: Henri Lefebvre and the right to the city. *Journal of Urban Affairs, 36*(1), 141–154.

Traganou, J. (2022). The paradox of the commons: The spatial politics of prefiguration in the case of Christiania Freetown. In L. Monticelli (Eds.), *The future is now: An introduction to prefigurative politics* (pp. 144–160). Palgrave Macmillan.

Yates, L. (2015). Rethinking prefiguration: Alternatives, micropolitics and goals in social movements. *Social Movement Studies, 14*(1), 1–21.

Yates, L., & De Moor, J. (2022). The concept of prefigurative politics in studies of social movements: Progress and caveats. *Social Movement Studies, 21*(1–2), 1–19.

Open Access This chapter is licensed under the terms of the Creative Commons Attribution-NonCommercial-NoDerivatives 4.0 International License (http://creativecommons.org/licenses/by-nc-nd/4.0/), which permits any noncommercial use, sharing, distribution and reproduction in any medium or format, as long as you give appropriate credit to the original author(s) and the source, provide a link to the Creative Commons license and indicate if you modified the licensed material. You do not have permission under this license to share adapted material derived from this chapter or parts of it.

The images or other third party material in this chapter are included in the chapter's Creative Commons license, unless indicated otherwise in a credit line to the material. If material is not included in the chapter's Creative Commons license and your intended use is not permitted by statutory regulation or exceeds the permitted use, you will need to obtain permission directly from the copyright holder.

14

Prefiguring Conservation, Peace, and Self-Determination in the Salween Peace Park in Karen State, Myanmar

Zali Fung and Sheila Htoo

1   Introduction

Prefigurative politics—collective efforts to imagine and enact more just and sustainable futures in the present—are receiving increased attention in human geography and related disciplines. Through prefigurative politics, movements and communities are not only challenging the extractive capitalist logics that are driving the climate crisis and rising inequality, but enacting alternative economies, spaces, and ways of living from the bottom-up (Jeffrey & Dyson, 2021). Rather than examine prefigurative politics in urban Global North contexts, we respond to calls for a "new approach to prefigurative politics that looks beyond the movements

Z. Fung (✉)
Institute of Geography and Sustainability, University of Lausanne, Lausanne, Switzerland
e-mail: zali.fung@unil.ch

S. Htoo
Faculty of Environmental and Urban Change, York University, Toronto, Canada

© The Author(s) 2026   **211**
S. Sareen and S. Juhola (eds.), *Societal Transitions to Sustainability*, Palgrave Studies in Environmental Transformation, Transition and Accountability,
https://doi.org/10.1007/978-3-032-07395-2_14

usually studied" (Dyson & Jeffrey, 2018, 574). We examine how prefiguration unfolds and persists under conditions of protracted conflict and contested political authority in Southeast Asia's borderlands.

Our case is the Salween Peace Park (SPP), an Indigenous Karen community-led peace and conservation initiative established in a Karen-controlled area of Karen State, Myanmar (Burma) and along the Thai-Myanmar border. Karen State has experienced conflict between the Karen National Union (KNU) and Myanmar government for over seven decades. The SPP is not only about conservation and peace but enacting Karen communities' ongoing demands for self-determination over livelihoods, resources, and territories. It is also about cultural protection and sustaining knowledge and resource-based livelihood practices that have been passed down over decades and generations. Here, prefiguration is not only about enacting desired futures now but also maintaining and sustaining lives, livelihoods, and culture in the present, reflecting Muñoz-Sueiro and Kallis' (2024) notion of "postfiguration."

We also explore factors that limit the transformative potential and spatial expansion of the SPP. This includes ongoing state violence by the Myanmar military, which has intensified since the 2021 military coup d'état, the denial of Karen self-determination by the Myanmar state, and proposals to develop destructive hydropower dams on the Salween River inside the SPP. Overall, the SPP both challenges and presents an alternative to top-down militarised hydro-development and extraction and embodies a conservation model that embraces, rather than excludes, people and resource-based livelihoods.

2   Prefigurative Politics

Prefigurative politics not only challenge oppressive power structures but encompass efforts to imagine, organise, and enact alternative futures in the present (Jeffrey & Dyson, 2021, 644). Rather than predetermined, the contents of prefiguration tend to be forged through negotiation, struggle, improvisation, creation, and experimentation (Ibid, 646). Scholars have used this framing to show how progressive social movements challenge rapacious capitalism and create alternative and more caring economies, spaces, and social relations. Much of this work focuses on organised social movements in urban, Global North contexts (but see Dyson & Jeffrey, 2018; Tadros, 2015).

Reflecting on the Occupy Movement—a quintessential example of prefigurative politics—and a range of factory and university occupations, Vasudevan (2015, 316) positions "occupation as a political process that prefigures and materialises the social order which it seeks to enact." Occupy activists enacted "the urban futures to which they aspired. Occupy protest camps in major cities often contained clinics, kitchens, healthcare ... and democratic-decision making structures that activists presented as prefigurative" (Jeffrey & Dyson, 2021, 644–645). Prefiguration thus attends to the social reproduction of movements and life, rather than economic production alone (Monticelli, 2021, 99). Yet London's Occupy Movement also reproduced gender and class divides, for instance, as women took on more social reproductive labour (Halvorsen, 2017). Rather than producing a physical (urban) space (see Asara & Kallis, 2023), Dyson and Jeffrey (2018) find that strengthening social networks is key to the "everyday" prefiguration of migrant youth in north India.

Beyond radical urban movements, localised efforts to transform food systems and ways of living include small-scale organic and urban farms, and cooperative and autonomous housing (Brandt & Larsen, 2025; Stojcheska et al., 2025). However, localised initiatives such as ecovillages can become "exclusive enclaves that reproduce class and racial divides" (Jeffrey & Dyson, 2021, 651) while activists have critiqued "micropolitical" endeavours for failing to abolish capitalist relations (Naegler, 2018, 508). We see both localised efforts and large-scale movements that are attuned to power and social difference as key to prefiguring more just and sustainable transformations. While prefigurative politics focus on enacting desires futures, Muñoz-Sueiro and Kallis (2024) show how traditional knowledges and practices with pre-capitalist origins and deep historical roots in rural Spain can "postfigure" degrowth alternatives.

Limited work examines prefiguration in contexts of contested political authority and conflict. While not framed as prefigurative, McConnell (2009) details how Tibetan activists in India have established a government-in-exile that performs "state-like functions" including elections, taxation, and healthcare and education provision. This draws parallels with Myanmar's ethnic resistance organisations that provide "state-like" services (Loong, 2025). These prefigurative actions bolster claims and perpetuate a movement for Tibetan sovereignty, even if this outcome is unlikely in the present (geo)political climate. Such cases of "proleptic prefigurative politics" are important because "acting 'as if' can challenge dominant claims about there being no alternative to existing structures of power" (Jeffrey & Dyson, 2021, 648–649).

In the SPP, prefiguring peace and conservation is inseparable from demands for self-determination. Self-determination is not only a struggle for political authority but also a "practice of resistance, prefigurative politics, and epistemic emancipation" (Constantinou et al., 2024, 2). Self-determination is a *relational* claim that is shaped by historical and contemporary struggles and "future-oriented hopes" and aspirations (Loong et al., 2023, 2). This highlights how self-determination struggles can be prefigurative and enrol multiple temporalities. Scholars have noted that prefigurative politics are central to the SPP (Cole, 2020; Constantinou et al., 2024; Loong et al., 2023). We use this case to expand the notion of prefigurative politics and its temporal frames.

3   Setting the Context: Karen State and the Salween Peace Park

The KNU is a political organisation with an armed wing that has struggled for Karen autonomy and self-determination over land and natural resources in Karen State for over seven decades (Bright, 2019, 71). The KNU commenced armed struggle in 1949, following failed attempts to negotiate with the colonial British and then Burmese governments for a separate state (Loong, 2023, 139).

The SPP is in Mutraw District, a largely KNU-controlled stronghold that has primarily remained outside central state control, in northern Karen State, southeastern Myanmar. In the 1990s, the Myanmar military increased attacks in KNU-controlled areas, which entailed human rights abuses and scorched-earth policies (Loong, 2023, 139). These violent campaigns have forcibly displaced more than 80% of Mutraw's population (Karen Rivers Watch [KRW], 2004). In KNU-controlled areas, the KNU collects taxes and provides health, education, and social services.

Following political and economic reforms led by Myanmar's military government in 2011, the National League for Democracy won general elections in 2015 and was Myanmar's ruling party until the 1 February 2021 coup. The KNU signed a bilateral ceasefire with the Myanmar government in 2012 and a so-called nationwide ceasefire in 2015. This generated friction between Karen groups that supported the ceasefire and those who were cautious about engaging in state-led peace processes, including in Mutraw District (Htoo, forthcoming). The ceasefires reduced, but did not eliminate, armed conflict (Loong et al., 2023, 9). The military violated the ceasefires through violence, failure to demilitarise, and by attempting to expand military roads and construct hydropower dams. This reflects Wood's (2011) notion of "ceasefire capitalism," where a temporary cessation in hostilities enabled military-state extraction within Myanmar's resource-rich, borderland "ceasefire zones."

The 2021 coup brought renewed conflict and state violence to Karen State and nationwide. Since this time, the military has intensified aerial bomb attacks inside the SPP, killing and injuring dozens, destroying homes and infrastructure, and forcing thousands to flee (SPP, 2021). This highlights the context in which the SPP emerged and persists.

3.1   The Salween Peace Park

The SPP was established in 2018, following decades-long efforts by Karen communities and civil society organisations (CSOs) to protect their lands, forests, and waterways. The Salween River, which flows through Tibet, China, Myanmar, and Thailand, forms the eastern boundary of the SPP and the international river border with Thailand. "Amorphous" dams and diversions have been (re)proposed, and resisted, for over four decades in the Basin, and remain unbuilt (Fung, 2024a). The Hatgyi Dam is proposed in the southernmost part of the SPP and would be developed by Burmese, Thai, and Chinese actors. Communities and CSOs are concerned that the dam would displace communities, and negatively impact fisheries, and riverine ecologies and livelihoods (Sein Twa et al., 2021). Attempts to construct this dam have exacerbated conflict around the dam site (Bright, 2019).

The SPP encompasses around 440 villages, cultivated land, wildlife sanctuaries, community and reserved forests, and 303 *Kaw* or customary territories (see Fig. 1). Each *Kaw* is "a physical place, a unit of land administration, and a social system" that encompasses customs, practices, and cosmologies that govern socioecological relations (Sein Twa et al., 2021). The boundaries of *Kaw* have been maintained for generations through social relations and kin-mediated access protocols. Residents practice subsistence agriculture, including irrigated rice paddy and diverse upland swidden agriculture that is rotated with forest fallow. They also hunt, gather, and fish in the forests and rivers.

The SPP is governed by a Charter that was drafted by Karen leaders and CSOs in consultation with residents and approved via a referendum (Constantinou et al., 2024, 7). According to the Charter, the SPP is a "grassroots, people-centred alternative" to top-down hydro-development and extraction (SPPSC & KNU, 2018, 4). The SPP embodies an "Indigenous Karen environmental ethic that integrates sustainable livelihoods, nature protection, and democratic governance … the Charter enshrines the right of Indigenous Karen people to self-determination over how to manage and govern their natural resources and lands" (Ibid). Section 5.1 of the Charter outlines how land tenure is established, regulated, and negotiated, based on community-led land management and the *Kaw* system (Ibid, 35–37). Governance is decentralised, with decision-making often undertaken at the *Kaw* scale, while an elected General Assembly is responsible for overall coordination (Sein Twa et al., 2021).

Given fraught peace processes and an ongoing coup, Karen residents and CSOs are prefiguring their own visions of community-led democracy, peace, and conservation. As the KNU and the Karen Environmental and Social Action Network (KESAN) explain: "the new Myanmar government [NLD] has promised to lead the country toward a devolved, federal democracy. The Karen are not waiting idly for this: the [SPP] is federal democracy in action" (2017, 3). For Loong et al., (2023, 10),

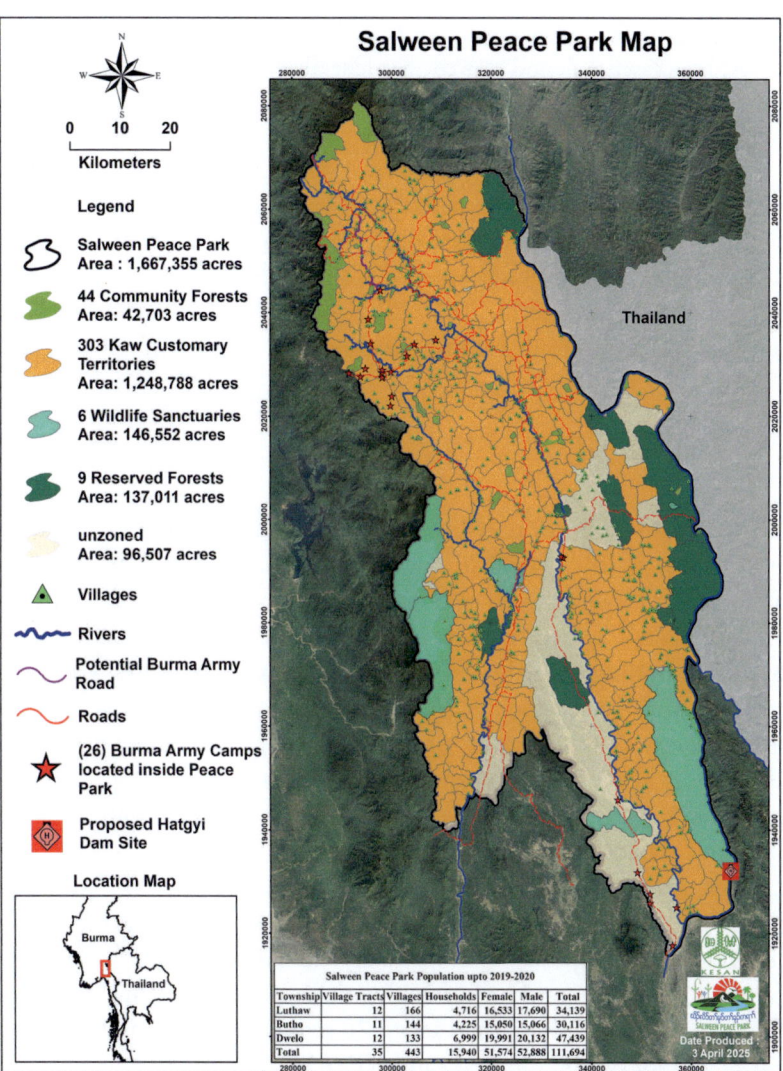

Fig. 1 Salween Peace Park map. *Source* KESAN (use permitted by email to authors, 20 May 2025)

Prefiguration is … key to the SPP … [for] the KNU and Karen civilians, the ceasefire period proved that federalism could not be built from the top-down … Through the Peace Park, the KNU sought to build federalism from the bottom-up.

Overall, the SPP embodies and spatialises Karen communities' visions of peace, conservation, and self-determination. Rather than a "fortress conservation" that excludes and displaces people from conservation zones (Brockington, 2002), the SPP positions the well-being of Karen communities and resource-based livelihoods as integral to conservation.

4   Methods

Zali draws on empirical fieldwork conducted in the Thai-Myanmar borderlands of the Salween Basin since 2021, including 80 semi-structured interviews with residents affected by proposed dams and diversions, civil society actors, and Thai state actors. Interviews were conducted in English, or in Thai or Karen with an interpreter. Zali speaks conversational Thai language. As a foreign researcher, Zali collaborates with Salween civil society actors, who shape the direction, questions, fieldwork, and outputs of research. Inspired by feminist research ethics, Zali aims to produce knowledge that is shaped by and useful to participants and collaborators (see Fung, 2024b).

Sheila draws on empirical fieldwork conducted in the SPP and Thailand. From December 2017 to March 2018, Sheila interned with KESAN in Chiang Mai and worked with KESAN staff leading SPP consultation processes. During fieldwork in Karen communities, Sheila conducted ethnographic observation; 20 semi-structured interviews with central and local KNU leaders, and KESAN staff; and five focus groups with Mutraw village tract and township leaders, Karen leaders, park rangers, villagers, and KESAN staff. Interviews were conducted in Sgaw Karen language, aside from three conducted in English with KESAN staff and a consultant. Sheila's position as a Karen and former refugee, who was born along the Salween River and grew up in multiple refugee camps along the Thai-Myanmar border, enabled her to build rapport with interlocutors. KNU leaders trusted Sheila because several family members previously worked with these leaders in Mutraw District. Sheila's situatedness enabled her to navigate travel obstacles and security challenges, including multiple Myanmar military posts and KNU-Mutraw security checkpoints.

All interviewees provided informed consent to participate in the research, and we have used pseudonyms to protect confidentiality.

5   Enacting Peace, Conservation, and Self-Determination in the Salween Peace Park

The SPP not only challenges military violence and state and corporate resource extraction and territorialisation but enacts community visions of peace, conservation, and self-determination in the present. According to Saw Paul Sein Twa, a Karen activist and co-founder and director of KESAN, the SPP represents a "vision to fulfil three core aspirations of the Karen Indigenous people, integrated into the Peace Park's design and implementation" (Sein Twa et al., 2021). These three pillars are peace and self-determination, environmental integrity, and cultural survival.

The vision of peace embodied by the SPP is not only an *absence* of conflict but also the *presence* of justice and self-determination for Karen peoples (Htoo, forthcoming). For Sein Twa (2024), "peace means self-determination ... [and] conservation. Peace means revitalisation of our culture and Indigenous knowledge." The SPP opens a democratic space for Karen people to create and implement their vision of peace and conservation *before* extractive and infrastructure projects materialise. In doing so, it creates "more democratic and equitable relations between the KNU, civil society, and ordinary civilians," including by allocating decision-making power to community representatives (Loong, 2025, 1).

The SPP protects some of the world's last remaining teak forests along the Salween River and several endangered species (KESAN, 2015). Land management and conservation are guided by rules, rituals, ceremonies, taboos, and reciprocal relations with more-than-human beings

(Paul et al., 2023). Rather than view people and resource-based liveli-hoods as detrimental to conservation, the SPP positions environmental, social, and cultural integrity as interconnected. As General Pu Traw, a KNU Mutraw District spokesperson, explained, Karen territories are governed by well-established rules about where to cultivate, and how to protect wildlife and manage forests and waterways sustainably, and "so our traditional Karen way of life is interdependent, intertwined with the environment" (Interview, January 2018). This knowledge is passed down from older to younger generations and integrated into SPP governance.

Fish and aquatic biodiversity are high in the Basin, but remain under-studied by international scientists (IUCN, 2024). Recently, researchers identified 170 fish species, 60 of which are endemic, in the Salween River, and community-led research documented 90 species of fish down-stream from the Hatgyi Dam site (Sein Twa et al., 2021). In the SPP, fisheries are managed by community fish conservation zones (Paul et al., 2023). Fish sustain riverine livelihoods, as a Thai academic explained: "the fish in the river are like the supermarket for local people. They catch them and this provides food for families" (Interview, February 2022). This highlights what is at stake for livelihoods and ecologies if the Hatgyi Dam is built.

When asked about the Hatgyi Dam, Karen leaders and residents emphasised that "ceasefire capitalism" in the form of hydropower devel-opment is not "genuine" peace, but rather "a symbol [of] violence and a tool used by the Burma Army to occupy our territory" (KRW, 2019). For Pu Traw:

> Mega-dams are not 'for peace' but rather for the economy only. This is what the military regime has planned … to pursue their interests and maintain power. This is not peace. By peace, we mean benefits for our civilians … They must feel confident to approve of peace and participate in the process. As of now, this is coming one-sidedly from the Burma military and it is done in our territory, so it contradicts our peace-building efforts/process. … We are not saying that we do not do 'development', but we will do it … in the most beneficial way to our people … These mega-projects are just one-sided. Residents, administrators, and leaders are not informed or consulted in any way. There is no consideration

of their interests and potential benefits. This approach is destroying the peace we are building. (Interview, January 2018)

The Hatgyi Dam is resisted by Karen communities on both sides of the river border. For two decades, communities and CSOs have gathered along the Salween River on the International Day of Action for Rivers to protest dams and diversions (shown in Fig. 2; see also Fung & Lamb, 2023). Two CSO employees (Interview, March 2022) explained that a key reason that proponents have been unable to construct dams, including Hatgyi, is because they are proposed inside "Karen territory [and] the SPP," highlighting how prefigurative actions are perceived to shape development outcomes. The SPP instead envisions communities powered by small hydropower and decentralised solar and livelihoods based on sustainable forest management, agroforestry, and organic farming (KNU and KESAN, 2017, 3).

Fig. 2 Karen residents protest against dams along the Salween River in 2023. The banner reads: "No dam! We need peace, equality, and justice," reflecting the prefigurative politics of the SPP. *Source* KESAN (use permitted by email to author Htoo, 25 May 2025)

However, conflict and displacement have severed the ties of many Karen people to their land and cultural practices, including in Sheila's experience. As Pu Traw explained:

> Since the Burmese military came to our territories … [Karen people] had to leave their villages, abandon their farmlands … peace that existed in Karen ways of life and communities was interrupted and destroyed by the Burmese military. They destroyed peace in our ancestral time and until this day.

Pu Traw highlights the factors limiting the Park's emancipatory potential and spatial expansion. State violence is "ever present" in the SPP, as the multiple Myanmar military camps shown in Figure B8.1 attest (Loong, 2025, 9). Because the SPP can only be enacted in "areas where the KNU has kept the Burmese state out, the [SPP] encompasses only parts of *Kawthoolei*" (Constantinou et al., 2024, 7). *Kawthoolei* refers to a "Karen homeland" that the KNU and some Karen residents have sought to establish since independence. Many Karen people in the SPP mobilise an Indigenous identity in their claims for self-determination. However, the Myanmar government has long suppressed Karen identity in government- or mixed-controlled areas (Loong et al., 2023). In a landscape fragmented by war, some Karen residents are less able to participate in the prefigurative politics of the SPP and broader claims to Karen self-determination.

Nevertheless, the SPP embodies and spatialises Karen communities' visions for peace, conservation, and self-determination on their own terms. In doing so, the SPP presents an alternative to top-down, militarised, extractive development or "fortress conservation" that excludes people and livelihoods.

6   Conclusion

Our chapter makes two key contributions to understandings of prefigurative politics. First, rather than theorise from urban Global North cases, we show how prefigurative politics unfold and persist amidst protracted conflict and state violence in Karen State. In the SPP, sustainability and peace are inseparable from Karen people's ongoing struggles for self-determination amidst seven decades of war.

Second, existing work on prefigurative politics emphasises how people collectively imagine and enact more just and sustainable *futures* in the *present*. Our conceptualisation of prefigurative politics accounts for a multitude of temporalities, and how the past, present, and future intersect (see Muñoz-Sueiro & Kallis, 2024). In the SPP, prefigurative politics are not only about enacting desired futures now but are profoundly shaped by historical and contemporary struggles over livelihoods, resources, and self-determination. The SPP sustains knowledge, resource-based livelihoods, and customary land management practices with deep cultural and historical roots.

Following critical political ecologists, our contribution demonstrates that conservation is anything but politically neutral (Brockington, 2002). Rather than exclude people from conservation zones, the SPP integrates Karen communities and resource-based livelihoods into its vision of conservation. Overall, the SPP illustrates how communities and diverse actors can collectively imagine and enact more just and sustainable futures in the present under difficult circumstances. Such cases offer hope in a time of multiple and intersecting crises, including climate change, and rising inequality, and authoritarianism.

Acknowledgements We sincerely thank the Salween residents and civil society actors who shared their time, knowledge, and stories with us. Special thanks to KESAN for their support, and especially Casper Palmano for reviewing this chapter. We are grateful to Siddharth Sareen and Sirkku K. Juhola for bringing together scholars to discuss prefigurative politics at a workshop in Oslo, Norway and collating this book. Finally, we thank workshop members for their critical and constructive feedback.

Competing Interests The authors have no conflicts of interest to declare that are relevant to the content of this chapter.

References

Asara, V., & Kallis, G. (2023). The prefigurative politics of social movements and their processual production of space: The case of the indignados movement. *Environment and Planning C: Politics and Space, 41*, 56–76. https://doi.org/10.1177/23996544221115279

Bright, S. J. (2019). Rites, rights, and water justice in Karen State: A case study of community-based water governance and the Hatgyi Dam. In C. Middleton & V. Lamb (Eds.), *Knowing the Salween river: Resource politics of a contested transboundary river* (pp. 71–86). Springer.

Brandt, J., & Larsen, J. L. (2025). Inhabitation as prefigurative politics and source of political transformations. In S. Sareen & S. Juhola (Eds.), *Societal transitions to sustainability: The prefigurative politics of present transformation* [this book]. Palgrave Macmillan.

Brockington, D. (2002). *Fortress conservation: The preservation of the Mkomazi game reserve*. Indiana University Press.

Cole, T. (2020). *Possessed earth: Ownership and power in the Salween peace park of Southeast Myanmar* (PhD dissertation). Stockholm University.

Constantinou, C. M., McConnell, F., Dirik, D., Regassa, A., Loong, S., & Kuokkanen, R. (2024). Reimagining self-determination: Relational, decolonial, and intersectional perspectives. *Political Geography, 118*, 103112. https://doi.org/10.1016/j.polgeo.2024.103112

Dyson, J., & Jeffrey, C. (2018). Everyday prefiguration: Youth and social agency in India. *Transactions of the Institute of British Geographers, 43*(4), 573–585. https://doi.org/10.1111/tran.12245

Fung, Z. (2024a). Amorphous infrastructure: Contesting the proposed Yuam River water diversion project in the Salween River Basin. *Environment and Planning E: Nature and Space, 7*(6), 2386–2412. https://doi.org/10.1177/25148486241288366

Fung, Z. (2024b). Navigating research ethics during COVID-19 in authoritarian Southeast Asia: A feminist approach to combining online and in-person qualitative methods. *Australian Geographer, 55*(4), 541–561. https://doi.org/10.1080/00049182.2024.2406059

Fung, Z., & Lamb, V. (2023). Dams, diversions, and development: Slow resistance and authoritarian rule in the Salween River Basin. *Antipode, 55*(6), 1662–1685. https://doi.org/10.1111/anti.12939

Halvorsen, S. (2017). Spatial dialectics and the geography of social movements: The case of Occupy London. *Transactions of the Institute of British Geographers, 42*(3), 409–421. https://doi.org/10.1111/tran.12179

Htoo, S. forthcoming. *The Salween Peace Park: Building 'Peace' from Grassroots Indigenous-led Conservation Movement in Karen State of Southeast Myanmar/Burma* (PhD dissertation). York University.

IUCN. (2024). *IUCN and Maejo University train civil society organisations on fish identification in Thailand river basin, 26 July.* Retrieved May 21, 2025, from https://iucn.org/story/202407/iucn-and-maejo-university-train-civil-society-organisations-fish-identification

Jeffrey, C., & Dyson, J. (2021). Geographies of the future: Prefigurative politics'. *Progress in Human Geography, 45*, 641–658. https://doi.org/10.1177/0309132520926569

KRW. (2019, March 14). International Day of Action for Rivers 2019: Large dams are driving conflict, not empowering people to build peace. *Progressive Voice.* Retrieved May 21, 2025, from https://progressivevoicemyanmar.org/2019/03/14/international-day-of-action-for-rivers-2019-large-dams-are-driving-conflict-not-empowering-people-to-build-peace/

KRW. (2004). *Damming at gunpoint: Burma army atrocities pave the way for Salween Dams in Karen State.* KRW.

KESAN. (2015). *A working concept of the Salween peace park* (Unpublished working paper).

KNU & KESAN. (2017). *Salween peace park: A vision for an indigenous Karen landscape of human-nature harmony in Southeast Myanmar.* Retrieved May 21, 2025, from https://prize.equatorinitiative.org/wp-content/uploads/formidable/6/Salween-Peace-Park-Briefer-2018.pdf

Loong, S. (2025). More-than-rebel territory: War, resistance, and relations in the Salween Peace Park. *Annals of the American Association of Geographers.* https://doi.org/10.1080/24694452.2025.2478262

Loong, S. (2023). In Myanmar, Generation Z goes to war'. *Current History, 122*(843), 137–142. https://doi.org/10.1525/curh.2023.122.843.137

Loong, S., Manby, A., & McConnell, F. (2023). Rethinking self-determination: Colonial and relational geographies in Asia.' *Territory, Politics, Governance,* 1–19. https://doi.org/10.1080/21622671.2023.2232410.

McConnell, F. (2009). De facto, displaced, tacit: The sovereign articulations of the Tibetan government-in-exile. *Political Geography, 28*, 343–352. https://doi.org/10.1016/j.polgeo.2009.04.001

Monticelli, L. (2021). On the necessity of prefigurative politics. *Thesis Eleven, 167*, 99–118. https://doi.org/10.1177/07255136211056992

Muñoz-Sueiro, L., & Kallis, G. (2024). Postfiguring degrowth: How traditional popular culture challenges growth-oriented common senses'. *Journal of Political Ecology, 31*(1), 951–971. https://doi.org/10.2458/jpe.5708

Naegler, L. (2018). 'Goldman-Sachs doesn't care if you raise chickens': The challenges of resistance prefiguration. *Social Movement Studies, 17*(5), 507–523. https://doi.org/10.1080/14742837.2018.1495074

Paul, A., Moo, S. S. B., & Roth, R. (2023). Water and fish conservation by Karen communities: An Indigenous relational approach. *The Focus 94*. Retrieved May 21, 2025, from https://www.iias.asia/the-newsletter/article/water-and-fish-conservation-karen-communities-indigenous-relational-approach

SPP. (2021, April 3). *Salween Peace Park Under Attack! Burmese military violence undermines Indigenous Karen conservation for peace*. Retrieved May 21, 2025, from https://www.karenpeace.org/report/briefing-salween-peace-park-under-attack/

SPPSC (SPP Steering Committee) & KNU. (2018). *Charter of the Salween Peace Park*. Mutraw District: SPPSC and KNU. Retrieved May 21, 2025, from https://kesan.asia/wp-content/uploads/2018/12/SPP-Charter-Eng.pdf

Sein Twa, P. (2024). 'Peace means self-determination': An interview with Paul Sein Twa. Interview by Lital Khaikin, 1 September. Retrieved May 21, 2025, from https://newint.org/indigenous-peoples/2024/peace-means-self-determination-interview-paul-sein-twa

Sein Twa, P., Fogerite, J., & Palmano, C. (2021). Hkolo Tamutaku K'rer: The Salween Peace Park in Burma/Myanmar. In I. C. C. A. Consortium (Ed.), *Territories of life: 2021 Report* (pp. 111–120). ICCA Consortium.

Stojcheska, A. M., Kotevska, A., & Tuna, E. (2025). Present transformations in food systems: Local initiatives towards sustainable and just futures in post-socialist context. In Sareen, S., & Juhola, S. (Eds.) *Societal transitions to sustainability: The prefigurative politics of present transformation* [this book]. Palgrave Macmillan.

Tadros, M. (2015). Contentious and prefigurative politics: Vigilante groups' struggle against sexual violence in Egypt (2011–2013). *Development and Change, 46*(6), 1345–1368. https://doi.org/10.1111/dech.12210

Vasudevan, A. (2015). The autonomous city: Towards a critical geography of occupation. *Progress in Human Geography, 39*, 316–337. https://doi.org/10.1177/0309132514531470

Woods, K. (2011). Ceasefire capitalism: Military-private partnerships, resource concessions and military-state building in the Burma-China borderlands.

The Journal of Peasant Studies, 38(4), 747–770. https://doi.org/10.1080/030 66150.2011.607699

Open Access This chapter is licensed under the terms of the Creative Commons Attribution-NonCommercial-NoDerivatives 4.0 International License (http://creativecommons.org/licenses/by-nc-nd/4.0/), which permits any noncommercial use, sharing, distribution and reproduction in any medium or format, as long as you give appropriate credit to the original author(s) and the source, provide a link to the Creative Commons license and indicate if you modified the licensed material. You do not have permission under this license to share adapted material derived from this chapter or parts of it.

The images or other third party material in this chapter are included in the chapter's Creative Commons license, unless indicated otherwise in a credit line to the material. If material is not included in the chapter's Creative Commons license and your intended use is not permitted by statutory regulation or exceeds the permitted use, you will need to obtain permission directly from the copyright holder.

Part III

Sectoral Movements

15

Climate Change as a Crisis of Recognition: Alternative Climate Imaginaries for Food Sovereignty in Brazil

Juliana E. Gonçalves

1   Introduction

Climate change is increasingly exposing global inequalities, where those who contribute the least to socio-environmental degradation disproportionately bear the impacts of climate change (Dodman et al., 2022; Sultana, 2022). Despite this systemic understanding grounded in critical scholarship, institutional responses to the climate crisis remain largely fixated on short-term, technology-based incremental improvements within existing systems (Loorbach, 2022). Such an approach not only fails to address the systemic causes of climate change and its structural consequences but also reinforces existing social inequalities and dispossession across gender, race, and class (Sultana, 2022). The persistent focus on risk management and system optimisation furthermore reflects a fundamental institutional lock-in, where strategies to address

J. E. Gonçalves (✉)
Department of Urbanism, Faculty of Architecture and the Built Environment, Delft University of Technology, Delft, The Netherlands
e-mail: J.E.Goncalves@tudelft.nl

© The Author(s) 2026
S. Sareen and S. Juhola (eds.), *Societal Transitions to Sustainability*, Palgrave Studies in Environmental Transformation, Transition and Accountability, https://doi.org/10.1007/978-3-032-07395-2_15

the climate crisis are constrained by the need to minimise disruption to the status quo (Loorbach, 2022).

In response to this apparent inertia, the last few years have seen a growing literature on imagination and imaginaries in relation to climate change, usually within a narrative of "crisis of imagination." Within this narrative, there is a distinction between those who understand imagination as a faculty of the individual mind and those who view the imaginary as a social construct and collective capacity. The former tends to focus on *expert* imagination (e.g. Hajer & Versteeg, 2019; Zevenbergen et al., 2024). While this direction can hold promise if tied to critical approaches that challenge ideological control, cultural hegemony, capitalist domination, colonial education, and cultural erasure (Fanon, 1952; Miraftab, 2017; Mbembe, 2019; Thiong'o, 1986), there is a risk of fostering top-down expert practices that (continue) to marginalise potentially disruptive alternatives.

This article centres on the second understanding—imaginaries as a political collective capacity. In this view, I argue that a crisis of imagination is also a crisis of recognition—a failure to recognise the transformative potential of alternative imaginaries that resist "on the margins" of dominant systems and narratives. Paraphrasing Esteva (2009), those most marginalised by dominant climate imaginaries are the ones increasingly dedicated to marginalising these imaginaries. They do so by creating the symbolic and material conditions for alternative imaginaries to emerge. These groups envision futures where values like ecological balance, social justice, and collective well-being take precedence over individualism and economic growth. Their collective and social practices of resistance encode claims to the normative—what it should be—as well as the possible—what it could be—forming alternative imaginaries that are "promissory, deterministic and performative" (Whiteley et al., 2016).

In the context of food systems, dominant imaginaries respond to climate change through efficiency-based, market-driven, and technology-centred discourses and strategies (Guthman & Fairbairn, 2023). In Brazil, such discourses and practices align with how agribusiness operates (Fernandes, 2024), exacerbated by Brazil's peripheral position in the global food system, which makes the country highly dependent on global demands. The historical legacy of colonialism has also led to concentrated land ownership, precarious labour conditions, and ongoing conflicts with indigenous communities (Porto, 2025). At the same time, farmer movements advocating for land redistribution have been active for decades, with new movements emerging more recently. One example is Teia dos Povos (translated as "Web of Peoples"), a network of communities, territories, peoples, and political organisations across rural and urban Brazil, committed to formulating paths for collective emancipation through food sovereignty. Teia dos Povos differs from traditional land redistribution movements by enacting a distinct food imaginary that promotes new relationships to land, nature, and people.

However, the dominant logic of food as a commodity often overshadows these alternative approaches. Discourses and marketing strategies continue to highlight agribusiness' role in driving the national economy (Fernandes, 2024). The focus on the economic value of the agrobusiness over land security and food sovereignty illustrates the how alternative imaginaries not fitting into onto-epistemology of the dominant order are consequently considered economically unfeasible or not scalable. This chapter therefore argues that alternative climate imaginaries, particularly those grounded in relational ontologies and prefigurative politics, offer transformative pathways to address the systemic roots of climate change but need to be investigated through another lens.

To fully understand alternative imaginaries, we must understand how they challenge dominant onto-epistemologies through their alternative ways of being as well as how they materialise and fixate in everyday life and practices. To do so, the chapter combines the concepts of climate imaginaries and prefigurative politics to understand the transformative potential of alternative imaginaries from a political ontology perspective. Through a grounded theoretical analysis of primary source material produced by Teia dos Povos, the chapter demonstrates how alternative climate imaginaries disrupt dominant ontologies, opening space through prefigurative politics for more just and ecological practices here and now.

2   Theoretical Background

2.1   Climate Imaginaries and Their Ontologies

The concept of "climate imaginaries" was introduced by Levy and Spicer (2013) as socio-semiotic systems that structure a field around a set of shared understandings of the climate. Accordingly, climate imaginaries imply a "particular mode of organising production and consumption and a prioritisation of environmental and cultural values" (ibid). Through this understanding, they examine the struggle among NGOs, businesses, and state agencies over four core climate imaginaries: "fossil fuels forever", "climate apocalypse", "techno-market", and "sustainable lifestyles". Their analysis focused on global discourses around climate change. Since then, imaginaries of climate change have been explored largely through a focus on discursive and symbolic inquiries, as recently highlighted in a special issue in the Geoforum Journal (Machen et al., 2023). Such discursive approach, however, overlooks how discourse and materiality are imbricated.

Following a materialistic approach, Celermajer et al. (2024) have put forward three globally dominant imaginaries of a climate changed future, which they call "business as usual", "techno-fix" and "apocalypse" and to which they contrast the alternative imaginaries from communities in India and Australia. From this understanding, they propose a more holistic definition of climate imaginaries, where "imaginaries are created, sustained, and transformed through debates, the generation of discourses, and through cultural practices and products like stories or films" (ibid). They furthermore argue that "the ways in which people live, the forms of material flows in which they are involved, and the concrete relations with other humans and the more-than-human that comprise their forms of life are not mere reflections of existing imaginaries, but prefigurative sources of those imaginaries" (ibid). This definition thus moves beyond the realm of economic production and consumption proposed earlier by Levy and Spicer (2013), highlighting the need for an ontologically approach to understanding alternative climate imaginaries.

When thinking about the climate crisis and political and cultural struggles that seek to create alternative imaginaries, it is necessary first to dwell into the cultural and philosophical tradition from which dominant imaginaries emerge and operate within. Following the political ontology approach delineate by Escobar (2018), this tradition follows a modern onto-epistemology rooted in four fundamental beliefs. The *belief in the individual* is linked to the notion that we exist as separate individuals and has been the most enduring and naturalised belief in Western modern society, which explains the "cultural war against relational ways of being" (Escobar, 2018, p. 83). The *belief in the real* is related to a rationalistic view on the world that prioritises reason, logic, and evidence as the primary sources of knowledge. It underlines both human mastery of nature, aligning with patriarchal structures that subordinate both women and nature, as well as the colonial notion of universalism (ibid). The *belief in science* relates to the hegemony of modern science and technology. It was with modernity that societies became pervasive with expert knowledge and discourses, which in turn transformed them. Given its monopoly over knowledge, modern science cannot even enter in dialogue with other forms of knowledge (ibid). The *belief in the economy* is a belief in a "almost-entity" called the economy, a separate domain of thought and action, supported by the science of economics and grounded in the principles of free markets, limited government intervention, and individual economic freedom.

With the consolidation of "the economy" in the eighteenth century, the four beliefs have been shaping the cultural history and everyday life primarily but not only in the West, particularly after globalisation. These beliefs have also shaped dominant climate imaginaries. Take the "techno-market" or "techno-fix" imaginary as example. This imaginary is committed to a narrative of "green" growth and progress with humans retaining an instrumental and extractivist relationship with the more-than-human. It foregrounds the narrative that climate change is amenable to technological solutions, produced by modern science and distributed through free market capitalism. Such imaginary heavily relies on geoengineering and the large deployment of renewable energy technologies, including solar energy and electric cars, while largely ignoring the socio-environmental impacts these technologies have (Kraaijvanger et al., 2023; Stock & Sovacool, 2023; Yenneti et al., 2016). The four beliefs entail an ontological dualism: the individual *vs* a relational understanding of the self; the "real" Western world *vs* pluriverse worlds; modern science *vs* "other" knowledges; and the economy *vs* alternatives to capitalism. Following this, alternative climate imaginaries seek to create conditions for alternative more-than-human relations, alternative worldviews, alternative knowledge, and alternative economies to emerge (Celermajer et al., 2024).

2.2   Prefiguring Alternative Climate Imaginaries

The ontological framework described above connects to the holistic approach to prefigurative politics as proposed by Monticelli (2022): "Prefigurative politics aims to imagine, produce and reproduce—materially—new collective subjects and subjectivities, new democratic modes of participation and new decision-making processes—in other words, new forms of life." (p. 24). This definition aligns closely with the definition of climate imaginaries proposed by Celermajer et al. (2024). Monticelli further presents three interrelated features to understand social change as conceived in prefigurative politics: the mechanism of change; the relationship to (state) power; and the temporality of change. Through mechanisms of interstitial erosion and embodiment, prefigurative politics grounds change in material reality and experimentation, highlighting alternative modes of social reproduction and presentation of life. The relation to (state) power of prefigurative politics is characterised by self-determination, emancipation, and empowerment through the creation of alternative material conditions that enable other ways of living. Finally, prefigurative politics entails social change that happens in the present but develops processually, immanently, and slowly, involving thus a different kind of temporality. This means that prefigurative politics cannot be understood through the same lenses as representative and contentious politics. Like alternative climate imaginaries, many scholars and critics of prefigurative politics have limited the understanding of prefigurative politics to the realm of economic production and reproduction, failing to grasp their holistic potential beyond economics (see Monticelli's critique, ibid, p. 24).

While Monticelli (2022) defines a framework to differentiate prefigurative politics from representative and contentious politics, Yates (2014) looks at the compound of five identifiable processes within prefigurative politics: collective experimentation; the imagining, production and circulation of political meanings; the creating of new and future-oriented social norms or "conduct"; the consolidation of these political meanings and social norms in material environments or social orders; and the diffusion and contamination of ideas, messages and goals to wider networks and groups. As such, "prefiguration involves combining the imaginative construction of 'alternatives', within either mobilisation-related or everyday activities, with some strategic attempt to ensure their future political relevance" (ibid). To these "constructive" processes, one more process can be added, which is the deconstruction of capitalist relations (Feola, 2026 this book).

The theoretical framework discussed above is illustrated in Fig. 1 as a conceptual model to understand prefigurative politics from an ontological perspective. At the centre of the diagram lies the alternative climate imaginary which emerges from *within* the dominant imaginary, that is, through interstitial erosion (Monticelli, 2022). Prefiguration here entails destructive processes (Feola, 2026, this book) by the rejection of the four fundamental beliefs pervasive in dominant imaginaries (Escobar, 2018) as well as constructive processes through experimentation, meaning making, encoding, consolidation, and diffusion (Yates, 2014). In the following section, I employ this conceptual framework to illustrate how alternative climate imaginaries manifest through these dynamics with Teia dos Povos as empirical context.

3 Alternative Climate Imaginaries for Food Sovereignty in Brazil

Teia dos Povos is a network of communities, territories, peoples, and political organisations across rural and urban Brazil, with the aim of formulating paths to collective emancipation through food sovereignty. Teia dos Povos emerged in 2012 in the first Jornada da Agroecologia (Journey of Agroecology), a local conference in the state of Bahia, which

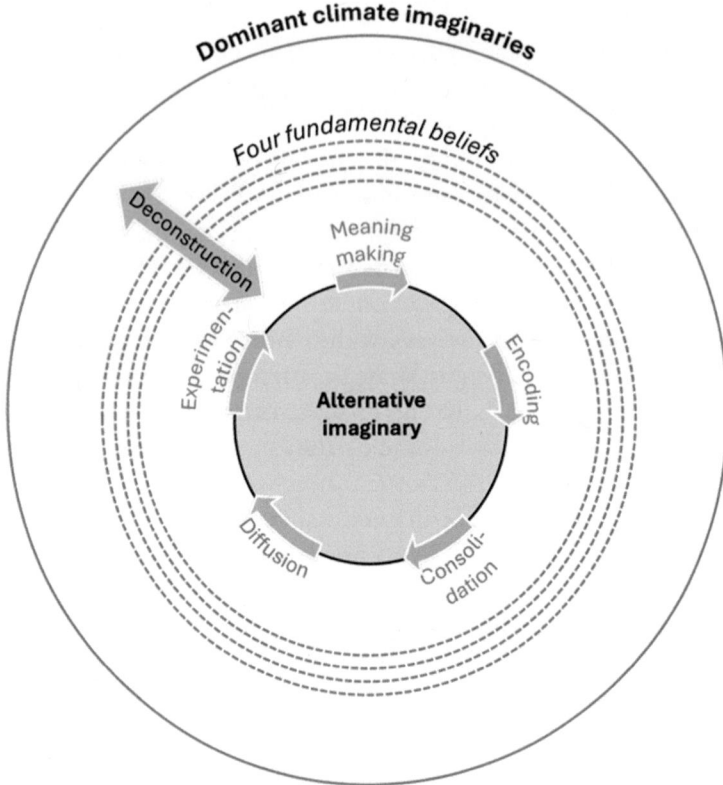

Fig. 1 Conceptual model of prefiguration from a political ontology perspective. *Source* Author.

brings together thousands of members from both the countryside and the city every two years. Organisationally, Teia dos Povos operates through two main elements: "Nucleos de Base" (freely translated as territorial core groups), which are territories where autonomy is built and defended, and "Elos da Teia" (freely translated to links of the web), which refer to agroecology collectives, intellectuals, academic research groups, students, associations, cooperatives and other groups that are not territorialised in their action. Nucleos de Base and Elos da Teia in each region are clustered into Regional Groups, providing an appropriate forum for decision-making that considers the regional biome, socio-environmental conflicts and own characteristics, which particularly relevant in the Brazilian context, given the continental size of the country. Finally, there are Divisions, which are specific to actions in each area, and composed of people from Nucleos de Base and Elos da Teia as well as individual activists that contribute to the movement. One example is the Communications Division that takes care of the website, social networks, communications, and contacts outside the territories.

Teia dos Povos functions as a decentralised *web* of communities connected by shared struggles and solidarities. It draws on Indigenous and Afro-diasporic traditions to frame climate action as an ontological struggle—a fight not only for material survival but also for the recognition of ways of life that colonialism and modernity seeks to erase (Ferreira & Felício, 2021). Teia dos Povos challenges the *belief in the individual* by its very solidarity approach that seeks to build a popular alliance between black, Indigenous, and farming communities over territorial struggles in Brazil, while also connecting to urban struggles. This comes from the recognition of interdependencies between these groups and the environment. In particular, the Indigenous peoples of Teia dos Povos have a deep connection to ancestrality, building on cosmovisions (Jourdan, 2024) and deeply relational forms of personhood. In addition, in bringing together different people and struggles, Teia dos Povos recognises that "no one needs to lower the flag of their struggle to form an alliance" (Teia dos Povos n.d., freely translated). In this way, Teia dos Povos prefigures a different way of movement building, one that ensures a plural movement instead of a monolithic unity that forces unity around a single cause.

Teia dos Povos breaks with the notion of what is *real*, affirming that a territory is not merely a demarcated area, something to be occupied, managed and owned by people. Instead, a territory is a place full of symbols of belonging based on the abundance of life (Ferreira & Felício, 2021). Through the occupation of inactive land and processes of "retomada" (taking back) of indigenous lands and the subsequent regeneration of these territories, Teia dos Povos anchors social change in material reality and alternative forms of socio-ecological reproduction. This relation of self-determination and emancipation from state power further exposes how the state has always worked in favour of colonial white elites (Porto, 2025).

Teia dos Povos foregrounds an anti-capitalist system centred on food sovereignty through agroecological practices. The use of native seeds connects their practices to traditional and ancestral wisdom, following an agricultural practice that differs from the current hegemonic agribusiness. A claim for food sovereignty is particularly prominent in Latin America, given the impact of industrial agriculture on farming livelihoods. By cultivating native crops, restoring degraded lands, and resisting agribusiness expansion, Teia dos Povos embodies a counter-hegemonic approach to climate resilience. Lima (2020) furthermore highlights that these practices are not merely about food production but are acts of resistance against land dispossession and environmental degradation. Other practices include care practices to foster individual and collective health, crowdsourcing initiatives to fund specific projects as well as the organisation collective action moments (called mutirões) focusing on specific activities, e.g. the cleaning of settlement buildings, the preparation of the land, or building of sustainable water infrastructure.

Teia dos Povos also has a clearly defined relationship with *science* through the Elos da Teia, which include intellectual and academic groups, among others. Although many of these scholars take critical and decolonial approaches, the direction of this relationship is towards supporting the Nucleos de Base: "those engaged in territorial struggles have *priority* over those engaged in the reflections about the struggles" (Teia dos Povos, n.d., freely translated, emphasis added). Nevertheless, in biennial Jornadas, academic sessions are held, covering topics such as popular and situated education, agroecology and climate change, culture, art and ancestrality, territorial conflicts, communication, water and energy autonomies, among others. In addition, Teia dos Povos supports education and political formation through its Escola de Formação (freely translated to training or hands-on school), where participants engage in knowledge-sharing rooted in Indigenous, Afro-Brazilian, and peasant worldviews. This pedagogical approach challenges the extractive logic of Western academia and private education and instead promotes co-learning and intergenerational knowledge exchange.

Teia dos Povos therefore rejects the four fundamental modern beliefs through *deconstructive* process, which materialise, for instance, as land occupation and retake practices, but also actively *constructs* an alternative climate imaginary based on socio-ecological experimentation and the politicising of food systems within the context of climate change, with clear attempts to ensure their relevance by consolidating and diffusing their new socio-ecological relations and practices to wider networks through Jornadas and pre-Jornadas as well as a dedicated Communication Division.

4 Conclusion

Asserting that a crisis of imagination in the context of climate change is also a crisis of recognition, this chapter draws on the notions of climate imaginaries and prefigurative politics to demonstrate the transformative potential of alternative climate imaginaries through a political ontology perspective. Based on a grounded theoretical analysis of primary source material produced by Teia dos Povos, a *web* of communities, territories, peoples and political organisations across rural and urban Brazil, the chapter shows that alternative climate imaginaries deconstruct dominant beliefs while also creating ontological alternatives to dominant imaginaries. With a relational, plural, situated, and autonomous imaginary for food sovereignty in Brazil, Teia dos Povos holistically challenges dominant climate imaginaries through prefigurative politics.

Competing Interests The author has no conflicts of interest to declare that are relevant to the content of this chapter.

References

Celermajer, D., Cardoso, M., Gowers, J., Indukuri, D., Khanna, P., Nair, R., Wright, G. et al. (2024). Climate imaginaries as praxis. *Environment and Planning E: Nature and Space, 7*(3), 1015–1033. https://doi.org/10.1177/25148486241230186

Davoudi, S., & Machen, R. (2022). Climate imaginaries and the mattering of the medium. *Geoforum, 137*, 203–212. https://doi.org/10.1016/j.geoforum. 2021.11.003OUCI

Dodman, D., Hayward, B., Pelling, M., Castán Broto, V., Chow, W., Chu, E., Dawson, R., Khirfan, L., McPhearson, T., Prakash, A., Zheng, Y., & Ziervogel, G. (2022). Cities, settlements and key infrastructure. In *Climate Change 2022: Impacts, adaptation and vulnerability. Contribution of working group II to the sixth assessment report of the intergovernmental panel on climate change* (pp. 907–1040). Cambridge University Press.

Escobar, A. (2018). *Designs for the pluriverse: Radical interdependence, autonomy, and the making of worlds*. Duke University Press.

Esteva, G. (2009). What is development? (Unpublished manuscript). Universidad de la Tierra, Oaxaca.

Fanon, F. (1952). *Black skin, white masks*. Pluto Press.

Fernandes, S. (2024, 9 de setembro). *Para combater as queimadas, Brasil precisa se libertar do agro*. Intercept Brasil. https://www.intercept.com.br/2024/09/09/para-combater-as-queimadas-brasil-precisa-se-libertar-do-agro/

Ferreira, J., & Felício, E. (2021). *Por terra e território: Caminhos da revolução dos povos no Brasil*.

Feola, G. (2026). Broadening the Understanding of Deconstruction in Prefigurative Social Spaces. In Sareen, S., & Juhola, S. (Eds.), *Societal transitions to sustainability: The prefigurative politics of present transformation* (this book). Palgrave Macmillan.

Guthman, J., & Fairbairn, M. (2023). Decoupling from land or extending the view: Divergent spatial imaginaries of agri-food tech. *Geographical Review, 115*(1–2), 35–48. https://doiorg.tudelft.idm.oclc.org/10.1080/00167428.2023.2261283

Hajer, M., & Versteeg, W. (2019). Imagining the post-fossil city: Why is it so difficult to think of new possible worlds? *Territory, Politics, Governance, 7*(2), 122–134. https://doi.org/10.1080/21622671.2018.1510339

Jourdan, C. (2024). Cosmologias do múltiplo e formas de vida anticoloniais. *Revista Tapuia*. https://doi.org/10.29327/2128853.2.4-13

Kraaijvanger, C. W., Verma, T., Doorn, N., & Goncalves, J. E. (2023). Does the sun shine for all? Revealing socio-spatial inequalities in the transition to solar energy in The Hague, The Netherlands. *Energy Research & Social Science, 104*, Article 103245. https://doi.org/10.1177/13505084134898

Levy, D. L., & Spicer, A. (2013). Contested imaginaries and the cultural political economy of climate change. *Organization, 20*(5), 659–678. https://doi.org/10.1177/1350508413489816

Lima, N. R. de. (2020). Articulação e autonomia para os povos em movimento: Reflexões sobre a construção da Teia dos Povos. In *Estado e sociedade sob olhares interdisciplinares: Experiências participativas, disputas narrativas, território e democracia.*

Loorbach, D. A. (2022). Designing radical transitions: A plea for a new governance culture to empower deep transformative change. *City, Territory and Architecture, 9*(1), 30. https://doi.org/10.1186/s40410-022-00176-z

Machen, R., Davoudi, S., & Brooks, E. (2023). Climate imaginaries and their mediums. *Geoforum, 138*, Article 103672. https://doi.org/10.1016/j.geoforum.2021.11.003

Mbembe, A. (2019). *Necropolitics.* Duke University Press.

Miraftab, F. (2017). Insurgent practices and decolonization of future (s). In *The Routledge handbook of planning theory* (pp. 276–288). Routledge.

Monticelli, L. (2022). Prefigurative politics within, despite and beyond contemporary capitalism. In *The future is now* (pp. 15–31). Bristol University Press.

Porto, R. N. (2025). Decomposing the law, composting the collectives: Indigenous struggles for lands and the making of life beyond rights. *Law and Critique, 1*–32. https://doi.org/10.1007/s10978-024-09411-7

Sovacool, B. K. (2023). Expanding carbon removal to the Global South: Thematic concerns on systems, justice, and climate governance. *Energy and Climate Change, 4*, 100103.

Sovacool, B. K., Dunlap, A. A., & Novaković, B. (2024) When decarbonization reinforces colonization: Complex energy injustice and solar energy development in the California Desert. *Annals of the American Association of Geographers, 1*–32. https://doi.org/10.1080/24694452.2024.2433040

Stock, R., & Sovacool, B. K. (2023). Left in the dark: Colonial racial capitalism and solar energy transitions in India. *Energy Research & Social Science, 105*, Article 103285. https://doi.org/10.1016/j.erss.2023.103285

Sultana, F. (2022). Critical climate justice. *The Geographical Journal, 188*(1), 118–124. https://doi.org/10.1111/geoj.12417

Teia dos Povos. (n.d.). Sobre. https://teiadospovos.org/sobre/

Whiteley, A., Chiang, A., & Einsiedel, E. (2016). Climate change imaginaries? Examining expectation narratives in cli-fi novels. *Bulletin of Science, Technology & Society, 36*(1), 28–37. https://doi.org/10.1177/0270467615622845

Yates, L. (2014). Rethinking prefiguration: Alternatives, micropolitics and goals in social movements. *Social Movement Studies, 14*(1), 1–21. https://doi.org/10.1080/14742837.2013.870883

Yenneti, K., Day, R., & Golubchikov, O. (2016). Spatial justice and the land politics of renewables: Dispossessing vulnerable communities through solar energy mega-projects. *Geoforum, 76*, 90–99. https://doi.org/10.1016/j.geoforum.2016.09.004

Zevenbergen, C., Harteveld, M. G., Bloemen, P., van Ham, M., van den Doel, W., Hertogh, M. H., & Tromp, E. (2024). Uniting imagination and evidence by design to navigate climate survival in urbanizing deltas. *NPJ Ocean Sustainability, 3*(1), 53. https://doi.org/10.1038/s44183-024-00094-2

Open Access This chapter is licensed under the terms of the Creative Commons Attribution-NonCommercial-NoDerivatives 4.0 International License (http://creativecommons.org/licenses/by-nc-nd/4.0/), which permits any noncommercial use, sharing, distribution and reproduction in any medium or format, as long as you give appropriate credit to the original author(s) and the source, provide a link to the Creative Commons license and indicate if you modified the licensed material. You do not have permission under this license to share adapted material derived from this chapter or parts of it.

The images or other third party material in this chapter are included in the chapter's Creative Commons license, unless indicated otherwise in a credit line to the material. If material is not included in the chapter's Creative Commons license and your intended use is not permitted by statutory regulation or exceeds the permitted use, you will need to obtain permission directly from the copyright holder.

16

The Potential Role of the Law Clinics in Prefigurative Law and Politics: Environmental Law Clinics as a Case Study

Zerrin Savaşan

1   Introduction

While the concept and practice of prefiguration have so far attracted significant attention in political theory, its intersection with law remains largely unexplored. This article seeks to bridge this gap by examining how law can embody and foster desired future social and legal arrangements in the present. This underexplored area reveals the potential for law not only to reflect societal change but also to pioneer it. Focusing on a specific context, it investigates the potential of environmental law clinics (ELCs) as crucial sites for cultivating prefiguration within law and politics.

Environmental Law Clinics (ELCs) are programs within law schools where students work on real-world environmental law cases and projects

Z. Savaşan (✉)
Department of International Relations, Sub-Department of International Law, Faculty of Economics and Administrative Sciences, Selçuk University, Konya, Türkiye
e-mail: szerrin@selcuk.edu.tr

© The Author(s) 2026
S. Sareen and S. Juhola (eds.), *Societal Transitions to Sustainability*, Palgrave Studies in Environmental Transformation, Transition and Accountability,
https://doi.org/10.1007/978-3-032-07395-2_16

under the supervision of experienced attorneys. Prefigurative law, as understood here, is a theory of law that moves beyond the traditional view that law merely reflects existing social norms. Instead, it argues that law should act as a catalyst for social transformation, embedding desirable future realities within the realm of existing law. In the context of ELCs, this refers to legal work that goes beyond immediate problem-solving and environmental outcomes. It aims to model, enforce, and promote the kinds of legal and social systems that environmental activists envision for the future. For example, an ELC that helps a community develop a community-owned solar project is an example of a model for decentralised renewable energy. Similarly, students drafting a local regulation prioritising green building practices prefigure more sustainable development patterns. Even litigating a case to stop a polluting project can establish a precedent for integrating climate change impacts into environmental reviews. By combining legal expertise with a deep commitment to social and ecological justice, challenging traditional legal approaches, empowering communities, and advocating for systemic change, ELCs possess the potential to play a pivotal role in providing and developing prefiguration in law and politics, thereby demonstrating its necessity and feasibility.

In this context, this study specifically seeks to answer the questions of whether law can truly be prefigurative and what role ELCs can play in fostering this prefiguration in the legal and policy landscape. For this purpose, it establishes ELCs as strategic sites for prefigurative law strategies. Before delving into details, it clarifies the conceptual and theoretical underpinnings of prefigurative law and its application within the ELC context. Following a brief overview of the methodology, it focuses on the central question: how can these strategies be best implemented in ELCs? To address this, it examines ELCs through the lens of three interconnected sub-fields: 1. Legal Education (3Ps-Pedagogical- Professional and Public education); 2. Legal assistance (social change lawyering models) and 3. Advocacy and social mobilisation (movements, advocacy, and activism). It also suggests a model for ELCs embodying ideal environmental practices. Finally, based on its findings, it discusses how these clinics operate to navigate change, providing an overview of the benefits and limitations of this approach for improving prefigurational environmental law and politics. Examining these questions through the lens of ELCs presents a promising and largely unexplored area of research.

2 Methodology

This study adopts a qualitative research approach, primarily relying on conceptual analysis and literature review within prefigurative politics, legal theory, and clinical legal education. Its arguments are mainly developed through theoretical synthesis and deductive reasoning. The study's limitations include the scarcity of direct empirical research on the prefigurative impact of ELCs.

This original analytical framework, which categorises the prefigurative functions of ELCs into legal education, legal assistance, and advocacy/social mobilisation, was developed by the author to address the need for a structured analysis of this novel intersection and aims to offer a fresh perspective to the existing literature. The purpose of this categorisation is to emphasise the distinct yet interconnected ways in which ELCs can contribute to embodying and promoting future legal and social norms in the present, e.g. while legal education prioritises long-term capacity building, legal assistance aims direct problem-solving. Advocacy and social mobilsation, on the other hand, seeks to establish a foundation for bottom-up change and transformation.

Furthermore, the conceptualisation of 'ideal environmental practices' was elaborated through: (1) A review of best practices identified in environmental law and policy literature; (2) An examination of publicly available information regarding the activities of various environmental law clinics; and (3) A synthesis of these findings to identify potential models for exemplary clinic operations. This framework was further refined through its application to the theoretical lens of prefiguration, considering how these practices could embody and promote desired future environmental and social norms within the current legal context.

3   Conceptual and Theoretical Framework

In general, prefiguration focuses on revising the means to achieve goals, often in grassroots activities that take shape in opposition to or apart from the existing order. It so involves rehearsing today the changes that will occur in the system in the future. Prefigurative law, then, can be defined as legal practices offering critiques of existing political struggles through the creation and enactment of legal drafts and reforms that embody desired future legal and social goals in the present. This concept draws upon insights from critical legal theory's questioning of law's neutrality and prefigurative politics' idea of embodying future social relations in present actions. It encourages lawyers and legal actors to engage in activities that embody the desired future social order, even within the constraints of the existing legal framework, and empowers communities to participate in shaping their own legal and social realities. The key point here is using the potential of law, not just as a tool for future change, but as a means to enact elements of that change in the here and now (Ashar, 2023; Cohen & Morgan, 2023; Cooper, 2020; Hesselink, 2024; Thorpe & Morgan, 2022).

When viewed through the lens of prefigurative law in the context of ELCs, it can be explained as reflecting the future into the present in an experimental way, where legal tools do not only serve as instrumental means to achieve a goal but also reflect desired goals. While there is not any specific study explicitly linking prefigurative law and ELCs, but relevant theories like critical legal theory (Gabel & Harris, 1982), collaborative governance approach (Ansell & Gash, 2008), and legal consciousness perspective (Chua & Engel, 2019), drawing on legal mobilisation theory (Vanhala, 2022), together with prefigurative politics (Törnberg, 2021), and scholarship on clinical legal education (CLE) (Levett, 2024), can arguably provide the theoretical building blocks to establish this connection and to understand the role of ELCs in enacting prefigurative legal practices. Indeed, the emphasis on using law, strategic litigation and other legal processes to advance social change offers the most direct and practical explanation of how ELCs contribute to prefigurative law. This is particularly because clinics can act as hubs for legal mobilisation (Epp, 1998), providing resources and expertise to support social movements working on environmental issues and increasing the effectiveness of environmental advocacy. They can help translate broad social demands into concrete legal claims and strategies, effectively bridging the gap between communities and the formal system. These aspects perfectly align with the prefigurative approach, which seeks to challenge existing legal frameworks, enact desired social and environmental changes, and advance broader environmental goals through legal action.

4   Environmental Law Clinic as a Site for Prefigurative Law Strategies

ELCs are programs within law schools where students, under the supervision of experienced attorneys, work on real-world environmental law cases and projects (Hamman & Witkowski Heaps, 2021). They can serve as a powerful tool for prefiguration through three main functions: Legal education (3Ps-Pedagogical-Professional and Public education); Legal assistance (social change lawyering models); and Advocacy and social mobilisation (movements, advocacy, and activism).

4.1   Legal Education

By promoting students' critical thinking about the law and its role in society, ELCs can train them in impact litigation strategies and involve them in real cases that challenge environmentally harmful practices, establish legal precedents, and push for stronger environmental regulations. For instance, a clinic engaging students in drafting a local ordinance for mandatory rainwater harvesting on new constructions not only provides practical legal skills but also prefigures sustainable water management policies. The prefigurative potential here is evident in the opportunities clinics provide for students to participate in collective action, express their opposition, and question existing power structures (Ashar, 2023). Thus, they can help develop a new generation of lawyers deeply committed to environmental justice and to building a more democratic and participatory society.

In addition to these pedagogical and professional goals, ELCs may also provide some form of legal information, advice, and sometimes legal services to the public (also known as street law or citizen law clinics) (Arthurs, 2015). By empowering the public with legal knowledge, these programs can help people understand their rights, navigate the legal system, become active participants in environmental decision-making processes, and advocate for themselves, fostering a more informed and engaged citizenship. Moreover, by raising awareness about environmental issues, encouraging participation in local environmental initiatives, and promoting environment-friendly and sustainable behaviours, lifestyles and consumption patterns, they can help foster a sense of responsibility for the environment and contribute to building a culture of environmental stewardship.

4.2   Legal Assistance

By integrating the tools of social change lawyering with a focus on prefigurative law, ELCs can play a vital role in shaping a desired future through present legal action. An effective ELC likely draws upon elements from various social change lawyering models, including collaborative (community-rebellious), movement, and catalytic lawyering (Cimini & Smith, 2023). However, community lawyering and movement lawyering are particularly well-aligned with the core functions of such clinics. For example, an ELC assisting a community in establishing a community-owned solar energy cooperative demonstrates a tangible model for decentralised and democratically controlled renewable energy systems. Indeed, community lawyering emphasises collaboration with a specific community to address their legal needs and empower them to participate in the relevant legal processes (Tokarz et al., 2008). ELCs driven by the movement lawyering model can also play a crucial role in supporting environmental movements to advance their goals through legal strategies. By providing legal expertise, representing activists, and pursuing impact litigation, these clinics contribute to broader efforts for environmental protection and social justice (Cummings, 2017). On the other hand, as mentioned, rebellious lawyering can also be useful to effectively challenge corporate or governmental actions harming the environment and local communities, as it can address power imbalances while empowering marginalised communities to use law for social change. Similarly, ELCs can engage in catalytic lawyering by selecting cases with broader implications, using litigation to raise public awareness, and advocating for policy reforms, since catalytic lawyering strategically employs law to create broader social change beyond individual cases.

Shortly, through social change lawyering methods, ELCs can provide legal service to community-based organisations working on environmental justice issues, as well as communities disproportionately affected by environmental injustices who may lack access to legal representation. This can empower these communities to fight against polluting industries or development projects, secure access to clean water and air, and influence decision-making processes that affect their environment, thereby addressing power imbalances and promoting access to justice.

4.3 Advocacy and Social Mobilisation

ELCs can also act as agents of social change by actively working in favour of environmental movements, environmental advocacy, and activism. They can empower communities affected by environmental issues to become more powerful and effective in organising protests and demonstrations, challenging environmental injustices through legal channels, and protecting them from legal repercussions for their activism (Ashar, 2008).

Environmental movements, encompassing a diverse array of individuals, organisations, and actions sharing the common goal of protecting the environment, e.g. Sunrise Movement, Gezi protests (as part of Occupy Movement) (Uncu, 2022), the Sierra Club, Greenpeace, and countless local groups working on related issues (Asara & Kallis, 2023), can greatly benefit the training, legal advice, and representation provided by ELCs. This might involve helping them navigate complex permitting processes, negotiate with polluters and use of mediation and other collaborative processes to resolve environmental disputes in a more equitable and participatory manner, or develop strategies for community-led environmental monitoring to hold polluters accountable and advocate for stronger environmental regulations, or even establish their own systems of environmental governance and management, incorporating traditional ecological knowledge, creating community-based conservation areas, and enacting local regulations to protect natural resources. Additionally, by fostering a sense of community and providing a space for dialogue between different community groups, environmental organisations, and other stakeholders working towards shared goals, or by providing resources and support for grassroots movements, ELCs can facilitate collaboration between them to amplify their collective voice and contribute to creating a stronger and more cohesive movement for environmental justice and systemic change. To illustrate, an ELC providing legal support to an environmental movement for the legal rights of nature foreshadows ecocentric legal frameworks. By advocating for the legal status and protection of ecosystems, it challenges the view that nature is just a property and seeks a legal system that recognises the intrinsic value of nature (Bastian et al., 2017; Rodríguez-Garavito, 2024).

Environmental advocacy involves taking specific steps to influence relevant policies to create a lasting impact on solving environmental problems. It can use a variety of platforms and tools, such as lobbying within existing political and legal systems, public awareness campaigns (such as media campaigns, social media campaigns, and public events), and research reports. (Jickling, 2003). For an ELC, this might involve working respectfully and inclusively with relevant community groups, NGOs, and government agencies to create a more conducive environment for changing public opinion and policy. For example, an organisation lobbying for more effective air pollution regulations or an advocate suing to protect an endangered species might be supported in this way. ELCs can also advocate for legislation by designing national or local regulations that can serve as models incorporating principles, such as supporting policies that prioritise environmental protection, sustainability, and environmental justice over development; working for a transition towards a circular economy; or for the right to a healthy environment or non-humans rights; and pressing for stricter environmental standards for industries. By responding to the challenges of the Anthropocene by creating hypothetical legal decisions from the future and imagining new legal concepts and approaches, the Anthropocene Judgments Project can be shown here as a powerful example of how preconceived legal arguments can influence the future development of environmental law (Rogers-Michelle Maloney, 2022; Goodrich & Zartaloudis, 2021).

Environmental activism emphasises more direct and often confrontational methods to bring about change. Activists may engage in a wide range of activities and diverse tactics (involving both legal and non-legal strategies, depending on the context and goals) aimed at raising public awareness and pressuring decision-makers to take action, like protests, demonstrations, civil disobedience, boycotts. (Rootes, 2003). An ELC helping a group of activists blockading a pipeline construction site or organising a large-scale march to demand action on climate change can be an example of a prefigurative approach.

In addition to leveraging these interconnected functions, an ELC can powerfully advocate for a better environmental future by also modeling ideal environmental practices. By embodying these practices in its own operations, the clinic itself becomes a living example of environmental responsibility. This includes: Implementing a sustainable office (prioritising reusable materials, energy efficiency, paper reduction, sustainable procurement, waste reduction); Promoting sustainable transportation (encouraging sustainable commuting options, and offsetting travel emissions); Actively engaging with the community on environmental issues. Moreover, by prioritising clients committed to sustainability, the clinic collaborates with those already engaged in or willing to adopt environment-friendly practices. This could involve representing eco-businesses implementing innovative environmental solutions or communities developing local resilience strategies, and promoting best practices, such as assisting a small business with a waste reduction program or helping a farm transition to more sustainable agriculture. Finally, the clinic can showcase success stories by publicising positive examples of sustainable practices being implemented by both the clinic and its clients.

This can be achieved through developing case studies, public presentations, media outreach, its website, and social media. This 'walk the talk' approach can be a particularly effective way for ELCs to implement prefigurative strategies.

5   Navigating Legal and Social Transformation: Merits and Limits

This study focuses on the idea that the prefigurative law can be redirected through ELCs by environmental activists, legal scholars, and interested students who are dissatisfied with current legal frameworks and aim for more equitable and sustainable legal and social systems. The prefigurative functions of ELCs have the potential to drive change at various scales and directions. Indeed, agency can be supported through legal education, which raises awareness of alternative legal approaches for both lawyers and citizens, and through legal assistance, which provides marginalised communities with the tools needed to challenge environmental injustices. Additionally, structural change can be targeted through advocacy and social mobilisation aimed at encouraging systemic reforms and building broader public support for them. Therefore, the potential impact of prefigurative legal models is not only on communities directly affected by environmental injustices, but can also provide a broader social basis for their formation.

In short, prefigurative ELCs hold significant value because of their unique capacity to demonstrate the feasibility of alternative legal concepts and frameworks for broader legal reform, policy, and social change. However, they also face a number of limitations and challenges. In this context, in addition to some of the fundamental limitations pointed out for all law clinics (having limited resources to undertake complex cases, being dependent on student availability, and ensuring long-term sustainability through continuous effort (Levett, 2024)), some limitations specific to their prefigurative functions should also be mentioned.

Persuading existing mechanisms to adopt prefigurative legal arguments is one of them, since the formal legal system generally trusts and prefers to work with established norms and procedures, and is resistant to change. Given that these clinics are already generally perceived as a threat to existing legal and policy frameworks, and that their efforts may be constrained, it is conceivable that both the immediate impact and the long-term structural legal and social change that ELCs' initiatives will generate may be limited. Possible problems that may arise in the balanced distribution of ELCs' limited resources among their prefigurative functions, the practical complexities of implementing each of these functions, may lead to clinics being inadequate to question existing structures and their practical effectiveness may be questioned. Last but not least, these ELCs need to operate in a way that rejects top-down approaches and recognises the mutually constitutive relationship between law and society. This requires adherence to high ethical standards, being open to different perspectives, avoiding power-based biases, and being critical of their own activities and their impacts, all of which present significant challenges in practice.

6   Conclusion

As detailed throughout the article, the inherent potential of ELCs to contribute to prefigurative law and politics persists despite the challenges they face. Indeed, the knowledge and advocacy generated by ELCs can gradually influence legal consciousness, and prefigurative ELCs can pave the way for navigating the tension between legal rigidity and the urgent need for legal and social change. Consequently, this article has demonstrated that ELCs offer a unique avenue to operationalise prefigurative law through their core functions of legal education, legal assistance, and advocacy/social mobilisation. By doing so, they can achieve crucial prefigurative goals, such as challenging current discourses, perspectives, regulations, and institutions that perpetuate environmental injustice; and transforming them by building alternative ones that embody and foster a deeper understanding of the connection between environmental sustainability and social justice. It is hoped that this study, which reveals the prefigurative potential of ELCs, will serve as a starting point for further research that delves deeper into how this potential can manifest in different contexts and contribute to structural change.

Competing Interests   The author has no conflicts of interest to declare that are relevant to the content of this chapter.

References

Ansell, C., & Gash, A. (2008). Collaborative governance in theory and practice. *Journal of Public Administration Research and Theory, 18*(4), 543–571.

Arthurs, S. G. (2015). Street law: Creating tomorrow's citizens today. *Lewis & Clark Law Review, 19*(4), 925–954.

Asara, V., & Kallis, G. (2023). The prefigurative politics of social movements and their processual production of space: The case of the Indignados movement. *EPC: Politics and Space, 41*(1), 56–76.

Ashar, S. (2008). Law clinics and collective mobilization. *Clinical Law Review, 14*(2), 355–414.

Ashar, S. M. (2023). Pedagogy of prefiguration. *The Yale Law Journal Forum, 132*, 869–903.

Bastian, M., et al. (Eds.). (2017). *Participatory research in more-than-human worlds*. Routledge.

Chua, L. J., & Engel, D. M. (2019). Legal consciousness reconsidered. *Annual Review of Law and Social Science, 15*, 335–355.

Cimini, C. N., & Smith, D. (2023). Modalities of social change lawyering. *Lewis & Clark Law Review, 26*(4), 1035–1075.

Cohen, A. J., & Morgan, B. (2023). Prefigurative legality. *Law & Social Inquiry, 48*(3), 1053–1082.

Cooper, D. (2020). Towards an adventurous institutional politics: The prefigurative 'as if' and the reposing of what's real. *The Sociological Review, 68*(5), 893–916.

Cummings, S. L. (2017). Movement lawyering (UCLA School of Law, Public Law Research Paper No. 17-43).

Epp, C. M. (1998). *The rights revolution: Lawyers, advocates, and social change*. University of Chicago Press.

Gabel, P., & Harris, P. (1982). Building power and breaking images: Critical legal theory and the practice of law. *NYU Review of Law and Social Change, 11*, 369–411.

Goodrich, P., & Zartaloudis, T. (Eds.). (2021). *The cabinet of imaginary laws*. Routledge.

Hamman, E., & Witkowski Heaps, J., et al. (2021). Environmental law clinics in Australia and the United States: A comparison of design and operation. In E. Hamman (Ed.), *Teaching and learning in environmental law: Pedagogy, methodology and best practice* (pp. 129–142). Edward Elgar.

Hesselink, M. (2024). Prefigurative private law. *Transformative Private Law*. https://transformativeprivatelaw.com/prefigurative-private-law/

Jickling, B. (2003). Environmental education and environmental advocacy: Revisited. *The Journal of Environmental Education, 34*(2), 20–27.

Levett, S. (2024). Clinic - Fit for practice? In M. Atkinson & B. Livings (Eds.), *Contemporary challenges in clinical legal education: Role, function and future directions* (pp. 185–201). Routledge.

Rodríguez-Garavito, C. (Ed.). (2024). *More than human rights, an ecology of law, thought and narrative for earthly flourishing*. Nyu Moth Project.

Rogers-Michelle Maloney, N. (2022). The Anthropocene Judgments Project: A thought experiment in future proofing the common law. *Alternative Law Journal, 47*(3), 173–178.

Rootes, C. (2003). The transformation of environmental activism: An introduction. In C. Rootes (Ed.), *Environmental protest in Western Europe* (pp. 1–19). Oxford University Press.

Thorpe, A., & Morgan, B. (2022). Prefigurative legality: Transforming municipal jurisdiction. *Urban Studies, 60*(11), 2096–2115.

Tokarz, K., et al. (2008). Conversations on "community lawyering: The newest (oldest) wave in clinical legal education." *Washington University Journal of Law & Policy, 28*, 359–400.

Törnberg, A. (2021). Prefigurative politics and social change: A typology drawing on transition studies. *Distinktion: Journal of Social Theory, 22*(1), 83–107.

Uncu, B. A. (2022). Embedding the prefigurations of the Gezi protests: The rhizomatic spread of new subjectivities and politicized identities. In Göle N (Ed.), *Public space democracy, performative, visual and normative dimensions of politics in a global age* (pp. 47–73). Routledge.

Vanhala, L. (2022). Environmental legal mobilization. *Annual Review of Law and Social Science, 18*, 101–117.

Open Access This chapter is licensed under the terms of the Creative Commons Attribution-NonCommercial-NoDerivatives 4.0 International License (http://creativecommons.org/licenses/by-nc-nd/4.0/), which permits any noncommercial use, sharing, distribution and reproduction in any medium or format, as long as you give appropriate credit to the original author(s) and the source, provide a link to the Creative Commons license and indicate if you modified the licensed material. You do not have permission under this license to share adapted material derived from this chapter or parts of it.

The images or other third party material in this chapter are included in the chapter's Creative Commons license, unless indicated otherwise in a credit line to the material. If material is not included in the chapter's Creative Commons license and your intended use is not permitted by statutory regulation or exceeds the permitted use, you will need to obtain permission directly from the copyright holder.

17

Prefiguring Healthcare: Community-Driven Digital Therapeutics and the Politics of Patient Innovation

João Rocha-Gomes

Patients don't merely envision better care; they help build it.

1   Introduction

In recent years a new form of patient activism has emerged in the digital health domain, where individuals and grassroots groups do not simply lobby for better care but create it themselves (Schermuly et al., 2021). This phenomenon can be understood through the lens of prefigurative politics—a mode of activism in which participants "embody and enact, within their activism, the socialities and practices they foster for broader society" (Moreira Fians, 2022). Unlike conventional patient

J. Rocha-Gomes (✉)
Department of Community Medicine, Health Information and Decision,
Faculty of Medicine, University of Porto, Porto, Portugal
e-mail: jngomes@med.up.pt

© The Author(s) 2026
S. Sareen and S. Juhola (eds.), *Societal Transitions to Sustainability*, Palgrave Studies
in Environmental Transformation, Transition and Accountability,
https://doi.org/10.1007/978-3-032-07395-2_17

advocacy, which might campaign for policy changes (Mackey & Schoen-feld, 2016), prefigurative projects attempt to build the desired healthcare system within the existing one (Monticelli, 2021). In this sense, commu-nity-driven digital therapeutics (DTx) projects—patient- and caregiver-developed software and devices for managing illness—serve as an alterna-tive model of health innovation (Tan et al., 2022). These projects often rely on open-source principles: participatory, non-commercial, and hori-zontal organisation (Moore & Adema, 2020); they mobilise digital tools and collective knowledge to address chronic health needs outside formal channels. Patient movements increasingly co-produce health innova-tions, illustrating how scientific development and societal participation are mutually constitutive processes (Freire & Sangiorgi, 2010). Digital platforms and data-sharing have given rise to new forms of collective health experimentation, where the personal becomes political and vice versa. The motto *"#WeAreNotWaiting"* embraced by people with Type 1 diabetes (T1D) developing DIY insulin delivery systems, captures the spirit of acting independently rather than waiting for industry or regu-latory solutions that may be too slow or too expensive (Arduser, 2018). Instead, citizens prefigure their vision of care through concrete action: hacking medical devices, writing code, and offering peer support to one another.

At its core, prefigurative politics involves practicing the future in the present (Swain, 2019). In health contexts, this means creating clinics, technologies, or social relations—empowerment, openness, affordability—that reflect desired values today. Community-driven DTx projects are in effect spaces for an alternative medicine (Ho et al., 2023). As Jansky notes of the *"#WeAreNotWaiting"* movement, its partic-ipants "engage in the development and usage of open-source closed-loop technology for the improvement of their chronic living" (Jansky & Langstrup, 2022). In other words, people with Type1-Diabetes (T1D) and their families have built their own artificial pancreas systems (APS) using software that anyone can read and modify. This grassroots inno-vation has evolved into a global network: a self-organised, digitally connected collective shaping both devices and social norms.

Here I trace the history and dynamics of the *#WeAreNotWaiting* community as a central case of prefigurative, community-driven DTx

(Dickson et al., 2022). I show how it began with a simple hack, grew into an international movement, and has begun to influence formal healthcare governance. In doing so I aim to paint a comprehensive picture of this phenomenon at the intersection of digital health and political theory: a world where people "do not wait" for permission to improve their own health.

2   The #*WeAreNotWaiting* Movement: Origins and Growth

The story of #*WeAreNotWaiting* begins in 2013 with Nightscout, one of the first patient-led digital health projects. John Costik, a father of a young child with T1D, created code to "hack" his son's continuous glucose monitor (CGM) so that its readings could be transmitted to the cloud and viewed on a smartphone (Kesavadev et al., 2020). This simple innovation allowed remote monitoring of blood glucose and alerted caregivers to highs and lows. It also embodied a bold principle: that device data should be freely accessible and shareable. Costik shared his code and instructions on a website and in a private Facebook group ("CGM in the Cloud"), inviting others to try it Lewis (2019). As he later recalled, the goal was to make the technology openly available to a community that needed it (Litchman et al., 2020). Nightscout thus set the stage for a patient collective to emerge around open-source diabetes tech: people quickly joined, translated the documentation into multiple languages, and improved the code together.

Nightscout soon proved to be a catalyst for more ambitious projects. In late 2015, a patient advocate named Dana Lewis (with programmer Scott Leibrand and others) took the next step: they used a Raspberry Pi computer to connect a CGM to an insulin pump, writing the first open-source algorithm for automated insulin dosing (Kesavadev et al., 2020). Their system, called OpenAPS, could predict blood sugar trends and automatically adjust insulin delivery overnight and between meals. Though relatively rudimentary compared to today's standards, OpenAPS worked—it closed the loop and kept glucose in target range much more consistently than before (Braune et al., 2022). Crucially, all source code was available on GitHub under permissive licenses: this meant anyone could review, copy, or modify the algorithm. Almost at once, other families and tech-savvy users began replicating OpenAPS setups, supported by detailed online guides and an active community forum.

Soon after, similar projects emerged on different platforms. In 2016 Pete Schwamb (a father of a T1D child) worked with Nate Racklyeft to develop Loop, an open-source iPhone app for closed-loop insulin control (Ware & Hovorka, 2022). By 2017 a parallel Android app, AndroidAPS, was released by European developers Milos Kozak and Adrian Tappe (Gawrecki et al., 2021). All these systems—OpenAPS, Loop, AndroidAPS—link a CGM and insulin pump to algorithms written by users. While commercial AP systems already existed or were in trials, the DIY versions often outpaced them in innovation: people used any available hardware (even out-of-warranty pumps) and shared tweaks rapidly online. They also focused on values neglected by industry, such as full transparency, endless customisability, and accessible use of existing devices (Braune et al., 2021).

By 2018 the movement had coalesced around the hashtag *#WeAreNotWaiting*—signifying a refusal to wait for slow regulatory approvals or pricey proprietary products. Online platforms (Facebook groups like "Looped", Slack channels, and Twitter) brought together thousands of participants. Indeed, an open-source consortium reported that " >10,000 individuals worldwide were using open-source AID [automated insulin delivery] systems, and the uptake continued to increase globally" (Cleal et al., 2025). In other words, long before companies introduced FDA-approved hybrid-closed-loop systems, a global community had already built and deployed its own artificial pancreas.

The *#WeAreNotWaiting* also community formalised its own evidence, with organised research consortia involving movement members as co-investigators (Heinemann & Lange, 2020). In a sense, these patients are writing their own peer-reviewed literature in real time. Informed by their own data streams, patients vet and document safety and efficacy for one another. Such "research in the wild" (Callon & Rabeharisoa, 2003) is rare in medicine but is a logical extension of the DIY ethos. They have run surveys and published some clinical studies demonstrating improved outcomes, finding overwhelmingly positive attitudes within the community. For instance, a 2020 Twitter study found that 82–85% of interactions in *#WeAreNotWaiting* and #OpenAPS were positive, focusing on media coverage, device showcases, user tips, and celebration of achievements (Litchman et al., 2020). The authors identified various participant "personas": tech-savvy "loopers", mission-driven parents, supportive providers, etc., all collaborating. Another study collected user stories from 383 participants across 29 countries, who reported large improvements in glycaemic control and quality of life after switching to DIY loops (Cleal et al., 2025). Although these accounts are not randomised trials, they point to meaningful health gains—for example, lower HbA1c and fewer hypoglycaemia events. Importantly, users consistently credit the *#WeAreNotWaiting* community itself for support: documentation, online forums, and personal coaching by experienced builders were cited as essential resources.

The technological side of the story is also fascinating and consequential. Though often called the "DIY Artificial Pancreas" movement, these community systems are early forms of digital therapeutics. By definition, digital therapeutics (DTx) are evidence-based interventions delivered via software to manage or treat health conditions (Crisafulli et al., 2022; Dang et al., 2020). The *#WeAreNotWaiting* tools fit this definition, however, unlike other DTx products, the community versions were developed outside formal R&D and clinical channels. This raises interesting questions: Can an unregulated, open-source software safely medicate people? In this case, the answer so far appears to be yes—or at least, the community judges it to be.

This body of practice exemplifies prefigurative politics. Community members "take these disease- and life-defining devices and use them to make change and express criticism of manufacturers and standard self-care regimes"; *#WeAreNotWaiting* users form a "recursive public" (Jansky & Langstrup, 2022): the technology loop (CGM → pump → algorithm → insulin) mirrors the social loop of the open-source community (tools → users → data → improvements). They are, in Kelty's terms, an online community sustained by a "shared concern" (chronic living with diabetes) and a commitment to self-determining, collective creation of technology (Kelty, 2020). In effect, every time a diabetic person sets up a Nightscout server or configures an OpenAPS rig, they enact the alternative of an open, horizontal, patient-led approach to healthcare—they live "another world" of health management in the present.

3   Theory and Practice: Device Activism and Patient Innovation

To situate this case, it helps to connect to two strands of theory. First is the literature on health activism and patient innovation, especially in Science and Technology Studies. Historically, marginalised patient groups—from HIV-positive activists in the 1980s to rare-disease advocates today—have at times moved from advocacy to direct involvement in research and care (Dunkle et al., 2010; Mbali, 2013). Scholars like Epstein and Rabeharisoa documented how "patient collectives" can reconfigure medical knowledge and norms either by demanding drug trials, or funding their own research (Epstein, 1995; Rabeharisoa, 2006). The *#WeAreNotWaiting* movement follows in this tradition but in a distinctly digital, gadget-oriented way. What makes it novel is that the space for action is not a university or hospital but kitchen tables and garages; the subjects are volunteers, and the "data" are one's own blood sugar readings. The movement practices "evidence-based activism", generating and applying knowledge even in the absence of formal trials.

Second, the concept of device activism has been proposed for precisely this phenomenon. Jansky suggests that for diabetics who depend on devices, those gadgets become political. In device activism, the objects of activism are simultaneously the tools. As one commentator puts it: participants engage with the devices in, on, and with their bodies, using them to push back on manufacturers' limitations (Jansky & Langstrup, 2022). This highlights how the *#WeAreNotWaiting* community is a matter of "material participation" as much as ideological politics. Their activism is fully embodied: it matters that an insulin pump or CGM is in use, because these become sites of contestation. By opening device data flows, by repurposing old pumps for new algorithms, by sharing networked monitoring of bodies—all this is a claim that technology should serve patients first, not just corporate profit. In practice, members document the (often opaque) inner workings of pumps and sensors, exposing companies' hidden data protocols and then rewriting them to serve communal goals. In doing so, they recreate the norms of technology development: instead of closed-source, black-box medical devices, they champion transparent, modifiable code and community review (Dexter, 2014).

These theoretical lenses help explain why the community operates the way it does—its social architecture is horizontal and participatory, with no central leader dictating a roadmap. Anyone can suggest a code tweak, and dispute is resolved in open forums. Decisions such as adding an "Advanced Meal Assist" feature to OpenAPS or supporting a new brand of pump come from community consensus. Funding is scarce and usually private (via personal donations or small grants), so for the most part projects rely on volunteer labour. This means the movement is decentralised, which aligns with anarchistic or "new left" ideals of prefigurative politics (Moreira Fians, 2022). In sum, *#WeAreNotWaiting* is a textbook case of a prefigurative, co-productive health movement—a global grassroots collective building its vision of healthcare today.

4   Regulatory and Commercial Tensions

No discussion of community-driven DTx would be complete without addressing the uneasy relationship with formal regulation and industry. By its nature, the *#WeAreNotWaiting* approach clashes with existing medical-device laws. In most jurisdictions, software that automatically manages insulin dosing would require approval as a medical device. Yet these systems are explicitly not FDA- or EMA-approved. The FDA first publicly acknowledged the phenomenon in 2019, issuing a warning after a user reported an overdose from an unauthorised algorithm (Caffrey, 2020). The Agency emphasised that "when patients combine devices that are not intended for use with other devices… new risks are introduced". In other words, regulators view DIY loops as a safety hazard outside their control.

This warning reflects a key tension. On one hand, the community's efforts are motivated by urgent patient need and have demonstrably helped many users. On the other hand, manufacturers and regulators worry about liability and generalisable safety. The regulatory stance has thus far been cautious: official bodies have mostly urged patients to be careful and report adverse events, while recognising that these users are "health savvy" and often report issues themselves (Synovitz & Larson, 2018). Interestingly, OpenAPS founders quickly issued a joint statement after the FDA notice, encouraging all users to submit event reports and work transparently. This illustrates a kind of hybridisation: the activists simultaneously defy and engage the regulatory system.

From the commercial perspective, the emergence of community loops has forced a reckoning. Insulin pump and CGM manufacturers initially remained silent, but eventually many developed their own automated systems (Akturk et al., 2022). In some cases, this was driven by patient demand: community activism made clear that customers wanted open data access and interoperability. For example, one nonprofit born in the movement, Tidepool, even partnered with industry: its "Tidepool Loop" project (started by two *#WeAreNotWaiting* fathers) is an FDA-regulated version of a DIY-style loop app (Kesavadev et al., 2020), showing that elements of the community's vision can be translated into the formal system.

Nevertheless, commercialisation remains a critical tension. Many in the movement are wary of for-profit motives and guarding the "open commons". There have been debates about whether the movement should stay nonprofit or allow venture capital. The founders of Tidepool (who began as nonprofit) explicitly emphasise that they fight for affordability and global access. Indeed, a fundamental ethic of *#WeAreNot-Waiting* is that no one should be excluded from these life-saving tools due to cost. This ties back to prefigurative goals: building a healthcare future with maximal inclusion and shared benefit.

Yet open-source health projects do face "digital divides". Setting up a DIY loop requires technical skill and dedication. The learning curve is steep: as one review noted, "the setting up of Nightscout and other algorithms is time-consuming and requires technical know-how… [and] can discourage patients" (Kesavadev et al., 2020). In other words, while the software is free, its usability is not trivial. This raises equity questions: those with computer skills—or access to people who have them—reap the benefits, while others may be left behind. It also raises sustainability questions: volunteers may burn out if the user base expands faster than support resources can handle. All in all, the community has grown largely through word of mouth and social media; if it ever became massively mainstream, new models (perhaps hybrid open-proprietary or corporate collaborations) might be needed to sustain support and training.

Regulatory authorities have also begun to adapt their frameworks in light of these developments. In policy circles, digital therapeutics are still often viewed under medical-device regulations (Wang et al., 2023), but the very existence of patients who essentially are the innovators is prompting discussion about new oversight models. For example, some have suggested that regulators could provide a "sandbox" for open-source developers, where safety standards are co-created with patient groups. The COVID pandemic's experience with rapid community-led ventilator projects has similarly encouraged agencies to consider flexible pathways for citizen innovation (Seow et al., 2021). Another realm is mental health and wellness: while commercial mental-health apps abound, there is also a growing interest in community-driven mental health tools. For example, some peer-support networks have built simple open platforms or chatbots for coping strategies. Organisations like Open Mental Health (OpenMH) are creating crowdsourced resource hubs and evidence-based guides, intended to be shared among communities without a corporate filter (Kornfield et al., 2022).

Of course, each case has unique features, but the critical tensions recur. Every project grapples with equity (who can participate and benefit?), regulation (who certifies or endorses these innovations?), commercialisation (do we welcome or resist corporate interest?), and sustainability (how to maintain volunteer-driven services?). These are the same themes that *#WeAreNotWaiting* must address as it evolves. The *#WeAreNotWaiting* community has even influenced public debates: in the UK and EU the question "should our devices be our own?" has been raised, and proposals for patient-centred certification have emerged, albeit slowly. The important point is that the activism has pushed these issues onto the agenda. The movement has not just built technology; it has forced policymakers to reckon with the idea that users want and can produce validated solutions.

5   Conclusion

The *#WeAreNotWaiting* case makes clear that community-driven DTx can play a powerful role in health innovation. By prefiguring an ideal and open, patient-centred care, these activists have created not only new tools but new ways of relating to disease and technology. Yet the journey is unfinished. Many questions remain open. How can these grassroots systems be evaluated and integrated into official care models? Should insurance ever pay for a DIY protocol? And importantly, can the values of the movement—inclusivity, openness, mutual aid—endure as it grows or intersects with markets? These unresolved issues mirror the larger challenge of prefigurative politics in a capitalist, bureaucratic world.

For now, what is undeniable is the energy and creativity this community has unleashed. As one participant noted, their mission has always been about "possibilities to think and create differently regarding personal healthcare" (Jansky & Langstrup, 2022). In other words, *#WeAreNotWaiting* and similar movements are themselves ongoing experiments in solidarity, in technological democracy, and in imagining a health system from below. Watching these experiments unfold offers a window into how healthcare might be organised in the future: a future where patients are not passive recipients, but active co-creators.

Competing Interests   The author has no conflicts of interest to declare that are relevant to the content of this chapter.

References

Akturk, H. K., Snell-Bergeon, J., & Shah, V. N. (2022). Efficacy and safety of tandem control IQ without user-initiated boluses in adults with uncontrolled type 1 diabetes. *Diabetes Technology & Therapeutics, 24*(10), 779–783.

Arduser, L. (2018). Impatient patients: A DIY usability approach in diabetes wearable technologies. *Communication Design Quarterly Review, 5*(4), 31–39.

Braune, K., Gajewska, K. A., Thieffry, A., Lewis, D. M., Froment, T., O'Donnell, S., Speight, J., Hendrieckx, C., Schipp, J., & Skinner, T. (2021). Why# WeAreNotWaiting—Motivations and self-reported outcomes among users of open-source automated insulin delivery systems: Multinational survey. *Journal of Medical Internet Research, 23*(6), Article e25409.

Braune, K., Krug, N., Knoll, C., Ballhausen, H., Thieffry, A., Chen, Y., O'Donnell, S., Raile, K., & Cleal, B. (2022). Emotional and physical health impact in children and adolescents and their caregivers using open-source automated insulin delivery: Qualitative analysis of lived experiences. *Journal of Medical Internet Research, 24*(7), Article e37120.

Caffrey, M. (2020). FDA issues warning on do-it-yourself artificial pancreas. AJMC. 2019.

Callon, M., & Rabeharisoa, V. (2003). Research "in the wild" and the shaping of new social identities. *Technology in Society, 25*(2), 193–204.

Cleal, B., Chen, Y., Wäldchen, M., Ballhausen, H., Cooper, D., O'Donnell, S., Knoll, C., Krug, N., Raile, K., & Ubben, T. (2025). Examining the emotional and physical health impact in users of open-source automated insulin delivery and sources of support: Qualitative analysis of patient narratives. *Journal of Medical Internet Research, 27*, Article e48406.

Crisafulli, S., Santoro, E., Recchia, G., & Trifirò, G. (2022). Digital therapeutics in perspective: From regulatory challenges to post-marketing surveillance. *Frontiers in Drug Safety and Regulation, 2*, Article 900946.

Dang, A., Arora, D., & Rane, P. (2020). Role of digital therapeutics and the changing future of healthcare. *Journal of Family Medicine and Primary Care, 9*(5), 2207–2213.

Dexter, M. H. (2014). *Open design and medical products: An open medical products methodology*. Sheffield Hallam University (United Kingdom).

Dickson, R., Bell, J., Dar, A., Downey, L., Moore, V., & Quigley, M. (2022). # WeAreNotWaiting DIY artificial pancreas systems and challenges for the law. *Diabetic Medicine, 39*(5), Article e14715.

Dunkle, M., Pines, W., & Saltonstall, P. L. (2010). Advocacy groups and their role in rare diseases research. In *Rare diseases epidemiology* (pp. 515–525).

Epstein, S. (1995). The construction of lay expertise: AIDS activism and the forging of credibility in the reform of clinical trials. *Science, Technology and Human Values, 20*(4), 408–437. https://doi.org/10.1177/016224399 502000402

Freire, K., & Sangiorgi, D. (2010). Service design and healthcare innovation: From consumption to co-production to co-creation. In *Proceedings of 2nd Service Design and Service Innovation conference, ServDes.*

Gawrecki, A., ozulinska-Ziolkiewicz, D., Michalak, M. A., Adamska, A., Michalak, M., Frackowiak, U., Flotynska, J., Pietrzak, M., Czapla, S., & Gehr, B. (2021). Safety and glycemic outcomes of do-it-yourself AndroidAPS hybrid closed-loop system in adults with type 1 diabetes. *PloS One, 16*(4), e0248965.

Heinemann, L., & Lange, K. (2020). Do It yourself" (DIY)-automated insulin delivery (AID) systems: Current status from a German point of view. *Journal of Diabetes Science and Technology, 14*(6), 1028–1034. https://doi.org/10.1177/1932296819889641

Ho, D., Sapanel, Y., & Blasiak, A. (2023). *Medicine without meds: Transforming patient care with digital therapies.* JHU Press.

Jansky, B., & Langstrup, H. (2022). Device activism and material participation in healthcare: retracing forms of engagement in the #WeAreNotWaiting movement for open-source closed-loop systems in type 1 diabetes self-care. *Biosocieties,* 1–25. https://doi.org/10.1057/s41292-022-00278-4

Kelty, C. M. (2020). *Two bits: The cultural significance of free software.* Duke University Press.

Kesavadev, J., Srinivasan, S., Saboo, B., Krishna, B. M., & Krishnan, G. (2020). The do-it-yourself artificial pancreas: a comprehensive review. *Diabetes Therapy, 11*(6), 1217–1235. https://doi.org/10.1007/s13300-020-00823-z

Kornfield, R., Mohr, D. C., Ranney, R., Lattie, E. G., Meyerhoff, J., Williams, J. J., & Reddy, M. (2022). Involving Crowdworkers with lived experience in content-development for push-based digital mental health tools: Lessons learned from crowdsourcing mental health messages. *Proceedings of the ACM on Human-Computer Interaction, 6*(CSCW1), 1–30.

Lewis, D. M. (2019). Automated insulin delivery. How artificial pancreas 'closed loop' systems can aid you in living with diabetes.

Litchman, M. L., Walker, H. R., Fitzgerald, C., Gomez Hoyos, M., Lewis, D., & Gee, P. M. (2020). Patient-driven diabetes technologies: Sentiment and personas of the# WeAreNotWaiting and# OpenAPS movements. *Journal of Diabetes Science and Technology, 14*(6), 990–999.

Mackey, T. K., & Schoenfeld, V. J. (2016). Going "social" to access experimental and potentially life-saving treatment: An assessment of the policy and online patient advocacy environment for expanded access. *BMC Medicine, 14*, 1–10.

Mbali, M. (2013). *South African AIDS activism and global health politics.* Springer.

Monticelli, L. (2021). On the necessity of prefigurative politics. *Thesis Eleven, 167*(1), 99–118.

Moore, S., & Adema, J. (2020). Community governance explored. COPIM.

Moreira Fians, G. (2022). Prefigurative politics. In *The Cambridge encyclopedia of anthropology.*

Rabeharisoa, V. (2006). From representation to mediation: The shaping of collective mobilization on muscular dystrophy in France. *Social Science and Medicine, 62*(3), 564–576. https://doi.org/10.1016/j.socscimed.2005.06.036

Schermuly, A. C., Petersen, A., & Anderson, A. (2021). 'I'm not an activist!': Digital self-advocacy in online patient communities. *Critical Public Health, 31*(2), 204–213.

Seow, H., McMillan, K., Civak, M., Bainbridge, D., van der Wal, A., Haanstra, C., Goldhar, J., & Winemaker, S. (2021). # Caremongering: A community-led social movement to address health and social needs during COVID-19. *PLoS ONE, 16*(1), Article e0245483.

Swain, D. (2019). Not not but not yet: Present and future in prefigurative politics. *Political Studies, 67*(1), 47–62.

Synovitz, L. B., & Larson, K. L. (2018). *Consumer health & integrative medicine: A holistic view of complementary and alternative medicine practices: A holistic view of complementary and alternative medicine practice.* Jones & Bartlett Learning.

Tan, R. K., Wu, D., Day, S., Zhao, Y., Larson, H. J., Sylvia, S., Tang, W., & Tucker, J. D. (2022). Digital approaches to enhancing community engagement in clinical trials. *NPJ Digital Medicine, 5*(1), 37.

Wang, C., Lee, C., & Shin, H. (2023). Digital therapeutics from bench to bedside. *NPJ Digital Medicine, 6*(1), 38.

Ware, J., & Hovorka, R. (2022). Closed-loop insulin delivery: Update on the state of the field and emerging technologies. *Expert Review of Medical Devices, 19*(11), 859–875.

Open Access This chapter is licensed under the terms of the Creative Commons Attribution-NonCommercial-NoDerivatives 4.0 International License (http://creativecommons.org/licenses/by-nc-nd/4.0/), which permits any noncommercial use, sharing, distribution and reproduction in any medium or format, as long as you give appropriate credit to the original author(s) and the source, provide a link to the Creative Commons license and indicate if you modified the licensed material. You do not have permission under this license to share adapted material derived from this chapter or parts of it.

The images or other third party material in this chapter are included in the chapter's Creative Commons license, unless indicated otherwise in a credit line to the material. If material is not included in the chapter's Creative Commons license and your intended use is not permitted by statutory regulation or exceeds the permitted use, you will need to obtain permission directly from the copyright holder.

18

Prefigurative Politics of Transformations in Land Use Decision-Making Systems in Europe

Simeon Vaňo, Julia Leventon, and Peter Mederly

1 Introduction

We live in the time of polycrisis spanning climate change, biodiversity loss, pollution and social injustice (Hooper et al., 2012; Vogel & O'Brien, 2022), the polycrises driven and caused by human agency (Díaz et al., 2019; Jaureguiberry et al., 2022; Richardson et al., 2023) including natural resource exploitation and land use alteration that take place on a large scale globally (Van Vliet et al., 2019). Scientists and politicians make urgent calls for transforming our society to enact system-wide shifts in views, structures, and practices across population and sectors and addressing human-nature relations (IPBES, 2024). Transformative

S. Vaňo (✉) · J. Leventon
Department of the Human Dimensions of Global Change, Global Change Research Institute of the Czech Academy of Sciences, Prague, Czechia
e-mail: vano.s@czechglobe.cz

S. Vaňo · P. Mederly
Department of Ecology and Environmental Science, Constantine Philosopher University in Nitra, Nitra, Slovakia

© The Author(s) 2026
S. Sareen and S. Juhola (eds.), *Societal Transitions to Sustainability*, Palgrave Studies in Environmental Transformation, Transition and Accountability,
https://doi.org/10.1007/978-3-032-07395-2_18

change was suggested as a mosaic of actions, through new strategies, mechanisms and roles of actors at multiple scales and levels of decision-making, with governments playing a key facilitating role (Gurung et al., 2024). Starting from this premise, the chapter explores prefigurative politics of possible present transformations in land use decision-making that, like prefigurative politics, adopts a future-oriented perspective on the "here and now".

Prefigurative politics refers to modes of action and social processes that embody the intended future society in the present time. It is an ongoing political practice that embodies social experimentation critiquing the status quo (Cornish et al., 2016), social relations, organisation, decision-making and culture envisioned as ultimate goals, aligning means with ends (Boggs, 1977; Swain, 2019) and fostering participatory democracy to challenge current social organisation to prefigure and build a more egalitarian society (Breines, 1980). Academic discourse has traditionally examined prefigurative politics primarily through the lens of emancipatory movements, such as anarchist and feminist movements, or alternative social-economic models like self-sufficient communities. This chapter adopts a systems perspective to identify and anchor transformative practices within existing social institutions, thereby addressing broader structural barriers that are often neglected (Feola et al., 2021). It illuminates both emerging opportunities and the inherently strategic nature of prefigurative politics (Swain, 2019), emphasising the need for its integration into decision-making and policy to achieve wider social change. It recognises that political and social transformation occurs through gradual, iterative processes in the present moment, while acknowledging that tensions and conflicts may prove just as transformative as points of consensus.

Land use is an important lens for understanding what prefiguration could enact transformative change. Shifts in land uses, such as changes in agricultural practices, natural resource management or urban development, influences and is influenced by all components of sustainability and the polycrisis. Land use decision-making is also a rich setting for exploring prefigurative politics because of the range of sectors, policies, actors, and interests involved through multiple levels of formal and informal decision-making processes, technical, and/or procedural processes. When viewed from local levels, planners and land use decision-makers must navigate complex needs and wishes across their citizens' broader stakeholders and policies across a range of sectors and interests (Bassi et al., 2025). It therefore provides multiple spaces for participation, negotiation and deliberation (Leventon et al., 2022), softening up the rigid nature of decision-making and taking a closer look at place-based politics emerging from various contexts (Richter, 2022).

In this chapter, we explore prefiguration in land use through a trans-disciplinary approach that generates knowledge concerning multiple actors from multiple levels, building on both the evidence and different realities. More specifically, the chapter draws on research conducted in collaboration with actors from 12 cases in Europe (EU, Switzerland and Britain) that were involved in the PLUS Change project, funded by Horizon Europe. This research investigated political economies—policies and actors—of land use decision-making systems and identified inter-vention points that represent prefigurative politics as a momentum to catalyse transformation, to soften or change the constraining structures and interactions, address power imbalances and the system's integrity.

The background of this research is summarised in the following section (Sect. 2), explaining the analysis of political economies to consider intervention points (Sect. 3) as prefigurative politics creating conditions for change. We then conclude that by implementing these interventions at different levels, we could achieve more effective and coherent policies as well as stronger and meaningful actors' engagement and interactions through decision-making (Sect. 4).

2   Investigating Political Economies to Understand Prefiguration for Sustainability Transformations

The political economies of a phenomenon such as land use decision-making are historically shaped by complex political, institutional, economic, social, environmental, or cultural shifts, and power rela-tions unique to different places (Copestake & Williams, 2014; Jessop,

2001) that evolved into accepted modes of governance and decision-making (Whaley, 2018). In this chapter, political economy specifically addresses interactions between policies and actors (Vaňo et al., 2025) that shape land use decision-making systems and influence how change occurs within them, how decisions are made, implemented, or challenged; how systems become institutionalised; and what barriers may hinder change. Policies represent formal structures and constraints of the system, creating rules and norms for actions to take place, while actors are referred to as persons or groups either influenced or affected by policies, decisions, and the related processes (Hauck et al., 2014). By understanding who the actors are, we can more effectively prioritise their involvement in decision-making, as well as in research and practice (Reed et al., 2009). We therefore draw on the understanding of policies and actors that surround our practice cases to consider where interventions might prefigure transformation.

We conducted surveys and participatory workshops in collaboration with policymakers, planning and management authorities, practitioners, local stakeholders and citizens from 12 cases spanning rural (nature conservation, natural resource management), peri-urban and urban development in Europe (Table 1), from local to regional and cross-border level. Our research examined policy themes, actor interactions, and their roles as decision-makers, influencers, or implementers, while considering policy impacts on different groups and other external influences. The analysis revealed current trends and gaps and identified which decision-making levels hold the greatest potential for transformative change, what key actors and policies drive change, what strategies amplify their influence, and how actor relationships shape decision-making dynamics and policy outcomes.

A two-stage policy survey and workshop with stakeholders from 12 cases generated a collection of 81 policy documents (strategies, plans, legislation) at local, regional, and national scales. We then analysed these documents for policy aspects such as objectives, targets, measures, monitoring. We further considered 221 actors, their interactions and roles, and categorised them into actor groups as central/regional/local governments, professional agencies, natural resource managers, NGOs, private sector, and citizens.

Table 1 Cases included in the original research on political economies (n = 12). *Source* Authors

Case	Size (pop.)	Type	Designation
Zaanstreek-Waterland (NL)	2 580 km^2 (2,5 million)	Urban	Metropol-region: 32 municipalities, 2 provinces
Nitra City, SK	122 km^2 (100 000)	Urban	Municipality area—built-up area and hinterland (agriculture, forestry)
Flanders, BE	13 500 km^2 (6,5 mil)	Peri-Urban	Region, with 300 municipalities, 5 provinces
Kaigu wetland, LV	19,55 km^2 (3 878)	Rural	A natural habitat located in the administrative territory of Jelgava region
Parc Ela, CH	659 km^2 (5 700)	Rural	Nature Park incl. 6 municipalities
Lucca, IT	1 773 km^2 (387 876)	Rural	Province with 33 communes
Green Karst, SI	1 456 km^2 (53 092)	Rural	Region, with 6 municipalities
Three Countries Park, DE, BE, NL	3 500 km^2 (under 4 million)	Peri Urban	Cross-border partnership: 3 countries, 5 regions, 104 municipalities
South Moravia, CZ	7 188 km^2 (1,3 million)	Peri Urban	Region with 7 districts and ~ 700 municipalities
Surrey, UK	1 663 km^2 (1,2 million)	Peri Urban	County, incorporating 11 districts
Ile de France	12 000 km^2 (12,3 million)	Urban, Peri Urban and rural	Region, with 8 departments and 1270 municipalities
Mazovia Region	35 558 km^2 (5,5 million)	Urban, peri urban and rural	Region, with 42 counties (incl. 5 cities with county status) and 314 municipalities

The analysis revealed strong linkages between policy objectives, targets, and measures, indicating effective translation of intentions into actions. However, weaker connections between measures and monitoring mechanisms suggest underdeveloped accountability systems and raised concerns about implementation gaps and traceability. Policy objectives strongly emphasised sustainability and climate adaptation, with significant overlaps in nature-based solutions and biodiversity conservation, though climate mitigation often remained inconsistent or vague. Energy efficiency, technological innovation, and socioeconomic aspects received limited attention despite complementing environmental objectives, while equity and public participation were notably underemphasised.

The analysis also found that decision-making remains dominated by national and regional governments. Local authorities, despite some possessing stronger powers, are largely relegated to implementer roles with limited autonomy to develop context-specific solutions. Policy formation remains centralised, excluding other actors and reinforcing institutional rigidity and power imbalances. This may pressure local authorities managing trade-offs between competing interests coming from policies and local needs.

Further, scientific expertise appears underutilised, indicating a science-policy gap, while private and NGO sectors have limited formal decision-making roles despite their policy influence. Natural resource managers are rarely involved in governance, implementation remains authority-led rather than co-managed, and public involvement stays largely consultative lasting partnerships. Analysis of external influences revealed land-use dynamics as the most prominent trend, suggesting spatial pressures significantly shape policy outcomes. Similarly, community-driven grassroots initiatives and environmental movements increasingly influence decisions through local restoration efforts and sustainability advocacy. Finally, market forces and economic factors such as real estate investments, land prices fluctuations, incentives, or funding constraints were also indicated as significant drivers of land use decisions.

This analysis led to identification of a system of intervention points (Fig. 1)—that is, the key prefigurative actions or strategies that would shift the political economy to support meaningful change towards sustainability. They specifically address the results around which actors hold what power, where policy and expertise gaps are, and the range of actors that can be mobilised to contribute to sustainable land use. The intervention points were suggested to catalyse change through (1) power decentralising and enhancing multi-actor participation; (2) bridging policy gaps and aligning policies for greater synergy; (3) responding to external trends and bottom-up initiatives; and (4) strengthening policy implementation and accountability, as detailed below.

Fig. 1 System of intervention points for land use decision-making change. *Source* Authors. Alt text: System of intervention points for land use decision-making change

3 Intervention Points to Enact Transformations in Land Use Decision-Making Systems

Intervention 1: Enhancing Multi-actor Participation, Equity, and Decentralisation in Decision-Making. This intervention focuses on defining the role of actors to improve policy and decision-making processes and is characterised by actions such as (i) decentralising decision-making, (ii) addressing power asymmetries, (iii) strengthening public engagement through meaningful participation at multiple decision-making levels, (iv) enhancing private sector involvement, and (v) involving underrepresented actors. It identifies a need for a shift from centralised, top-down, governance models towards more participatory and decentralised models with a distribution of mandates, to open up decision-making to more relevant actors. Findings suggest that local governments could adopt stronger positions in decision-making on local and regional issues. For example, local and regional governments operate more as executive institutions, while re-funding and giving a greater mandate to the local level, could strengthen the focus on local needs. Addressing power imbalances especially among governmental bodies, the private sector, and civil society is necessary to create more inclusive and fair policy outcomes. Non-governmental sector and citizens could play an active role in shaping policies rather than remaining limited to consultative roles. Achieving meaningful engagement requires strengthening public participation, shifting towards collaborative decision-making processes with affected actors having a greater direct influence on current decisions that shape the future of the environment, infrastructure, and services. Prioritising social equity is crucial to addressing key challenges like displacement, affordability, and socio-economic disparities, ensuring that strategies reflect the needs of all communities. Emphasising governance mechanisms, such as decentralised, bottom-up, policymaking, and robust stakeholder engagement, could be essential to foster more inclusive, equitable, and effective policies. Furthermore, the findings suggest that private sector actors, particularly in environmentally focused policy areas, need to be more involved to ensure effective incentives and collective responsibility for sustainability goals, where major businesses, landowners or land managers become more involved in land use change that addresses nature and climate stewardship.

Intervention 2: Bridging Policy Gaps and Enhancing Cross-Sectoral and Cross-Scale Integration. This intervention focuses on the process of integration of sectoral objectives into policies at different scales in order to enhance synergies and is characterised by actions such as (i) integrating stronger social and environmental considerations, (ii) improving linkages between policies at different levels, (iii) enhancing the integration of economic and social dimensions, (iv) highlighting sustainability as a core priority, and (v) integrating technological and infrastructure innovation and energy efficiency. Lack of legal instruments to implement higher-level policy at the local level often creates problems. Addressing such gaps and disconnects would embed conservation efforts deeper into land use planning. Similarly, a stronger linkage between policies at different levels, as well as strategic planning policies and regulatory frameworks, is needed to enhance policy execution, as seemingly weak connections (and sometimes contradictory objectives) currently hinder the effectiveness of policy enforcement. Beyond environmental concerns, policy frameworks could more comprehensively incorporate economic and social dimensions, technological and infrastructure innovation, integrating equity, public participation, and economic resilience into land use governance. Lack of focus on long-term societal and economic transformations may indicate a need for more forward-thinking policy frameworks. Overall, sustainability could be emphasised more as a cross-cutting and broad theme that ties across policy sectors (e.g. housing, transport), rather than prioritising only its environmental dimensions.

Intervention 3: Responding to External Trends and Emerging Challenges. This intervention deals with the integration of external challenges (e.g. European trends and challenges) and is characterised by actions like (i) adapting policies to rapid land use changes and trends, (ii) scaling and integrating impactful bottom-up initiatives, (iii) engaging broader environmental movements and initiatives in policymaking, and (iv) balancing market forces with regulatory safeguards to enhance policy objectives. Policies are required to be flexible and adaptive to the rapid pace of land-use changes, including urban expansion, shifts in zoning, and evolving spatial planning needs. More dynamic governance approaches could be introduced to ensure that policies remain relevant and responsive. Additionally, bottom-up initiatives such as community led conservation efforts and local urban planning movements might be formally recognised and integrated into official land use governance frameworks. Economic market forces heavily influence land use decisions and could therefore be better balanced with regulatory safeguards to ensure that financial drivers do not undermine sustainability, needs of local actors or social equity. At the same time, environmental movements advocating for climate action and biodiversity protection could be actively engaged and perhaps brought into policymaking processes in a formalised way, ensuring that public demand for stronger environmental governance translates into legal and policy commitments.

Intervention 4: Strengthening Policy Implementation, Monitoring, and Accountability. This intervention addresses policy implementation including the responsibilities of actors and is characterised by actions like (i) developing robust and reliable monitoring and evaluation mechanisms, (ii) explicitly defining policy implementers, and (iii) establishing independent oversight bodies and controlling mechanisms. Policy effectiveness is often undermined by the lack of well-structured monitoring mechanisms. Establishing well-defined monitoring and evaluation frameworks with clear indicators, benchmarking and measurable targets could ensure greater accountability in policy execution. Additionally, policies could explicitly outline the roles and responsibilities of implementing bodies and controlling mechanisms to prevent execution insufficiencies. Independent oversight bodies may be established to provide transparency and accountability, ensuring that land use policies are not only well-intentioned but also effectively executed and enforced. Strengthening policy oversight through independent bodies would prevent unchecked influence from dominant stakeholders.

4   Conclusion and Future Avenues

While most contributions in this book focus on immediate, often radical movements highlighting important social experiments and place-based solutions, questions remain about their lasting impact on social structures and functioning. In response, this chapter deliberately adopts a system-level perspective on prefigurative politics, emphasising systemic interventions that rethink current political-economic structures, power balances, actor roles and strategies at multiple scales and levels of decision-making (Gurung et al., 2024). The chapter examines prefigurative politics through the lens of political economies of land use decision-making, where identified interventions in policies and actor interactions can be implemented as catalysts for transforming current decision-making structures and processes. These interventions emerged from analysing trends, gaps, and opportunities in policies and actor relationships across diverse European cases, seeking to enhance or restructure existing decision-making models.

The identified interventions provide critical entry points for transforming land use decision-making by addressing power relations, inclusiveness, policy coherence, and effectiveness through targeted changes to policies, actors, and their relationships. While formulated at a high level, these intervention points can be developed into more specific, actionable measures that serve as a "palette of options" for policymakers, and indeed any actors seeking change to land use decision-making and its outcomes, to strategise and govern land use changes. These interventions can be adapted to address context-specific challenges by creating new partnerships, aligning policies, and developing governance models tailored to different decision-making systems. Since each case represents a unique context requiring careful consideration, specified actions must respond directly to local and regional situations. This approach could be particularly relevant for the high-level governance entities, such as the European Union, which continuously seek to improve policymaking through enhanced coordination while respecting national and regional sovereignty and place-specific needs.

Finally, it is important to recognise that prefigurative politics, while inspired by and building upon local small-scale experiments in radical societal transformation, represents a long-term process rather than rapid change. Rebuilding society's structures, practices, and worldviews requires engaging deep systemic levers that need time and iterations to mature into new forms. Similarly, these interventions serve as seeds planted to gradually shift the system, with some taking root more quickly while others develop at a slower pace.

Competing Interests   The authors have no conflicts of interest to declare that are relevant to the content of this chapter.

References

Bassi, A., M., Andreasson, E., & Guzzetti, M. (2025). Report on the policy drivers of land use change. PLUS change project. (_WP3_T3.2_D3.2_ VIAA_v.8_ 31-1-2025).

Boggs, C. (1977). Marxism, prefigurative communism, and the problem of workers' control. *Radical America, 11*(6), 99–122.

Breines, W. (1980). Community and organization: The new left and Michels' "Iron Law." *Social Problems, 27*(4), 419–429. https://doi.org/10.2307/800170

Copestake, J., & Williams, R. (2014). Political-economy analysis, aid effectiveness and the art of development management. *Development Policy Review, 32*(1), 133–153. https://doi.org/10.1111/dpr.12047

Cornish, F., Haaken, J., Moskovitz, L., & Jackson, S. (2016). Rethinking prefigurative politics: Introduction to the special thematic section. *Journal of Social and Political Psychology, 4*(1), 114–127. https://doi.org/10.5964/jspp.v4i1.640

Díaz, S., Settele, J., Brondízio, E. S., Ngo, H. T., Agard, J., Arneth, A., Balvanera, P., Brauman, K. A., Butchart, S. H., Chan, K. M., Garibaldi, L. A., & Zayas, C. N. (2019). Pervasive human-driven decline of life on Earth points to the need for transformative change. *Science, 366*(6471). https://doi.org/10.1126/science.aax3100

Feola, G., Vincent, O., & Moore, D. (2021). (Un) making in sustainability transformation beyond capitalism. *Global Environmental Change, 69*, Article 102290. https://doi.org/10.1016/j.gloenvcha.2021.102290

Gurung, J., Leventon, J., Wickson, F., Dabezies, J., Olemako, T., Penca, J., Rajvanshi, A., Remans, R., Turnhout, E., Yoshida, Y., & Renaud, A. (2024). Transformative change and a sustainable world. In K. O'Brien, L. Garibaldi & A. Agrawal (Eds.), *Thematic assessment report*. IPBES secretariat. https://doi.org/10.5281/zenodo.11382238

Hauck, J., Saarikoski, H., Turkelboom, F., & Keune, H. (2014). Stakeholder analysis in ecosystem service decision-making and research. OpenNESS Ecosystem Service Reference Book. EC FP7 Grant Agreement (308428). https://doi.org/10.t1068/a32183

Hooper, D. U., Adair, E. C., Cardinale, B. J., Byrnes, J. E. K., Hungate, B. A., Matulich, K. L., Gonzalez, A., Duffy, J. E., Gamfeldt, L., & O'Connor, M. I. (2012). A global synthesis reveals biodiversity loss as a major driver of ecosystem change. *Nature, 486*(7401), 105–108. https://doi.org/10.1038/nature11118

Jaureguiberry, P., Titeux, N., Wiemers, M., Bowler, D. E., Coscieme, L., Golden, A. S., Guerra, C. A., Jacob, U., Takahashi, Y., Settele, J., Díaz, S., Molnár, Z., & Purvis, A. (2022). The direct drivers of recent global anthropogenic biodiversity loss. *Science Advances, 8*(45). https://doi.org/10.1126/sciadv.abm9982

Jessop, B. (2001). Institutional re (turns) and the strategic–relational approach. *Environment and Planning A, 33*(7), 1213–1235. https://doi.org/10.1068/a32183

Leventon, J., Suchá, L., Nohlová, B., Vaňo, S., & Harmáčková, Z. V. (2022). Participation as a pathway to pluralism: A critical view over diverse disciplines. In *Advances in ecological research* (pp. 175–199). https://doi.org/10.1016/bs.aecr.2022.04.006

IPBES. (2024). Summary for policymakers of the thematic assessment report. In K. O'Brien, L. Garibaldi, A. Agrawal, E. Bennett, O. Biggs, R. Calderón Contreras, E. Carr, N. Frantzeskaki, H. Gosnell, J. Gurung, S. Lambertucci, J. Leventon, C. Liao, V. Reyes García, L. Shannon, S. Villasante, F. Wickson, Y. Zinngrebe & L. Perianin (Eds.), *IPBES secretariat*, Bonn, Germany. https://doi.org/10.5281/zenodo.11382230

Reed, M. S., Graves, A., Dandy, N., Posthumus, H., Hubacek, K., Morris, J., Prell, C., Quinn, C. H., & Stringer, L. C. (2009). Who's in and why? A typology of stakeholder analysis methods for natural resource management. *Journal of Environmental Management, 90*(5), 1933–1949. https://doi.org/10.1016/j.jenvman.2009.01.001

Richardson, K., Steffen, W., Lucht, W., Bendtsen, J., Cornell, S. E., Donges, J. F., Drüke, M., Fetzer, I., & Rockström, J. (2023). Earth beyond six of nine

planetary boundaries. *Science Advances, 9*(37). https://doi.org/10.1126/sci
adv.adh2458

Richter, J. (2022). Thinking about "The room where it happens": Using place
to teach about alexander Hamilton and early America. In C. Northrop (Ed.),
The Hamilton phenomenon (pp. 185–202). Vernon Press.

Swain, D. (2019). Not not but not yet: Present and future in prefigurative
politics. *Political Studies, 67*(1), 47–62. https://doi.org/10.1177/003232171
7741233

Vaňo, S., Mederly, P., & Leventon, J. (2025). Intervention points for creating
land use policy and decision-making change. PLUS change project. (_
WP4_T4.1_D4.1_VIAA_v.3_ 31-3-2025).

Van Vliet, J. (2019). Direct and indirect loss of natural area from urban expansion. *Nature Sustainability, 2*(8), 755–763. https://doi.org/10.1038/s41893-
019-0340-0

Vogel, C., & O'Brien, K. (2022). Getting to the heart of transformation. *Sustainability Science, 17*(2), 653–659. https://doi.org/10.1007/s11
625-021-01016-8

Whaley, L. (2018). The critical institutional analysis and development (CIAD)
framework. *International Journal of the Commons, 12*(2). shttps://doi.org/
10.18352/ijc.848

Open Access This chapter is licensed under the terms of the Creative Commons Attribution-NonCommercial-NoDerivatives 4.0 International License (http://creativecommons.org/licenses/by-nc-nd/4.0/), which permits any noncommercial use, sharing, distribution and reproduction in any medium or format, as long as you give appropriate credit to the original author(s) and the source, provide a link to the Creative Commons license and indicate if you modified the licensed material. You do not have permission under this license to share adapted material derived from this chapter or parts of it.

The images or other third party material in this chapter are included in the chapter's Creative Commons license, unless indicated otherwise in a credit line to the material. If material is not included in the chapter's Creative Commons license and your intended use is not permitted by statutory regulation or exceeds the permitted use, you will need to obtain permission directly from the copyright holder.

19

More-Than-Human Politics of Short Food Supply Chains in Landscapes of Human-Plant Migration

Elisa T. Bertuzzo◉

1   The Multi-Level Transformations Brought by Human-Plant Co-Migration

In the early 2000s, Bangladeshi immigrants started growing vegetables of the native country, or *deshi sobji,* in different parts of Italy, of which they represent a recent but large community (Della Puppa, 2022). The *karala* (bitter gourd), *chichinga* (snake gourd), *lau* (bottle gourd), *sag* (spinach/green leaves), coriander, methi, green chilli, etc., revealed well-suited to climate and soil, and prospered. Encouraged by the positive results and the high demand from Asian consumers, who had hitherto relied on expensive, and less tasty, vegetables imported via the UK or produced in Dutch greenhouses, some growers moved from cultivating for private consumption to dedicated farming, experimenting with seed

---

All interviewees' names have been changed for anonymity.

E. T. Bertuzzo (✉)
Zentrum Moderner Orient (ZMO), Berlin, Germany
e-mail: et_bertuzzo@posteo.de

© The Author(s) 2026   **289**
S. Sareen and S. Juhola (eds.), *Societal Transitions to Sustainability,* Palgrave Studies in Environmental Transformation, Transition and Accountability,
https://doi.org/10.1007/978-3-032-07395-2_19

varieties brought from home on leased fields at the peripheries of cities or in the countryside (Siddique, 2020). Within a decade, Italy's first generation of independent migrant farmers established a supply chain providing a whole new range of vegetables at affordable prices and 'zero kilometres' (Fig. 1).

Fig. 1 Well-tended sobji plants near Cagliari, Sardinia. *Source* Author Elisa T. Bertuzzo 2022

I have been studying the phenomenon since 2021, through a posthuman framework that departing from anthropocentric agency concepts, highlights more-than-human intra-action (Barad, 2007). My intuitive fascination as an Italy-born scholar working on migration in Bangladesh and India, and relating to the yearning for familiar tastes due to my own experience of migration, morphed with time. I set off to reconstruct the trajectories of the co-mobile humans and plants through the mix of personal engagements, participant observation, attunements with the landscape, archival and lab research which characterises multispecies ethnography (Kirksey & Helmreich, 2010), aiming to corroborate research about the benefits of migration. However, those trajectories soon revealed circular, not linear or mono-directional: involving the re-investment of remittances and savings, return migration, as well as the planting of seeds of 'Italian' herbs and flowering plants in Bangladesh, they prompted enquiry into questions of connectivity, multilocality, conviviality, and human-nonhuman cooperation. Simultaneously, following the sobji in Italy's countryside directed attention to broader ecological processes and the problems of current food systems.

In 2021, the anti-pandemic measures governed everyday life in Italy. Operations in many sectors were slow, road communications curbed, and this affected major goods supply chains. Notwithstanding, the Bangladeshi-run groceries and vegetable stands around which I lingered to make contacts, were constantly well stocked. One salesman, I will call him Salam, struck me with his optimism: 'Nowadays, our shops are supplied more regularly than others, and our prices are competitive. So, Italian customers come and see this and that sag and ask me how to cook those. I explain and also tell them that karala, for example, purifies the liver and is good against diabetes. Soon I'll convince someone!'. Sellers throughout observed that Italian clients, hyper-reluctant towards the more exotic-looking items, were recently starting to purchase sag, which resembles spinach.

A few kilometres away from the market where I met Salam, in the hinterland of Venice, I set foot in my first sobji farm. It was mid-summer and the variety of lush green crops, many of them climbers, almost overwhelming. The farmers told me that the majority of Bangladeshi sobji were suited to the erratic climate, simultaneously withstanding wetness and periods of water scarcity, adding they were using fractions of the quantities of fertiliser required back home. Recognising parallels between the flat, humid, mosquito-prone landscape extending before us and Bangladesh's, it occurred to me that as the vegetables, also these growers should be able to intra-act with it best. The suggestion that climate change and migration can foster transformative human-nonhuman conjunctures (Tsing 2015, Tsing et al., 2018) resonates, and a case in point is the cultivation of *kulmi sag* (water spinach). This tasty leaf thrives in water and develops spontaneously in the monsoon season but now grows in orderly lines for most of the year in farms all over Italy. Someone understood to make use of the webs of pipes that many fields come with in this country to keep the plants continuously moistened.

In Rome's Tor Pignattara neighbourhood, a hub of the 'Bangla' community, I befriended Sayed, a wholesale dealer. The next Sunday, he took me along to one of his sobji suppliers in the Agro Pontino (Pontine Plain) area. The status of the farmhouse suggested the property had long been underused by humans. Anwar, who ran the farm along with his brother, confirmed my impression: 'The owners have never farmed themselves. Over generations, the male members of the family held positions in ministries. The current landlords are old, and their children live abroad, there's nobody to look after the land. We signed a two-year lease and started farming here three years ago; this year we started using the full 14 hectares. So far, they're happy with us and we're happy with them'.

On a revisit in spring 2025, the *gamcha*—multi-use thin cotton towels that in Bangladesh, are typically checked and colourful—hanging on a line by the house's south-looking side announced from afar the stability of this relationship. I recognised, and was in turn recognised by, three among the ten labourers who were relaxedly plucking coriander and methi on a small portion of the land, tilled for the new season. However, I also found some things to be changed. The most prominent was that half of the property was now occupied by solar panels. Incremental European Union (EU) subsidies were pushing the production of solar energy since the war in Ukraine and embargoes on Russian gas. The boss was now leasing additional plots from an aged farmer nearby, said the labourers.

2   The Politics of More-than-Human Cooperations

The previous section described humans and nonhumans in positions of marginality and/or disadvantage who are carving out space of their own, thereby improving the diets of thousands, arguably through successful intra-action. While physicist Karen Barad's contention, that 'individuals emerge through and as part of their entangled intra-relating' (2007: iv), foregrounds the idea that existence is not an individual affair, using 'cooperation' for 'intra-action' seems in order to highlight the politics of the phenomenon. Thus put, this section will deal with the transformative effects of the cooperations of the co-mobile plants and humans, in and with the landscapes to which they have been introduced, and it will be doing so from a political ecology angle.

But what is being transformed, in the first place? The changes are visible and multi-level. The access of migrant communities to fresh food has improved. Arguably, the sobji's introduction qualifies as what in economics is called disruptive innovation, giving 'a whole new population of consumers at the bottom of a market access to a product or service that was historically only accessible to consumers with a lot of money or a lot of skill' (Bower & Christensen, 1995: 45). The social relations in agriculture, usually controlled by agribusiness firms, local landowners, but also the mafia-like *caporalato* networks, are another site of change. Many labourers reflected that cultivating sobji among countrymen was both safer and more meaningful than working in conventional farms, while the owners recognised the advantages of controlling production and distribution, as opposed to selling to large firms. Under the usual norms and controls, the sobji could not have landed on Italian (and meanwhile many central European) markets as fast. Even with so far minimal interactions between labourers and local communities, the farms strengthen local economies in regions inhabited by ageing and dwindling populations. A positive ecological balance results from the fact that earlier, the veggies came with higher CO_2 emissions, due to their import via the UK or input-heavy cultivation in greenhouses.

2.1   Socio-Ecological Infrastructures

A political ecology reading underscores that the constant becoming at the sobji farms is not just 'happening'. Its actants might be less of following a plan than improvising, taking the chance, running risks, but their cooperations strictly depend on infrastructures or, I suggest, *socio-ecological infrastructures*. Infrastructures are understood to be physical and virtual, formal and informal supports facilitating cooperation and synchronisation on multiple levels, throughout geographically disjoined places, and requiring human and nonhuman actions (Larkin, 2013). 'Socio-ecological infrastructures' makes explicit that the co-ordinating and synchronising always ultimately implies exchanges and circulations of matter. For example, the infrastructure supplied by the biologically coordinated work of water, fungi, bugs and minerals in the soil intertwines with the infrastructures of irrigation, both at the scale of the region and that of the farm, where cultivators are adapting simple technologies to help growth—think of the kulmi sag—and respond to the specificities of climate and soil.

In nationwide WhatsApp- and imo-groups (imo being an audio/video calling and instant messaging software widely used by Bangladeshi migrants), the growers share their farming experiences and stay informed, e.g. on the constant changes in labour visa regulations. Sub-groups built along the lines of kinship, region of origin and political affiliation, are useful in case of emergencies or when recruiting labourers. Digital communication also offers elements of an infrastructure of care, where 'care' articulates moral and emotional support. While some of the men I interviewed have arrived with work visas (recently restricted to seasonal agricultural work), most spent months if not years in Eastern Europe, or in camps in Northern Africa, before entering Italy. Trauma and longing, barely expressed in person, are shared with travel companions and friends from times before emigration in the evening and rest time, through the mobile phone.

Zooming out from the farms, the distribution of the sobji relies on roads, kinship ties, the tools of digital and AI-supported communication and translation, etc. Actions like importing and growing seeds from Bangladesh, transferring money to relations, co-ordinating major purchases, land transaction, the construction of new houses from afar, involve migrant and expat networks as well as bio-geological supports, digital communication and social media, money transfer apps and banking networks, financing schemes, roads, railway, air routes, etc. Some of these infrastructures are presently disturbed due to climate change and over-extraction emanating from the power relations that have prevailed for centuries; others embody such relations in their design. Thus, politics imbue socio-ecological infrastructures, which are physical as well as virtual, technological as well as affective supports built on social relations and ecological transactions.

2.2 Transformative Foodscapes

Do these cooperations prefigure a transformation of food systems towards sustainable production, supply, processing, and consumption? The persistence of hunger in countries with a colonial past, many of which continue to be exploited as producers of raw materials, and nutrition-related illnesses worldwide should be a sufficient reason to press for that transformation. The globally aggravating climate and supply chain crises on the one hand, and the work done by anticolonial and Indigenous authors to foreground embodied food knowledge and 'food pluriverses' (Kothari, 2020) on the other, are boosting the search for alternatives that should value the relationship between food, place, and communities. The bottom line is that food systems ought to be transformed in a way that doesn't repeat the mistakes of current global food systems, forged by extractivist and colonial logics. Relocalisation is widely regarded as a strategy and scholars who identify the dominance of global players and industrial agriculture as the problem, recommend easing the access to land for different actors and adopting approaches like agroecology (Anderson et al., 2020; Demarais et al., 2010).

The sobji farmers in Italy took advantage of the comparatively easy access to farmland (through leases). The other side of the coin is a high level of land abandonment, a phenomenon that has political economy reasons. To climate change, a formidable challenge for agriculture, add the negative demographics and lasting emigration, urban sprawl, mining, transport, and commercial activities, soil exhaustion derived from industrial cultivation, the dependency on multinationals for pesticides, fertilisers, and seeds, as well as often counterproductive EU agricultural policies promoting specialisation and cash crops. Many of these problems are to do with capitalist practices and globalised structures, to which the sobji farms, though not existing outside of capitalism, offer a local emergent answer reliant on the interdependence between people and environments. The methods do not (yet) align with the principles of agroecology, but the knowledge of these farmers and the adaptable sobji could certainly help Italian agriculture face the growingly unpredictable precipitation timings and quantity. In a sector that prevalently relies on migrant labour also for the logistics, both near- and long-distance (Ambrosini, 2024), the Bangladeshi entrepreneurs running their own supply chain already embody the paradigm of food from farm to table.

The environmental effects of the sobji's introduction, unfolding in longer periods, require dedicated analysis. If a 'lack of commitment to a particular temporality is a hallmark of prefigurative politics, because it relates to something emergent' (Sareen & Juhola, 2025: this book), assessing human-plant cooperations and their impacts necessitates taking more-than-human temporalities into account, complicating things further. The intra-relationality of the socio-ecological processes sustaining life on the planet, cutting across spaces and species, also cuts across (human) times, i.e. past-present-future (Bertuzzo, 2025b). Awareness of more-than-human temporalities places the sobji on a continuum with maize, tomatoes, potatoes, a number of medical plants, etc.: long-naturalised migrant crops whose trespassings have crucially improved the diets, health, survival chances and quality of life in the 'old continent'. Yet today, conservation and policy discourses that distinguish 'native' and 'alien' on the basis of the distribution of species in the year 1492 are framing crops like the sobji as dangerous, selectively blocking their cultivation for fear of 'invasion'.

2.3   The Decolonised Temporality of Migrant Sobji

In bibliographic research, I have been able to spot one scientific publication on the 'ethnic farms' in Italy (Cristaldi & Leonardi, 2016); fieldwork was conducted in 2014 and apparently not followed up, in spite of the authors' remark that the new supply chain was promising. Given the wealth of research on different aspects of the immigration from Bangladesh in Italy, the low interest in the phenomenon is likely not due to access but to research funding issues. Several reports on confiscations and bans of 'Chinese' vegetable gardens in local newspapers and the sceptical or openly negative reactions to my stories about the sobji farms, always regarding the 'invasiveness' of the 'aliens', nurture the suspicion that the underlying problem is a refusal to see migrants and specifically, racialised, often ex-colonial subjects, as guardians of life-sustaining knowledge and innovative practices that Europe urgently needs.

The ways in which the migrant cultivators interface with the sobji, the environment, and climate change would be wrongly conceptualised as guardianship. Instead of pretensions of mastery, I am encountering pragmatism and eagerness to experiment. The kulmi sag case mentioned above is just one example of both, the often-joyful process of applying one's own, acquiring, and hybridising farming knowledges in a new country, and the promptness with which the innovation was shared, rather than claimed by one 'inventor'. The relationship to the planet this translates is diametrically opposed to the sense of entitlement prevalent in centuries of Western colonialism-extractivism. In that logic, the introduction of tomatoes, maize, potatoes, or medical plants epitomised Europe's domination. The introduction of the sobji owes to a cooperation of humans and nonhumans mobilised by devastating extractivist backlashes that is to do with care and I cannot but characterise as decolonial. Whose decolonisation? I anticipate the question, echoing a question which punctuated our workshop, 'Whose desirable futures?'. As held by Parreñas (2019), decolonisation 'is about experimentation. It's about a moment of deep uncertainty in which some kind of new way of living has to happen because the state of colonialism is no longer tenable'. Full emancipation from anthropocentric logics remains a horizon, but the sobji and their growers speak to an incomplete, unintentional, necessarily uncertain, and irreversible, decolonised temporality.

3   Configuring Enjoyable Mobile More-Than-Human Politics of Coexistence

How do the dynamics and infrastructures co-created by these mobile humans and plants in order to sustain life far away from home, prefigure desirable futures? The answer must be parted. The working definition of prefigurative politics as 'embodied strategies to render desirable futures with immediacy' (Sareen & Juhola, 2025: this book) does not include or exclude nonhumans. One could argue that the sobji, prospering in a different continent, are realising their plant desire for reproduction (Pollan, 2001), but environmental historian Jason Moore's point that 'the irreducibly dialectical relation between human and extra-human natures' (Moore, 2015: 5) necessitates a new language of world ecology kicks in. Barad's replacing 'agency' with 'intra-action', terminology like plants' 'purposive and transformative action' (Jones & Cloke, 2002) or 'symbiopolitics' (Helmreich, 2009), are steps. Through the concept of socio-ecological infrastructures, I emphasise that transformative food systems in particular and transformation in general, involving human-nonhuman cooperation and ecological transactions, are imbued with politics. By functioning, or disfunctioning together, *through*—not 'in spite of'—geographic distance and precarity, multiple actants in my fieldwork sketch a politics of more-than-human coexistence as locationally adapted, relational, and highly improvisational *attending (and sometimes, attending to) while going*.

For the human migrants, a general orientation towards a better future (this volume's understanding of desirable futures) should be the constant since departure from Bangladesh. Saying that, I on the one hand acknowledge a gap in perceptions of futurity and renderings of future possibilities between majority and minority world subjects that is to do with class and colonial difference, or with privilege and the lack thereof

(Appadurai, 2013). On the other hand, I aim to underscore the benefits and potentials (beyond labour!) which thousands of recently arrived migrants are offering to Italy and Europe. While in the political arena, particular interests are unsettling and delaying urgently needed measures to address the environmental crises, the emergent foodscape proves that change is rendered in everyday life. Amidst immediate embodiment and resistance (Bertuzzo, 2025a), nor the farmers nor the labourers I am talking to make political claims; for some it would be dangerous. Strategic thinking flows into navigating immanent constraints and pressures, gauging options. The aim and indeed, result, is (nonetheless) an expansion of possibilities.

For scholars and readers in transformation research, the question relates to the systemic effects of the experiment. The sobji farms have reduced food miles and contribute to food sovereignty by providing fresh and culturally appropriate vegetables to migrant communities, and being controlled by the migrant farmers, independently from local groups and large firms. While the cultivation methods are not (yet) socially and/or ecologically sound and circularity is only partially achieved, some are eyeing food processing. The risks of climate change and ecological factors like soil exhaustion or new plant pests, on the one hand, and lack of institutional support and the so far weak resonance in the 'host' community, on the other, appear to be the bigger limitations. Mutually performing ideas of hospitality and conviviality would help to inaugurate cooperations, e.g. with agroecology initiatives (there are several in Italy), building an inclusive food sovereignty movement. Institutional support could encourage the farmers to upscale, perhaps with the intermediation of NGOs specialising on migrant entrepreneurship, but also to adopt more ecological approaches.

This brings the reality of current politics of transformation to the fore. At the peak of the COVID-19 crisis, the already mentioned bans of 'Chinese' crops were concomitant with the state's selective channelling of EU funds towards 'Italian' farms cultivating 'exotics', such as mango or avocado. This was publicised as a climate change adaptation measure (Agenzia Nova, 2022). With 1.1 million farms, 53% of its population living in rural areas, one of Europe's largest agricultural and food processing sectors, and 50% of its agricultural land classified as constrained (DG AGRI, 2025), Italy—simultaneously a climate change hotspot—has an existential interest in making agriculture climate resilient. Biased policies are prone to failure and so are myopic ones. The sobji farms, off the authorities' radar so far, would be at stake should Italy decide to implement the EU's bio-surveillance policies, which since 2019 prescribe 'plants passports' coupled to QR codes for the tracking of introduced plant species. Would control be articulated in a constructive way, labour rights and safety at the worksites could be improved, the introducing and/or saving of seeds monitored, and the intra-acting of sobji and local ecologies adequately studied.

Feminist economists J.K. Gibson-Graham, writing about the 'marginal' economies that worldwide, 'are actually more prevalent, and account for more hours worked and/or more value produced, than the capitalist sector' (2008: 617), share how studying these forms of enterprise expanded their imaginaries of change. The smallest interventions could be conceived as potentially rendering structural change. Studying the dynamics around the sobji farms is enabling me to contribute to the ongoing work of conceptualising *and pressing* for transformation, acknowledging that change is about friction and resistance, conviviality and care, necessarily embedded in the present and reliant on hybrid practices, disruptive innovations, and human-nonhuman relationships. Such an understanding accompanies the call to leave Euro- and anthropocentric ambitions of control and mastery behind, and recognise unfinished, non-perfect, decentralised actions which cannot be bothered by borders, as foundations of coexistence in times of polycrisis. This configures an enjoyable mobile more-than-human politics of coexistence.

Competing Interests The author has no conflicts of interest to declare that are relevant to the content of this chapter.

References

Ambrosini, M. (2024). Immigrant work and the production of Italian agrifood. *Journal of Immigrant & Refugee Studies*, 1–16. https://doi.org/10.1080/15562948.2024.2424164

Anderson, C. R., et al. (2020). Agroecology now. *Agroecology and Sustainable Food Systems*, *44*(5), 561–565. https://doi.org/10.1080/21683565.2019.1709320

Appadurai, A. (2013). *The future as cultural fact*. Verso.

Barad, K. (2007). *Meeting the universe halfway*. Duke University Press.

Bertuzzo, E.T. (2025a). Re-articulating resistance in urban* environments. For edgy, plural, more-than-human good lives in an age of translocalisation. *Global Environment*, *18*(2), 412–431. https://doi.org/10.3828/whpge.638 37646622528

Bertuzzo, E.T. (2025b). Past, present, future recombinant urban-rural natures. *Urban Studies* [forthcoming]

Bower, J., & Christensen, C. M. (1995). Disruptive technologies: Catching the wave. *Harvard Business Review*, *73*(1), 43–53.

Agenzia Nova. (2022, December 3). Clima: nell'anno più caldo è boom di prodotti tropicali italiani. Agenzia Nova. Retrieved May 21, 2025, from https://www.agenzianova.com/news/clima-nellanno-piu-caldo-e-boom-di-prodotti-tropicali-italiani/

Cristaldi, F., & Leonardi, S. (2016). Tra importazioni e filiere corte. In L. Romagnoli (Ed.), *Spunti di ricerca per un mondo che cambia* (pp. 73–89). Edigeo.

Della Puppa, F. (2022, May 3). *A "bidesh" called Italy. Migration from Bangladesh to Italy, and beyond*. Bundeszentrale für politische Bildung. Retrieved May 21, 2025, from https://www.bpb.de/themen/migration-integration/laenderprofile/english-version-country-profiles/507928/a-bidesh-called-italy-migration-from-bangladesh-to-italy-and-beyond

Demarais, A.A. et al. (2010). *Food sovereignty. Reconnecting food, nature and community*. Fernwood Publishing.

DG AGRI, Directorate-General for Agriculture and Rural Development. (2025, March 17). At a glance: Italy's CAP strategic plan. Retrieved May 21, 2025, from https://agriculture.ec.europa.eu/cap-my-country/cap-strategic-plans/italy_en

Gibson-Graham, J. K. (2008). Diverse economies: Performative practices for 'other worlds.' *Progress in Human Geography, 32*(5), 613–632. https://doi.org/10.1177/0309132508090821

Helmreich, S. (2009). *Alien ocean: Anthropological voyages in microbial seas.* University of California Press.

Jones, O., & Cloke, P. (2002). *Tree cultures: The place of trees and trees in their place.* Berg Publishers.

Kirksey, S. E., & Helmreich, S. (2010). The emergence of multispecies ethnography. *Cultural Anthropology, 25*(4), 545–576. https://doi.org/10.1111/j.1548-1360.2010.01069.x

Kothari, A. (2020, September 30). *Do we have the stomach for it?* Global Working Group Beyond Development. Retrieved May 21, 2025, from https://beyonddevelopment.net/re-imagining-food/

Larkin, B. (2013). The politics and poetics of infrastructure. *Annual Review of Anthropology, 42*, 327–343. https://doi.org/10.1146/annurev-anthro-092412-155522

Moore, J. W. (2015). *Capitalism in the web of life: Ecology and the accumulation of capital.* Verso.

Parreñas, J. S. (2019, August 19). *Decolonizing extinction: An Interview with Juno Salazar Parreñas.* Engagement. Retrieved May 21, 2025, from https://aesengagement.wordpress.com/2019/08/19/decolonizing-extinction-an-interview-with-juno-salazar-parrenas/

Pollan, M. (2001). *The botany of desire: A plant's-eye view of the world.* Random house.

Sareen, S. & Juhola, S. (2025). The prefigurative politics of present transformation. In S. Sareen & S. Juhola (Eds.), *Societal transitions to sustainability: The prefigurative politics of present transformation* (this book). Palgrave Macmillan.

Siddique, S. B. (2020) *Taste matters! A qualitative study of eating traditional vegetables among Bangladeshi immigrants living in Venice, Italy* (Master Thesis, Università Ca Foscari, Italy).

Tsing, A. (2015). *The mushroom at the end of the world.* Princeton University Press.

Tsing, A. et al. (Eds.) (2018). *Arts of living on a damaged planet.* University of Minnesota Press.

Open Access This chapter is licensed under the terms of the Creative Commons Attribution-NonCommercial-NoDerivatives 4.0 International License (http://creativecommons.org/licenses/by-nc-nd/4.0/), which permits any noncommercial use, sharing, distribution and reproduction in any medium or format, as long as you give appropriate credit to the original author(s) and the source, provide a link to the Creative Commons license and indicate if you modified the licensed material. You do not have permission under this license to share adapted material derived from this chapter or parts of it.

The images or other third party material in this chapter are included in the chapter's Creative Commons license, unless indicated otherwise in a credit line to the material. If material is not included in the chapter's Creative Commons license and your intended use is not permitted by statutory regulation or exceeds the permitted use, you will need to obtain permission directly from the copyright holder.

The holistic agroecological approach to farming that integrates ecological principles, social justice, and community empowerment also plays a central role in prefigurative politics in agri-food systems. Agroecological practices do not just present technical solutions to agricultural problems but also challenge dominant agricultural paradigms by promoting practices that centre on the needs of local communities, environmental health, and food security (Altieri, 2002, Altieri et al., 2017). Agroecology, in this sense, is seen as a form of political action, aiming to demonstrate the viability of alternative, equitable, and sustainable food systems through practice. Some of the innovative agroecological approaches to agriculture include, but are not limited to, organic farming, climate-smart agriculture, biodynamic agriculture, sustainable intensification, and regenerative agriculture (Muhie, 2022).

The global scope of prefigurative politics in agri-food systems also cannot be ignored. Movements for food sovereignty and agroecology are not limited to specific countries but have a global reach, connecting struggles for justice across borders. For instance, Via Campesina, an international movement for food sovereignty, has become a key actor in promoting prefigurative politics globally by advocating for policies that address the inequities within the global food system (Borras, 2008). The ability to envision and act upon alternative food systems in different regions of the world reflects the power of prefigurative politics in the fight for global food justice.

This chapter investigates how local actors (specifically small-scale farmers engaged in cooperatives and innovative short supply chains) are reimagining food production, distribution, and consumption in alignment with principles of sustainability, sufficiency, and social justice (McGreevy et al., 2022). The dominant post-socialist food regime is increasingly shaped by agribusiness concentration, market liberalisation, and integration into global value chains (McGreevy et al., 2022; Smith & Jehlička, 2013). The transition from state socialism to market economies often reinforces neoliberal models of development and marginalised grassroots sustainability efforts. Despite being underappreciated in mainstream policy discourse, actions practices such as for instance food self-provisioning reflect what Smith and Jehlička (2013) call "quiet sustainability"; these practices are rooted in cultural traditions of mutual aid, task-sharing, and self-reliance, which persist despite shifts in political economy.

Two case studies—the National Federation of Farmers' e-Farm initiative and the First Organic Cooperative—serve as empirical examples to explore how local efforts can prefigure alternative food system models while addressing broader socio-environmental challenges in post-socialist South-East European countries such as North Macedonia. The methodological framework draws on interdisciplinary concepts including prefigurative politics, food sovereignty, sustainability, social justice, and local knowledge, particularly as they relate to post-socialist contexts. These concepts are situated within the multi-level perspective, which distinguishes between niche-level innovations (micro), dominant socio-technical regimes (meso), and broader socio-political and environmental landscapes (macro) (Geels, 2010; Törnberg, 2021). Within this framework, the two case studies are analysed as niche "free" spaces where actors experiment with alternatives to the binding agri-food regime, which is characterised by market concentration, institutional lock-ins, and socio-economic inequalities.

To deepen the analysis of how change is enacted within these niches, Yates's (2015) five interrelated prefigurative processes (experimentation, perspectives, conduct, consolidation, and diffusion) are used as analytical lenses. These processes allow for an examination of how local food initiatives not only challenge the dominant regime but also embed alternative values and practices in everyday life and organisational forms. An open-ended qualitative questionnaire was developed to guide data collection, structured around six thematic areas: key activities and practices; context and motivation; prefigurative politics and impact; barriers and challenges; local and systemic levels of change; and future directions and sustainability. In February 2025, in-depth interviews were conducted with representatives from the National Federation of Farmers (NFF) and the First Organic Cooperative (FOC). These interviews provided insight into how local actors engage in collective experimentation (e.g. testing new sales models), develop shared perspectives (e.g. sustainability and food justice discourses), establish new conduct (e.g. democratic governance, solidarity-based exchange), consolidate these practices into organisational and digital infrastructures (e.g. e-Farm platform, cooperative structures), and diffuse them to broader communities through public engagement and partnerships. Integrating a multi-level perspective with Yates's process-oriented approach to prefiguration, this methodological approach enabled a nuanced analysis of how local food initiatives operate as sites of both resistance and innovation, attempting to build viable alternatives and at the same time addressing constraints of post-socialist transitions and global food system dynamics.

2   Post-socialist Agri-Food Setting: North Macedonia Context

North Macedonia's agricultural sector remains a key pillar of the economy, supporting employment, rural livelihoods, and food security. Although it contributes about one billion EUR annually, its share of national gross value added (GVA) has declined from 11.7% in 2014 to 8.1% in 2023 (SSO, 2025), signalling long-term structural shifts and sectoral challenges.

A major turning point came with the dissolution of Yugoslavia in 1991. During the socialist period, agriculture in Macedonia was organised through agro-kombinats—large, vertically integrated agricultural-industrial complexes that combined farming, processing, and distribution under state or socially owned management, vis-à-vis a very small number of marginalised small-scale farmers (until 1988 the private legal ownership maximum was 10 hectares, Melmed-Sanjak et al., 1998). These structures played a dominant role in production, employment, and rural development. With the transition to a market economy, agro-kombinats were dismantled or privatised, often in ways that led to asset stripping, fragmentation of landholdings, and the weakening of local food infrastructures.

Following its independence from Yugoslavia in 1991, North Macedonia transitioned to a market-oriented economy. As an EU candidate since 2005, aligning national policies with EU agricultural and rural development standards has become a key strategic goal (Kotevska et al.,

2024). However, the country continues to face deep-rooted socio-economic vulnerabilities, particularly within the agri-food sector. It is dominated by 178,000 small family farms averaging just 1.8 hectares (SSO, 2017), with over 60% smaller than one hectare. These smallholders generate modest output (70% earn less than 4,000 EUR annually) and contribute just 22% of total agricultural production, while 30% of larger farms account for 78% of agricultural output. This concentration of production in larger, more economically viable farms highlights stark disparities in productivity and market access, with marginalisation of the small farmers in the agri-food systems. Cooperative organisation is weak, with only 44 functioning agricultural cooperatives and fewer than 800 members (MAFWE, 2021). Power asymmetries in the agri-food chain favour buyers and intermediaries, leaving smallholders with limited bargaining power. Demographic challenges further threaten sector sustainability: two-thirds of farmers are over 55 years old, and only 4% are under 35 (SSO, 2017), as younger generations are discouraged by from entering agriculture due to lower wages and a perceived lower quality of life in rural areas compared to other sectors (SWG, 2024). Climate change and environmental degradation, such as soil depletion, water scarcity, and biodiversity loss, amplified by the weak adaptive capacities compound these challenges (Martinovska Stojcheska et al., 2024). Together, these systemic issues reflect the fragility of the post-socialist agri-food system in North Macedonia. The case studies explored in this chapter reflect these broader dynamics. They illustrate how small-scale, grassroots food initiatives are emerging at the niche level to challenge prevailing structures and experiment with more sustainable, just, and resilient food system alternatives.

3   Local Initiatives towards Sustainable and Just Futures: Case Studies

A growing number of farmer-led initiatives in North Macedonia are reshaping local food systems through sustainable practices, collective action, and direct market engagement. We examined two such initiatives, the National Farmers Federation (NFF) and the Farmers Organic

Cooperative (FOC), analysing their organisational practices, challenges, and broader contributions to sustainable food systems. Operating within niche spaces, both initiatives are experimenting with new models that challenge the dominant agro-industrial regime and respond to shifting socio-political and environmental conditions.

NFF engages in lobbying, advocacy, education, and entrepreneurship support, specifically targeting rural development and related socio-economic issues. Since 2020, it has developed the e-Farm platform, a digital space aimed at strengthening farmers' business development and promoting locally produced food. FOC was formalised in 2019, emerging from a pre-existing informal network of organic farmers. A comparative overview of NFF and FOC summarising their key characteristics, activities, challenges, and strategic goals is given in Table 1.

The emergence of both initiatives was driven by profound socio-ecological transformations. NFF responded to rising land consolidation by large agricultural companies and diminishing opportunities for smallholders ("Large companies in agriculture are becoming more and more powerful and are buying more and more land, and on the other hand, the land of small farmers is increasingly sold or abandoned."). The values underpinning NFF's work include social justice, sustainability, tradition, and community empowerment. Social justice remains central, as the organisation aims to address inequalities in access to services, opportunities, and resources ("While many farmers have become economically successful, they remain vulnerable due to these ongoing injustices").

FOC operates with a similar commitment to sustainability, promoting organic production, biodiversity, and soil health. Its approach emphasises fair pricing, economic empowerment, and the preservation of local knowledge. The cooperative model facilitates shared knowledge and resources while prioritising sustainable and ecological farming practices. By enabling small-scale farmers to organise and act collectively, the cooperative strengthens members' negotiating power. Prior to its establishment, farmers often faced unfavourable conditions during individual negotiations with buyers. Centralised sales have since allowed FOC members to achieve higher prices and improved contractual terms. The launch of the "B Organic" brand enhanced product visibility and market access. This line includes a variety of processed and packaged organic

Table 1 Overview of key activities, structures, and goals of NFF and FOC. *Source* Own elaboration based on NFF and FOC interviews (February 2025)

Category	NFF	FOC
Key activities	Lobbying and advocacy; education and information; entrepreneurship support	Resource and knowledge sharing; organic farming; collective marketing and sales
Members/ Structure	7,620 members; over half are part of 34 organizations (local associations, cooperatives, small agricultural companies)	13 certified organic producers (68.5 ha of almonds, pomegranates, persimmons, plums, sour cherries, apricots, plus 1,800 beehives)
Support platforms	e-Farm platform (launched in 2020) to support business development, market access, and consumer education	"B Organic" brand; machinery ring for almond harvest; centralized marketing and branding system
Focus areas	Rural development, social justice, tradition, sustainability, community empowerment	Sustainability, fair pricing, local knowledge and tradition, community empowerment
Outputs/ products	Educational resources; digital platform for local food sales; events promoting local producers and sustainable practices	Processed and packaged organic goods: almonds, almond butter/cream, honey, dried apples/ persimmons, beans, vinegar
Impact	Facilitated direct links between producers and consumers; empowered farmers; raised awareness of sustainability and local food	Improved economic outcomes via centralized marketing; better pricing; reduced costs; increased visibility and access to national distribution
Challenges	Behavioural resistance; lack of government funding; entrenched agro-industrial lobbies; need for greater consumer awareness	Bureaucratic and legal hurdles; policy instability; limited institutional support; need for targeted financial support for cooperatives

(continued)

Table 1 (continued)

Category	NFF	FOC
Partnerships/ actors	Academia, extension services, chambers of commerce, sub-sectoral policy groups (limited engagement with consumer organizations)	Chamber of Organic Producers; Network of Agricultural Cooperatives; leadership of Rural Youth Union initiative
Future goals	Financial sustainability via membership fees; expanding e-Farm functionalities (online sales, training modules); long-term trust building with farmers and consumers	Scaling up through new partnerships; maintaining transparency and fair governance; preserving agriculture as core to local rural economies

items, such as almonds, nut spreads, honey, dried fruits, and vinegar, all distributed through a national network.

Both NFF and FOC also focus on educating and engaging consumers. The e-Farm platform fosters direct relationships between farmers and buyers by sharing personal stories and product details. These efforts aim to raise awareness about sustainable consumption and local food systems. FOC uses its branded products to similar effect, offering transparency and reinforcing consumer trust. These activities reflect elements of prefigurative politics, where existing practices model future alternatives to dominant systems. NFF promotes peer learning and collaboration among producers. For example, in Rosoman, a farmer expanded ajvar production through community participation, while Meden Istok led training sessions on sustainable beekeeping. Partnerships with local municipalities, such as one promoting rice cultivation in Česinovo-Obleševo, further illustrate efforts to foster regional agricultural development.

FOC has invested in infrastructure that improves efficiency and product quality, including machinery for almond harvesting, vacuum sealing, cold storage, and drying facilities. Collaboration with local suppliers ensured consistent access to high-quality inputs. Both initiatives challenge the conventional agro-industrial model by reducing dependence on intermediaries. They enable farmers to retain greater control over pricing and distribution by facilitating direct sales between producers and consumers. These approaches create more equitable trade conditions and reinforce a sustainable food system. Digital platforms and social media are used to enhance transparency, share production information, and celebrate community achievements.

Despite these advances, significant barriers remain. Shifting consumer habits and producer behaviour is a long-term endeavour that requires persistence and consistent outreach. Public scepticism and resistance to change, combined with institutional inertia, complicate progress. Entrenched lobby groups and oligopolistic market structures further hinder the adoption of alternative models. The e-Farm platform operates on a membership fee basis and receives no direct government funding. Local governments support the initiative primarily by providing space for farmer-consumer events. Regulatory frameworks, such as the Rule-books for direct sales (adopted in 2019–2020), have been essential in legitimising and enabling the platform's work.

FOC faces similar obstacles, particularly in navigating legal and bureaucratic systems. Unlike EU-based cooperatives, which often benefit from robust institutional support, FOC contends with inconsistent policies and administrative burdens. For instance, the line ministry of agriculture announced a call in 2023 which was then unexpectedly cancelled, disrupting planned investments. FOC continues to advocate for targeted policy measures that would enable small-scale producers to unite and market their products collectively.

In terms of systemic engagement, NFF collaborates with universities, agricultural extension services, and business chambers. It maintains regular dialogue with policymakers through participation in sub-sectoral working groups. However, partnerships with consumer organisations remain limited due to their low capacity. NFF emphasises the need for improved coordination among public institutions and private sector actors to advance food system reform.

FOC actively engages in cooperative development through its memberships and cooperation with the Chamber of Organic Producers, Network of Agricultural Cooperatives to improve national policy frameworks for small producers and leading an initiative to create a national Rural Youth Union, with the goal of empowering rural youth as leaders in community and agricultural development.

Looking forward, both organisations have outlined strategic goals for long-term sustainability. NFF seeks to further develop e-Farm through expanded e-commerce capabilities, educational modules, and enhanced member services. The initiative relies on financial self-sufficiency and places strong emphasis on maintaining high standards of quality and transparency. Building lasting trust with both producers and consumers is viewed as essential. Once lost, trust can be difficult to restore, making consistency and accountability central to the initiative's success. FOC is pursuing scale through partnerships with like-minded producers and a continued focus on democratic governance. The cooperative underscores that personal interests must not outweigh collective goals. Transparency, fairness, and mutual trust are identified as the cornerstones of a resilient and effective organisation. Both NFF and FOC view agriculture not only as a vital economic sector but also as a foundation for social equity and environmental sustainability, contributing to preserving rural livelihoods and reinforcing food sovereignty in North Macedonia.

4 Challenging Dominant Agro-Industrial Paradigms in Post-Socialist Food Systems

Local initiatives can challenge dominant agro-industrial paradigms and propose alternative models that engage with both local knowledge and broader sustainability goals. The role of prefigurative politics in these movements is critical: local food producers are not merely reacting to current system failures but actively shaping the future of food systems. Free social spaces at micro level nurture early innovations, but major societal change only occurs when these align with large-scale disruptions (Törnberg, 2021).

NFF has adopted a strategy to directly implement the changes they seek by placing and humanising the farmer, facilitating contextualised direct contact with consumers. FOC has formalised a network of local organic producers, promoting resource and knowledge sharing while encouraging youth engagement in agriculture. Building on Yates' (2015) framework, experimentation is recognised in both initiatives' efforts to develop new ways of organising production and market relations. NFF's creation of the e-Farm platform and FOC's implementation of centralised marketing and on-farm processing infrastructure demonstrate how practical innovations are being tested to improve farmers' autonomy, income stability, and consumer engagement. These experiments reflect a response to existing challenges, but also a proactive construction of alternatives. Political perspectives are embedded in each initiative's broader vision for food system transformation. Through their communications, branding, and educational efforts, both initiatives articulate coherent ideological narratives that critique the dominant food regime and offer grounded visions of sustainability and equity. New social conduct is being established through internal practices of cooperation, mutual accountability, and democratic decision-making. NFF's emphasis on peer learning and community-driven development, and FOC's transparent governance and member-first principles, both indicate evolving norms that challenge hierarchical, profit-driven structures. Consolidation is evident in the embedding of these norms and values into physical and institutional infrastructures. Although consolidation often remains localised and constrained by structural limitations, these materialisations increase the visibility and viability of alternative models. Finally, diffusion is occurring through both formal and informal channels, with efforts to extend influence beyond immediate members. Consumer outreach, public events, and the circulation of branded goods serve to normalise sustainable consumption and widen the reach of prefigurative food practices. While the initiatives impact is currently small, these models have potential to be replicated and lead to broader change. It remains to equitably address and balance the trade-offs among multiple goals, while maintaining persistent commitment to long-term transformation.

Competing Interests The authors have no conflicts of interest to declare that are relevant to the content of this chapter.

References

Altieri, M. A. (2002). Agroecology: The science of natural resource manage-ment for poor farmers in marginal environments. *Agriculture, ecosystems & environment 93*(1–3), 1–24. https://doi.org/10.1016/S0167-8809(02)000 85-3

Altieri, M. A., Nicholls, C., & Montalba, R. (2017). Technological approaches to sustainable agriculture at a crossroads: An agroecological perspective. *Sustainability, 9*(3), 349. https://doi.org/10.3390/su9030349

Borras, S. M. J. R. (2008). La Vía Campesina and its global campaign for agrarian reform. *Journal of Agrarian Change, 8*, 258–289. https://doi.org/10.1111/j.1471-0366.2008.00170.x

Geels, F. W. (2010). Ontologies, socio-technical transitions (to sustainability), and the multi-level perspective. *Research Policy, 39*(4), 495–510. https://doi.org/10.1016/j.respol.2010.01.022

Koensler, A. (2020). Prefigurative politics in practice: Concrete utopias in Italy's food sovereignty activism. *Mobilization: An International Quarterly, 25*(1), 133–150. https://doi.org/10.17813/1086-671-25-1-133

Kotevska, A., Martinovska Stojcheska, A., & Erjavec, E. (Eds.) (2024). Euro-pean integration and agriculture in the western Balkans: Current trends and challenges. Skopje: Regional Rural Development Standing Working Group in South-East Europe (SWG).

Leach, D. K. (2013). Prefigurative politics. In D. A. Snow, D. Della Porta, P. G. Klandermans & D. McAdam (Eds.), *The Wiley-Blackwell encyclopedia of social and political movements* (pp. 1004–1006). Wiley-Blackwell. https://doi.org/10.1002/9780470674871.wbespm167

Maeckelbergh, M. (2009). *The will of the many: How the alterglobalisation movement is changing the face of democracy.* Pluto Press.

MAFWE. (2021). National Strategy for Agriculture and Rural Development (2021–2027). Ministry of Agriculture, Forestry and Water Economy of the Republic of North Macedonia.

Martinovska Stojcheska, A., Mukaetov, D., Tanaskovikj, V., Chukaliev, O., & Andonov, S. (2024). Climate change adaptation in agriculture – Status and prospects in Western Balkan economies. EU4Green/SWG. https://doi.org/10.13140/RG.2.2.11495.56483

McGreevy, S. R., Rupprecht, C. D., Niles, D., Wiek, A., Carolan, M., Kallis, G., Kantamaturapoj, K., Mangnus, A., Jehlička, P., Taherzadeh, O., Sahakian, M., & Tachikawa, M. (2022). Sustainable agrifood systems for a post-growth world. *Nature Sustainability,* 5(12),1011–1017. https://doi.org/10.1038/s41893-022-00933-5

Melmed-Sanjak, J., Bloch, P. C., & Hanson, R. (1998). Project for the analysis of land tenure and agricultural productivity in the Republic of Macedonia. Land Tenure Center, University of Wisconsin-Madison, 147p.

McMichael, P. (2006). Peasant prospects in the neoliberal age. *New Political Economy, 11*(3), 407–418. https://doi.org/10.1080/13563460600841041

Muhie, S. H. (2022). Novel approaches and practices to sustainable agriculture. *Journal of Agriculture and Food Research, 10,* 100446. https://doi.org/10.1016/j.jafr.2022.100446

Smith, J., & Jehlička, P. (2013). Quiet sustainability: Fertile lessons from Europe's productive gardeners. *Journal of Rural Studies, 32,* 148–157. https://doi.org/10.1016/j.jrurstud.2013.05.002

SSO. (2017). Structure and typology of agricultural holdings 2016. State Statistical Office, Republic of North Macedonia.

SSO, 2025SSO. (2025). MakStat database. State Statistical Office, Republic of North Macedonia.

SWG. (2024). Study on the situation and the needs of Rural Youth in the Western Balkan countries and territories (Synthesis report). Regional Rural Development Standing Working Group.

Thompson, P. B. (2010). *The agrarian vision: Sustainability and environmental ethics.* University Press of Kentucky. https://doi.org/10.5810/kentucky/9780813125879.001.0001

Törnberg, A. (2021). Prefigurative politics and social change: A typology drawing on transition studies. *Distinktion: Journal of Social Theory, 22*(1), 83–107. https://doi.org/10.1080/1600910X.2020.1856161

Vizuete, B., Oteros-Rozas, E., & García-Llorente, M. (2024). Role of the neo-rural phenomenon and the new peasantry in agroecological transitions: A literature review. *Agriculture and Human Values, 41*(3), 1277–1297. https://doi.org/10.1007/s10460-023-10537-0

Wittman, H., Desmarais, A., & Wiebe, N. (2010). The origins and potential of food sovereignty. In A. A, Desmarais, N. Wiebe, H. Wittman, (Eds.), *Food sovereignty: Reconnecting food, nature and community* (pp. 1–14). Food First Books.

Yates, L. (2015). Rethinking prefiguration: Alternatives, micropolitics and goals in social movements. *Social Movement Studies, 14*(1), 1–21. https://doi.org/10.1080/14742837.2013.870883

Open Access This chapter is licensed under the terms of the Creative Commons Attribution-NonCommercial-NoDerivatives 4.0 International License (http://creativecommons.org/licenses/by-nc-nd/4.0/), which permits any noncommercial use, sharing, distribution and reproduction in any medium or format, as long as you give appropriate credit to the original author(s) and the source, provide a link to the Creative Commons license and indicate if you modified the licensed material. You do not have permission under this license to share adapted material derived from this chapter or parts of it.

The images or other third party material in this chapter are included in the chapter's Creative Commons license, unless indicated otherwise in a credit line to the material. If material is not included in the chapter's Creative Commons license and your intended use is not permitted by statutory regulation or exceeds the permitted use, you will need to obtain permission directly from the copyright holder.

21

Place Attachment and Quality of Life as Catalysts for Sustainable Rural Transformation: Addressing Depopulation and Promoting Socioecological Resilience

Aleksandra S. Dragin[ID], Aleksandra Tešin[ID],
Maja Mijatov Ladičorbić[ID], Zrinka Zadel[ID],
Kristina Košić[ID], Juan Manuel Amezcua-Ogáyar[ID],
Alberto Calahorro-López[ID], and Tamara Surla[ID]

1 Introduction

The European Commission identifies rural regions on the basis of urban–rural typology. The classification of regions is determined by identifying the population in rural grid cells (all cells outside urban clusters) and their proportion. Therefore, predominantly rural regions are defined as those in which more than half of the population lives in rural grid cells (Eurostat, 2024). Rural depopulation has become a pressing issue in many parts of the world, where migration of young people from rural to urban areas exacerbates demographic imbalances (Li et al., 2019).

This study explores how place attachment and quality of life influence young people's decisions to stay in rural areas of Serbia and Croatia, highlighting place identity as a key factor and framing the findings within the concept of prefigurative politics to show how enhancing local well-being can support rural sustainability in line with global transformation goals.

A. S. Dragin (✉) · A. Tešin · M. Mijatov Ladičorbić · K. Košić · T. Surla
Department of Geography, Tourism and Hotel Management, Faculty of
Sciences, University of Novi Sad, Novi Sad, Serbia
e-mail: sadragin@gmail.com

© The Author(s) 2026
S. Sareen and S. Juhola (eds.), *Societal Transitions to Sustainability*, Palgrave Studies
in Environmental Transformation, Transition and Accountability,
https://doi.org/10.1007/978-3-032-07395-2_21

Between 2015 and 2020, the population of predominantly rural regions across the EU declined by an average of 0.1% per year, while the population of predominantly urban regions grew by an average of 0.4% per year (Eurostat, 2023).

According to the 2022 census, Serbia (see Fig. 1) has about 6.7 million people, but 95% of its settlements have experienced population decline since 2011. Of the 4,721 settlements, one in ten has no children, 55 have no women, and 545 lack women of reproductive age (20–34). In 718 settlements, the average age of women exceeds 60, and in nearly 2,900 settlements, the elderly outnumber youth under 20 by more than double. This decline can largely be attributed to a long-term decrease in the number of women of childbearing age: 400,000 fewer than 20 years ago, including 200,000 fewer women aged 20–34. Even with an increase in birth rates, reversing this demographic trend would require long-term strategies. Nearly one-third of Serbia's population is now over 60 (RZA, 2022).

Croatia is facing similar challenges. With a population of 3.9 million in 2021 and an average age of 44.3, the country's rural areas are also aging rapidly. In 2020, over 2,900 Croatian settlements recorded no births, and nearly 2,000 had fewer than 100 residents. In 46% of settlements, people aged 80+ outnumber children under five (CBS, 2024).

These trends are especially evident in rural areas, where depopulation has reached critical levels. Youth migration to cities or abroad creates serious demographic imbalances (Tešin et al., 2024). Young people are vital to the future of rural areas, forming the foundation of sustainable demographic, economic, and social development. Ensuring adequate living conditions increases their likelihood of staying and contributing to rural growth. Addressing these challenges requires targeted policies

A. S. Dragin · J. M. Amezcua-Ogáyar · A. Calahorro-López
Department of Business Organization, Marketing and Sociology, Faculty of Social Sciences and Law, University of Jaén, Jaén, Spain

Z. Zadel
Faculty of Tourism and Hospitality Management, University of Rijeka, Rijeka, Croatia

Fig. 1 Location of the two case study countries (Serbia and Croatia). *Source* Authors' rendition. Alt text: Location of the two case study countries (Serbia and Croatia)

and rural development strategies to retain young people and support community revitalisation.

While socio-economic factors are often cited as key drivers of rural out-migration, less attention is given to the emotional and psychological aspects influencing decisions to stay or leave. For young people, attachment to rural places can inspire prefigurative political action, with emotional bonds shaping social possibilities. Such connections often guide individuals' involvement in decision-making, development, and environmental stewardship (Darabaneanu et al., 2024), reflecting their values and everyday practices. These emotional ties act as both catalyst and motivation for societal transformation. Among them, place attachment, the emotional bond between individuals and their environment, plays a crucial role in shaping belonging (Tešin et al., 2024). This link between place attachment and quality of life is increasingly seen as a lever for sustainable rural transformation that may counter rural depopulation.

Place attachment has been widely studied in fields such as environmental psychology, sociology, and urban studies, but its application to rural depopulation remains underexplored (Tešin et al., 2024). In particular, how place identity (the emotional connection to one's environment) and place dependence (reliance on a place for daily needs) affect quality of life in rural areas remains an area for further investigation.

2   Literature Review

Place attachment refers to the emotional and cognitive connections individuals form with their socio-physical environments (Scannell & Gifford, 2017). Over time, people develop a strong bond with a place, fostering a sense of identity, belonging, and affection. This attachment is complex, involving not only the bond to the physical and social environment but also functional attachment—a dependence on specific living conditions that may be difficult to replicate elsewhere (Williams & Vaske, 2003).

Two key dimensions of place attachment are widely studied: place identity and place dependence. Place identity refers to how a place influences an individual's sense of self, shaping character, values, and beliefs (Counted, 2019). Place dependence, on the other hand, reflects the functional connection between a person and a place, based on how well the place meets their needs. This often results in reluctance to leave, as the place supports necessary activities and needs. The stronger this dependence, the more the place becomes part of an individual's identity (Anton & Lawrence, 2014).

This study builds on these two dimensions, using items adapted from Williams and Vaske (2015). It also introduces a third dimension, place bonding, developed by the project team. Previous literature indicates that connections to a place also include social elements, such as relationships with family, friends, and community (Moore, 2021). The place bonding dimension incorporates these social factors and other elements identified for further exploration. It includes five items that explore how friends, family, safety, financial convenience, and overall security influence one's decision to live in a place and develop a sense of connection to it. This dimension was piloted by the authors in 2024 and published in the journal *Land* (Tešin et al., 2024).

The primary goal of the *Quality of Life* instrument is to explore factors affecting perceptions of life quality in rural border areas. It incorporates both objective indicators, like living standards, and subjective indicators, such as personal judgments about life circumstances and happiness (Guillen-Royo & Velazco, 2012). Higher levels of subjective well-being correlate with greater life satisfaction, fewer negative emotions, and reduced likelihood of relocation.

Understanding residents' needs and attitudes is key for rural development, particularly for younger generations, who are vital to sustainable demographic, social, and economic growth. Providing satisfactory living conditions is essential to ensuring their stay and contribution to rural communities. The questionnaire will use a methodology combining approaches from the EU, WHO, and EU-SILC (Dolan et al., 2011).

This study builds upon previous work by the authors, all stemming from a unique research project conducted in Serbia and Croatia, further discussed in the methodology section. Key findings from earlier studies, which form the foundation for this research, will be presented in the following text.

Tešin et al. (2024) analysed place attachment and its impact on young people's decisions to stay or leave, identifying three key dimensions:

Place Dependence has a moderate mean score (M = 3.309), suggesting a functional but not overly dominant connection to place. Place Identity was the strongest dimension in their study (M = 3.867), indicating that personal identity is strongly tied to one's place of residence. Place bonding—while important (M = 3.767), showed less consistency across respondents.

Tešin et al. (2024) also explored residents' overall quality of life, identifying two key satisfaction factors:

Satisfaction with Infrastructure Quality includes satisfaction with essential services like roads, medical services, public safety, and prices. Satisfaction with pricing and service quality had the highest factor loadings (0.803, 0.682), highlighting the importance of economic and service factors in shaping residents' perceptions (Diener et al., 2000).

Satisfaction with Culture and Education reflects how well cultural and educational needs are met. Educational institutions play a crucial role in shaping quality-of-life satisfaction, while cultural amenities, though important, were slightly less influential.

The varying reliability and factor loadings indicate the complexity of place attachment. Additionally, satisfaction with infrastructure, economic factors, and education plays a key role in residents' perceptions of their environment.

This paper addresses these issues through a case study of young residents in rural Serbia and Croatia. By examining how place attachment and quality of life impact rural youth's decisions to stay or migrate, this research aims to uncover strategies to promote socioecological resilience and contribute to more sustainable rural development, while also exploring differences in their perceptions. These are two neighbouring countries—Croatia, an EU member, and Serbia, which is on its path to EU membership. According to the latest census, both countries are experiencing significant migration of young people from rural areas to urban centres (RZA, 2022; CBS, 2024).

Understanding the attitudes of young members of local communities is crucial. As a result, youth involvement in local communities has become an increasingly important focus for researchers (Ivasciuc & Ispas, 2023), driven by young people's growing awareness of sustainable development and strong concern for the future (Dragin et al., 2024).

3   Methods

The first part of the questionnaire included respondents' socio-demographic characteristics. The sample consisted of 299 respondents from Serbia (150) and Croatia (149), aged 18 to 30 (M = 20.16). Most were women (72.9%). The majority had a high school education (77.3%), followed by a bachelor's degree (19.4%), a master's degree (1.3%), and primary school (1.7%). Only 11.7% were employed. Most respondents were single (60.9%) or in a relationship (36.5%), with 2.3% married. Half (51.5%) reported an average income, while 27.1% had above-average and 21.1% below-average financial status.

The second part measured place attachment among young rural residents. A scale of 18 items was used. The study modified 12 items from Williams and Vaske (2003), categorised into two dimensions: place dependence and place identity. The authors supplemented the second dimension with additional item (*When I spend time in my place, I feel fulfilled and tranquil*), and introduced a third dimension, place bonding. It consisted of 5 items designed to investigate the impact that friends, family, safety, financial privileges, and overall security have on the decision to live in a specific location. The third part intended to determine young inhabitants' attitudes towards the quality of living in rural areas. The study employed Đerčan et al.'s (2021) instrument, which included 18 statements describing the standard of living in rural areas, such as the quality of infrastructure, roads, public services, prices, schools, etc. All the statements were rated on a five-point Likert (1—strongly disagree; 5—strongly agree).

The study was carried out in Serbia and Croatia, between December 2023 and May 2024. The research was conducted among young resident (18+) of rural areas in both countries. Specialised marketing research agencies distributed the questionnaires.

4   Results and Discussion

Since the study involved participants from Serbia and Croatia, it was assessed whether there were any disparities in their responses based on their country of origin. For these purposes, an independent samples t-test was performed. Place attachment was first examined, and the results revealed significant differences in all three factors (see Table 1). Moreover, respondents from Croatia reported higher levels of place dependence and place bonding compared to those from Serbia. On the other hand, Serbian residents expressed a stronger place identity than Croatian citizens.

The findings also revealed statistically significant differences in respondents' perceptions of the quality of life depending on their country of origin. According to the results presented in Table 1, young people of rural Croatia showed greater levels of quality-of-life satisfaction than

Table 1 T-test results for place attachment and quality-of-life factors. *Source* Authors

Factors	Mean		T-value
	Serbia(n = 150)	Croatia(n = 149)	
T-test results: place attachment factors			
Place dependence	2.68	3.94	−14.204*
Place identity	3.95	3.78	1.991*
Place bonding	3.59	3.94	−3.963*
T-test results: quality-of-life factors			
Satisfaction with the quality of infrastructure	3.25	3.91	−9.002*
Satisfaction with culture and education	3.30	3.81	−5.512*

Note '*': $p < 0.05$

those from Serbia. They are more satisfied with the quality of infrastructure, as well as the cultural and educational aspects of their rural residence.

Higher levels of place dependence and bonding in Croatia indicate that emotional and functional ties to place are stronger among rural youth in Croatia than Serbia. In contrast, higher levels of place identity in Serbia suggest that personal identity may play a more significant role in rural attachment, possibly reflecting a stronger emotional connection to rural way of life.

These findings have important implications for rural development policies. Enhancing quality of life, such as improving infrastructure and expanding educational and cultural opportunities, could strengthen place attachment and incentivise young people to stay in rural areas. Moreover, fostering social ties through community engagement and supporting local identity could further mitigate rural exodus.

Place attachment is more than just personal sentiment—it can be a catalyst for broader socioecological resilience and sustainable rural transformation. By linking local attachments to global sustainability goals like circular economy and degrowth, rural areas can capitalise on their unique qualities, such as local food production (Martinovska Stojcheska et al., 2025), renewable energy, and community-led initiatives. Thus, improving place attachment through targeted interventions that enhance quality of life can contribute to broader shift towards sustainability in rural communities.

Findings of this study aim to inform development of policies and strategies that not only address challenges of rural depopulation but also promote social, economic, and environmental sustainability of rural communities. In this regard, it is important to emphasise key role that place attachment and quality of life play in shaping sustainability of rural areas. Results of this study provide valuable insights into emotional and psychological connections young people have to their rural environments, which can be leveraged to inform more holistic and context-specific policy frameworks.

As research highlights, place attachment is a multifaceted concept, encompassing emotional, functional, and relational connections to one's environment. To promote long-term social, economic, and environmental sustainability, rural development policies must go beyond addressing basic infrastructure and service provision. They should also foster sense of belonging, identity, and dependence on rural place. For example, as seen in study, higher levels of place dependence and bonding among rural youth in Croatia suggest that stronger social ties and functional connections to their environment may be an important factor in their decision to stay. This insight could encourage policies that prioritise community building, participatory decision-making, and localised development initiatives, which align with desires and needs of rural residents. This study builds on broader discourse of rural transformation, specifically focusing on prefigurative politics of sustainable rural development.

Development of spa tourism, particularly in rural settings, can contribute to transformative processes by enhancing local community attachment and improving quality of life through better infrastructure and employment opportunities (Bacsi et al., 2012; Cristian-Constantin et al., 2015). Spa tourism in villages has emerged as key driver of rural development in many countries. In Portugal, for instance, health and wellness tourism is recognised as strategic development sector that revitalises and builds upon country's centuries-old traditions (Pereira et al.,

2023). Spa tourism in countries such as Hungary, Serbia, and Croatia has long-standing tradition rooted in idea that these services should be accessible to everyone, regardless of economic status. This inclusive approach has been supported by national healthcare programs and government subsidies. As a result, local populations have grown accustomed to visiting spa centres, either based on medical prescriptions or personal choice, as regular part of their lifestyle and well-being. This is supported by the idea that this region is rich in thermal and mineral springs, and it is up to rural population to recognise and harness potential of these natural resources. In Hungary, concept of developing spa centres has improved quality of life of local population: road infrastructure has been developed, new jobs have been created, accommodation services have been provided in homes of local population, as well as sale of their products. In villages in Serbia and Croatia with thermal mineral springs, this transformation happens sporadically. Bacsi et al. (2012) identify spa tourism as significant driver of tourism growth in Hungary. Beyond economic benefits, tourism can also contribute to social and cultural development. This includes increased interaction with visitors, enhanced quality of life, expanded opportunities for recreational activities, and potential improvements in overall health and well-being. Arrival of visitors often leads local population to develop deeper appreciation for environment and stronger attachment to community (Mijatov Ladičorbić et al., 2024).

Development of spas has been shown to reduce tourism seasonality and promote local development in rural areas (Gonçalves & Guerra, 2019). It also provides platform for intergenerational dialogue, especially with youth, who are often more environmentally conscious and supportive of sustainable practices (Dragin et al., 2022a, 2022b, 2024). Understanding their perceptions is key for fostering positive socio-environmental outcomes, as their engagement in tourism initiatives can ensure they remain in rural areas, helping curb depopulation and preserving natural and cultural heritage.

In that way, integrating sustainable tourism practices, such as health spas, can align local development with global sustainability objectives, ultimately contributing to balanced, long-term rural transformation.

From an economic perspective, policies aimed at creating employment opportunities, supporting local entrepreneurship, and promoting sustainable agricultural practices could capitalise on local identities and foster place attachment. Rural areas could act as hubs for sustainable practices that tie local economic development to global environmental goals, encouraging young people to see their communities as integral parts of sustainable future.

On the environmental front, place attachment can play a significant role in strengthening young residents' engagement with eco-friendly practices, such as the production of locally grown, eco-friendly food, as well as geographically designated products and similar items. If local communities are emotionally and practically invested in the well-being of their environment, they may be more likely to adopt sustainable practices in areas such as waste management, energy consumption, and land use. Thus, integrating sustainability into the fabric of rural identity and attachment could enhance resilience of these communities in the face of global environmental challenges.

Finally, this study underscores the importance of viewing rural transformation not solely through an economic or infrastructural lens, but as a process deeply intertwined with emotional and relational bonds that people have with their environment. By incorporating place attachment and quality of life into rural development strategies, policymakers can create more effective, sustainable interventions that address both the functional needs of rural residents and their emotional connections to their communities. This nuanced approach offers a promising pathway for the adaptation and thriving of rural areas within the context of global transformations, ensuring that depopulation does not lead to the decline of these regions, but serves as an opportunity for innovative, resilient, and sustainable development.

5   Conclusion

This study provides insights into the complex and multidimensional nature of place attachment and quality-of-life satisfaction. It offers a methodological approach for understanding how individuals relate to their environment and the factors that shape their sense of well-being. By examining both the psychological and practical aspects of place attachment, this research contributes to a deeper understanding of how people connect with their surroundings, which is essential for improving rural spaces and fostering stronger community bonds.

The T-test results suggest that Croatians tend to have higher levels of place dependence and bonding, while Serbians report slightly higher levels of place identity. These findings could have implications for understanding how people in different countries relate to their environments, important for urban planning, community building, and social policies aiming to strengthen people's connection to their places.

The T-test results also indicate that Croatia fares better than Serbia in terms of quality-of-life satisfaction related to infrastructure and culture/education.

This study highlights the importance of place attachment and quality of life in addressing rural depopulation and promoting sustainable rural transformation. By understanding how these factors influence young people's decisions, policymakers and planners can better support the development of rural areas that meet both functional and emotional needs. Findings from Serbia and Croatia suggest that improving infrastructure, enhancing educational and cultural opportunities, and fostering social ties can strengthen place attachment and contribute to more resilient, sustainable rural communities.

The integration of spa tourism into rural areas can contribute to sustainable development if the local community's perceptions are addressed and aligned with local needs and global sustainability objectives. This research emphasises the importance of involving the local population, particularly youth, in decision-making processes to ensure the success and longevity of tourism initiatives in rural areas.

Future studies should investigate how different dimensions of place attachment interact with socio-economic factors, such as employment opportunities and environmental conditions, to shape rural depopulation trends. Additionally, examining the role of prefigurative politics—imagining and implementing future sustainable practices today—can illuminate how rural areas can not only survive but thrive in the face of global challenges.

6   Competing Interests

The authors have no conflicts of interest to declare that are relevant to the content of this chapter.

Acknowledgements This research is part of a project approved by the Autonomous Province of Vojvodina, Provincial Secretariat for Higher Education and Scientific-Research Activity, Program 0201, with the project title "Research of the entrepreneurial potentials among the local population for using the thermo-mineral water resources of Vojvodina", registration number: 003018958 2024 09418 003 000 000 001-01. We also acknowledge the financial support of the Ministry of Science, Technological Development and Innovation of the Republic of Serbia (Grants No. 451-03-137/2025-03/ 200125 & 451-03-136/2025-03/ 200125).

References

Anton, C. E., & Lawrence, C. (2014). Home is where the heart is: The effect of place of residence on place attachment and community participation. *Journal of Environmental Psychology, 40*, 451–461.

Bacsi, Z., Kovács, E., & Lőke, Z. (2012). Spa successes and challenges in Transdanubia, Hungary-Results of a survey in three spa towns. *Deturope, 4*(1), 27–47.

CBS. (2024). Available online. Retrieved 05, 2024, from. https://dzs.gov.hr/vij esti/objavljeni-konacni-rezultati-popisa-2021/1270

Counted, V. (2019). Sense of place attitudes and quality of life outcomes among African residents in a multicultural Australian society. *Journal of Community Psychology, 47*, 338–355.

Cristian-Constantin, D., Radu-Daniel, P., Daniel, P., Georgiana, C. L., & Igor, S. (2015). The role of SPA tourism in the development of local economies from Romania. *Procedia Economics and Finance, 23*, 1573–1577.

Darabaneanu, D., Maci, D., & Oprea, I. M. (2024). Influence of environmental perception on place attachment in Romanian Rural Areas. *Sustainability, 16*, 1106.

Diener, E. (2000). Subjective well-being: The science of happiness and a proposal for a national index. *American Psychologist, 55*(1), 34.

Dolan, P., Layard, R., & Metcalfe, R. (2011). *Measuring subjective well-being for public policy: Recommendations on measures.* Centre for Economic Performance, London School of Economics and Political Science.

Dragin, A. S., Majstorović, N., Jančić, B., Mijatov, M. B., & Stojanović, V. (2022a). Clusters of Generation Z and travel risks perception: Constraining vs. push–pull factors. In P. Mohanty, A. Sharma, J. Kennell, & A. Hassan (Eds.), *The Emerald handbook of destination recovery in tourism and hospitality* (pp. 375–395). Bingley.

Dragin, A. S., Mijatov, M. B., Ivanović, O. M., Vuković, A. J., Džigurski, A. I., Košić, K., Knežević, M. N., Tomić, S., Stankov, U., Vujičić, M. D., Stojanović, V., Bibić, L. I., Đerčan, B., & Stoiljković, A. (2022b) Entrepreneurial intention of students (managers in training): Personal and family characteristics. *Sustainability, 14,* 7345.

Dragin, A. S., Surla, T., Mijatov Ladičorbić, M., Jovanović, T., Zadel, Z., Nedeljković-Knežević, M., Tešin, A., Amezcua-Ogáyar, J. M., Calahorro-López, A., Košić, K., Stojanović, V., Ivkov-Džigurski, A., Pavlović, D., & Vasić, Ž. (2024). Exploring the link between openness and entrepreneurial capacity in young people: Building resilient and sustainable rural territories. *Land, 13*(11), 1827.

Đerčan, B., Bubalo Živković, M., Gatarić, D., Lukić, T., Dragin, A., Pivarski, B. K., Lutovac, M., Kuzman, B., Puškarić, A., Banjac, M., Grubor, B., & Simović, O. (2021). Experienced well-being in the rural areas of the Srem Region (Serbia): Perceptions of the local community. *Sustainability, 14,* 248.

Eurostat. (2023). Predominantly rural regions experience depopulation. Retrieved August 03, 2024, from https://ec.europa.eu/eurostat/web/products-eurostat-news/w/ddn-20230117-2

Eurostat. (2024). Rural development methodology. Retrieved August 03, 2024, from https://ec.europa.eu/eurostat/web/rural-development/methodology

Gonçalves, E., & Guerra, R. J. D. C. (2019). O turismo de saúde e bem-estar como fator de desenvolvimento local: uma análise à oferta termal portuguesa. *PASOS. Revista de Turismo y Patrimonio Cultural, 17*(2), 453–472. https://doi.org/10.25145/j.pasos.2019.17.030

Guillen-Royo, M., & Velazco, J. (2012). Happy villages and unhappy slums? Understanding happiness determinants in Peru. In Happiness across cultures: Views of happiness and quality of life in non-western cultures (pp. 253–270). Springer Netherlands.

Ivasciuc, I. S., & Ispas, A. (2023). Exploring the motivations, abilities and opportunities of young entrepreneurs to engage in sustainable tourism business in the mountain area. *Sustainability, 15*(3), 1956.

Li, Y., Westlund, H., & Liu, Y. (2019). Why some rural areas decline while some others not: An overview of rural evolution in the world. *Journal of Rural Studies, 68*, 135–143.

Martinovska Stojcheska, A., Kotevska, A., & Tuna, E. (2025). Present transformations in food systems: Local initiatives towards sustainable and just futures in post-socialist context. In S. Sareen, & S. Juhola (Eds.), *Societal transitions to sustainability: The prefigurative politics of present transformation.* Palgrave Macmillan.

Mijatov Ladičorbić, M., Dragin, A. S., Surla, T., Tešin, A., Amezcua-Ogáyar, J. M., Calahorro-López, A., Stojanović, V., Zadel, Z., Košić, K., Ivanović, O. M., Džigurski, A. I., Vujičić, M. D., Knežević, M. N., Bibić, L. I., Tomić, S., & Anđelković, Ž. (2024). Towards healthy and sustainable human settlement: Understanding how local communities perceive and engage with spa tourism development initiatives in rural areas. *Land, 13*(11), 1817.

Moore, T. (2021). Planning for place: Place attachment and the founding of rural community land trusts. *Journal of Rural Studies, 83*, 21–29.

Pereira, R., Costa, V., & Gomes, H. (2023). Health and wellness tourism: An overview of thermal tourism in Portugal. *JTSW, 11*(3), 136–147.

RZA. (2022). *Population census in Serbia.* RZS

Scannell, L., & Gifford, R. (2017). The experienced psychological benefits of place attachment. *Journal of Environmental Psychology, 51*, 256–269.

Tešin, A., Dragin, A. S., Mijatov Ladičorbić, M., Jovanović, T., Zadel, Z., Surla, T., Košić, K., Amezcua-Ogáyar, J. M., Calahorro-López, A., Kuzman, B., & Stojanović, V. (2024). Quality of life and attachments to rural settlements: The basis for regeneration and socio-economic sustainability. *Land, 13*(9), 1364.

Williams, D. R., & Vaske, J. J. (2003). The measurement of place attachment: Validity and generali-zability of a psychometric approach. *Forestry Sciences, 49*, 830–840.

Open Access This chapter is licensed under the terms of the Creative Commons Attribution-NonCommercial-NoDerivatives 4.0 International License (http://creativecommons.org/licenses/by-nc-nd/4.0/), which permits any noncommercial use, sharing, distribution and reproduction in any medium or format, as long as you give appropriate credit to the original author(s) and the source, provide a link to the Creative Commons license and indicate if you modified the licensed material. You do not have permission under this license to share adapted material derived from this chapter or parts of it.

The images or other third party material in this chapter are included in the chapter's Creative Commons license, unless indicated otherwise in a credit line to the material. If material is not included in the chapter's Creative Commons license and your intended use is not permitted by statutory regulation or exceeds the permitted use, you will need to obtain permission directly from the copyright holder.

Part IV

Cross-Sectoral and Transdisciplinary Transitions

22

Energy at the Edge of the State. Extractivism, Agropastoralism, and the Making of Frontiers

Shayan Shokrgozar, Usman Ashraf, and David Singh

1 Introduction: Energy, Extractive Frontiers, and Rural Dispossession

The past decade has witnessed a large growth in the appropriation of land for extractive and developmental projects, dubbed the global land rush (Bluwstein & Cavanagh, 2023). Most of this change in land cover and land use has occurred in the majority world (ibid.), yet over three quarters of research on transition studies and cognate aspects is conducted in the minority world. The impetus for these changes in patterns of land use primarily stem from "developmental" ambitions, purportedly climate change mitigation goals, and national security justifications across the majority and minority world (Baka, 2017; Jamali, 2013; Levien, 2018;

S. Shokrgozar (✉)
Centre for Climate and Energy Transformation, University of Bergen, Bergen, Norway
e-mail: Shayan.Shokrgozar@uib.no

U. Ashraf
Global Development Studies (GDS), University of Helsinki, Helsinki, Finland

© The Author(s) 2026
S. Sareen and S. Juhola (eds.), *Societal Transitions to Sustainability*, Palgrave Studies in Environmental Transformation, Transition and Accountability, https://doi.org/10.1007/978-3-032-07395-2_22

Sharma, 2023). Emerging political ecology and agrarian studies scholarship on the intersections of capital-state-powerbroker has situated these changing patterns of land use and resource access within territorialisation (Peluso & Lund, 2011), and the foreclosure of alternatives (Shapiro & McNeish, 2021).

Extractive impulses encompass the material act of extracting natural resources (Nygren et al., 2022), representing a mode of accumulation—hallmark of contemporary "extractive capitalism" (Shapiro & McNeish, 2021), which has evolved into a broader way of organising life (Chagnon et al., 2022). A key dimension of (energy) extractivism, whether renewable or conventional, is frontier-making, a cyclical process that dismantles existing institutional structures, property regimes, and land claims to establish new territorial rules, property laws, and land acquisition frameworks (Rasmussen & Lund, 2018). This tabula rasa approach facilitates the transformation of resource frontiers into commodity frontiers, achieved through land occupation for extraction. Territorialisation thus plays a dual role: producing land both as a commodity and as a space where state authority, access, and exclusion regimes are enforced (Peluso & Lund, 2011).

A consistent thread in these socio-ecologically violent patterns is the central role of power in shaping certain futures while closing off alternatives (Stoddard et al., 2021), manifesting across institutional, relational, and socio-material realms (Sareen & Shokrgozar, 2022). Creating post-extractive futures requires both challenging power structures and nurturing alternatives. One approach to alternative future-making, particularly for marginalised voices, is prefigurative politics—"ways of knowing, acting, and occupying space and social networks that are not

D. Singh
Center for Social Humanities (CSH), New Delhi, India

Department of Food and Resource Economics (IFRO), University of Copenhagen, Copenhagen, Denmark

Center for South Asian and Himalayan Studies, EHESS, Paris, France lived in the shadow of dominant powers' temporal strategies" (Jeffrey & Dyson, 2021, 653).

In this chapter, by examining the development of thermal power in Sindh, solar power in Rajasthan, and wind power in Gujarat (see Fig. 1), we illustrate how, despite differing institutional, relational, and socio-material settings, an extractive resource politics is driving the emergence of similar patterns of enclosure across neighbouring regions of Pakistan and India. Section 2 explores extractive frontiers, territorialisation, and the key dimensions of enabling emancipatory futures. Section 3 introduces our case studies, positioning them within the capital-state-powerbroker nexus and the emergence of alternatives from these borderland spaces. Finally, Sect. 4 examines the implications of our findings for counter-majoritarian and anti-casteist transformative practices, asking: Can prefigurative politics enable alternative energy futures in agropastoral borderlands?

2   Extractive Frontiers, State Power, and Social Identity

The frontier is not a natural or an indigenous category, neither is it simply discovered at the edge of the state or in borderlands. The frontier was enrolled in sixteenth century Ireland when the British imposed new land laws, in the right of "discovery" and the notion of "vacant" lands in nineteenth century United States of America, and in the colonisation of Africa and South Asia's lands. In all cases, frontiers convey a sense of change, disruption and rupture with the past. The frontier is a two-sided landscape: on the one side, the unruliness, the backwardness, the disorder, the wildness, the emptiness; and on the other, the symmetry, the order, the abundance, the civilised, making the frontier "empty of people, histories and claims, but full of potential for new and improved use" (Li, 2014, 592). This binary construction of the frontier establishes legitimate land uses and users whose claims are legalised and sanctioned by the state as rights, while it denies those same rights to others as they become criminalised, depicted as "poachers" or "squatters" (Peluso & Lund, 2011).

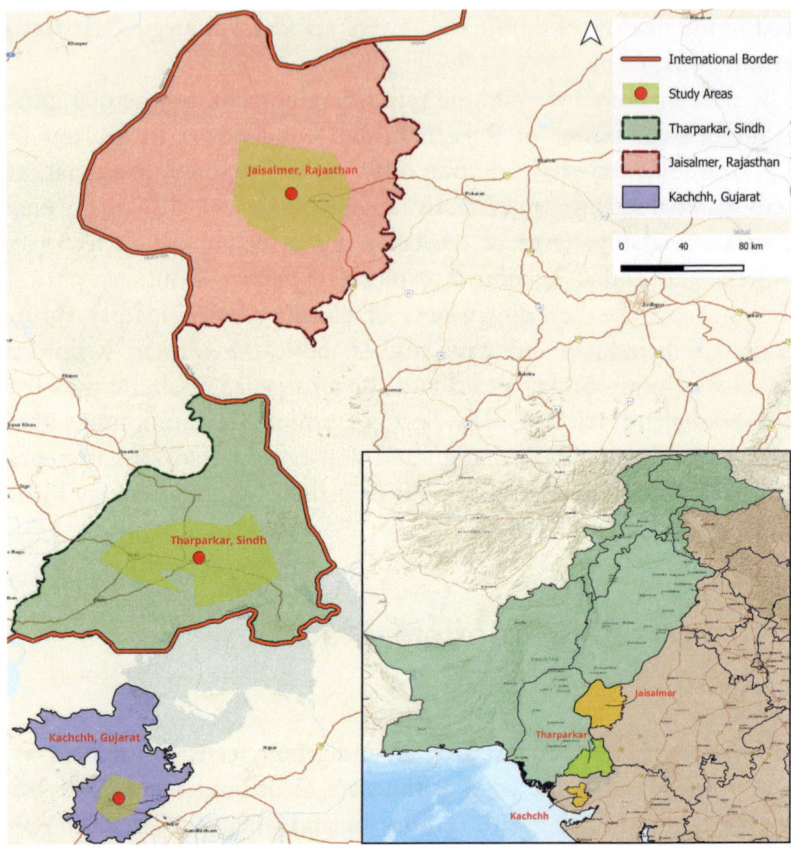

Fig. 1 Study areas (*Source* Author Usman Ashraf's rendition). The data used to create shape files/maps is from the publicly available GADM database (https://www.gadm.org), version 2.5, July 2015

Something new emerges out of the frontier, a reconfiguration of the geography of resource control, or what many scholars have called territorialisation or territorial expansion (Rasmussen & Lund, 2018). Indeed, frontiers have become privileged sites where (non-)state territorial control and authority over resources and people are being constantly redefined, reinvented and (de)institutionalised. Elden (2010, 810) defines territory as a historically constituted political technology of rule aimed at rendering space "owned, distributed, mapped, calculated, bordered and controlled." For this reason, territorialisation is associated with the establishment of boundary-making, the mapping and zoning of space through institutional and legal arrangements such as land cover categories, cadastral maps, measurement standards, property legislations. For historical reasons, this process was mostly the prerogative of the state and expressed within national boundaries, borderlands being sites par excellence of what has been called state (or internal) territorialisation. Under neoliberalism, nonetheless, the state continues to play an important role as an enabler of capital accumulation, what Levien (2018) in the context of Rajasthan has describes as a "land broker state," facilitating the transfer of usage rights from agropastoralist communities to extractive industries.

However, these global processes interact with local lives, sparking "political reactions from below" (Borras & Franco, 2013), often centred around contestations over land access and control over resources (Nygren et al., 2022). As Nygren et al (2022) note, these struggles often take place in "resource spaces," involving heterogenous social groups for their land, livelihood, territories, environment, and culture with varying degree of success (Kröger, 2021). Conflicts not only arise against state and corporation but also within and between communities (Borras & Franco, 2013). These conflicts explain the variegated political reactions from below and in some cases, its absence (ibid.). These responses are often shaped by intersecting social identities—such as gender, ethnicity, race, caste, and religion—which overlap with class dynamics to produce exclusionary relations. Kröger (2021), for example, highlights how caste relations underlie the violence and exclusion inherent in extractive capitalism in India. Thus, the extractive assemblage is not solely composed of state and capital but also includes local power brokers who play a significant role in sustaining these exclusionary practices. Resultantly, Tharparkar has become "a zone of exception" (Khan, 2020)—in which property regime, political jurisdiction, rights and social contracts are set aside.

Nevertheless, the changing regimes of extractivism and territorialisation demonstrate how the present and the future are always in the making (Bluwstein & De Rosa, 2024; Jeffrey & Dyson, 2021). Yet despite incumbent powers striving to advance what they perceive a "pragmatic" future—employing various legitimation practices to dismiss alternatives as pipedreams—the future remains open to prefiguration (Stoddard et al., 2021). While dominant narratives of developmentalism and modernisation propelled much of the imaginaries and praxes for place making, anxieties about their threats to the ecosphere and ethnospehere are propelling others. These transformations, however, often come with sociotechnical implications, marked by aggressive and undemocratic changes to landscapes, alongside bleak prospects for Indigenous and agropastoral futures (Paprocki & McCarthy, 2024), leading to opposition and resistance (Brisbois & Cantoni, 2025).

Simultaneously, myriad movements and initiatives have emerged that are advancing anti-majoritarian, and in the case of Pakistan and India anti-casteist agendas, seeking to enable more emancipatory and socioecologically sustainable futures. These efforts represent marginalised communities coming together to prefigure alternative "social relations, political structures and cultural practices" to form "counter-future" making (Jeffrey & Dyson, 2021, 641). Changing social relations brought through dispossessory processes that (re)marginalise peoples require (re)negotiating pervasive inequalities and employing emancipatory space. As Naess and colleagues (2025, 2) observe, intersectional subjectivities are a potent tool in "examining vulnerability, institutions and possibilities for emancipatory change." Specific to the context of energy, these practices can include the rise of energy communities, which decentralise energy production through open and voluntary networks that often produce energy on a small-scale and distribute its socioeconomic benefits (Sareen et al., 2024). The growth and advancement of such ECs can be based on place specific epistemic values and ontological lifeworlds and thus rooted in a commitment to plurality.

3   Case Study

3.1   Sindh

Tharparkar, situated in south-east of Pakistan, is part of the great Indian desert bordering with Rajasthan, India. In 2015, the Thar coal mining project was included into China's infamous Belt and Road Initiative (BRI). Thus, Thar coal mining project became part of Pakistan's national development, energy security, and foreign investment narrative. Thar coal's proximity to the Indian border has strengthened militarisation of development process in the region. In the discourse of militarisation, Thar became a zone of exception in which the rights of Thari people were set aside and the Thari came to be seen as "expendable" (Khan, 2020).

The expulsion from land in Tharparkar benefitted a small upper caste land-owning elite who were able to secure better land compensation. However, about 70% of the enclosed land for mining is common land from where landless people derive their livelihood. The enclosure of commons resulted in loss of land and alienation of the landless classes without legal claims for compensation. The politics of land compensation in a multi-ethnic, religious, and class society inadvertently entangled with politics of class, caste, and religion and their consequential effect on unequal land distribution.

The intersections of class, caste, and religious politics surrounding landownership have generated multiple, and at times competing, forms of mobilisation. For example, the class interest of land-owning Muslims was to increase compensation for land. While the landless Hindu minority groups' mobilisation centred around access to commons. Correlatively, the activism around land compensation alienated other castes and classes because of unequal effects of compensation and identity politics. The landless lower class-caste Hindu minority Bheel, Kolhi, and Meghwars faced dual process of marginalisation that Lerche and Shah (2018) call "conjugated oppression"—a feature of capitalist expansion in caste-based societies. First, the Bheel, Kolhi, and Meghwar communities face social marginalisation and oppression due to their designation as "untouchables" within the caste hierarchy (Khan, 2020). Second, the expansion of capitalist land relations at the resource frontier has economically dispossessed and marginalised them by displacing them from their customary usufruct rights to common and forest lands.

Counter intuitively, the Thari did overcome social structures of exclusion through organising mass mobilisation. The Gorano movement, as it came to be known, by Thari people used a wide repertoire of activism including sit-in, hunger strike, direct action, social media activism, and legal actions. The first wave of protests fell apart after a year owing to caste-class and religious politics. After the first setbacks, the Gorano movement proactively used prefigurative politics to counter the caste-class-religious divisions. For example, one of such modalities was inter caste-class alliance building by involving historical harmonious relationship between different religious groups which has been the hallmark of Tharparkar. The protest organising committee was subsequently constituted to include multiple caste-class and religious groups. This move ensured inclusion of issues of different communities which subsequently ensured participation of those communities in protests. Gorano movement was able to create inter and intra class-caste and religious alliances transform conflict into prefigurative politics of a multi-ethnic pluralist society.

Another tactic used by the Gorano protest movement was Rajoni Kath (people's assemblies). These assemblies were used by activists to encourage cross caste-class knowledge production of the issues, to encourage solidarity and politicisation of polity. For example, on one Rajoni Kath some Muslims were seen praying in the premises of a Hindu Temple. In discussion that followed and on social media posts, this was highlighted as an example of the pluralist society that Tharparkar is. In conclusion, the Thari movement created a multicaste-class and religiously diverse movement that emulated the pluralist society that the movement imagined.

3.2 Rajasthan

The product of combining about two dozen princely states, chiefships, and a British district—first into the Indian Union and later the Government of India—Rajasthan is known for its palaces, forts, and natural beauty. With an economy primarily based on agriculture, Rajasthan contrasts with its business-oriented neighbour Gujarat and is often seen as financially constrained, facing "high-energy sector debt, austerity measures in the public sector, high rates of energy poverty and an acute concern with affordability" (Sareen et al., 2023, 142). The state has historically relied on thermal power, making it vulnerable to protests and supply chain disruptions tied to coal mining and transport. To attract national and international investments and become more energy self-reliant, Rajasthan began promoting unconventional energy sources as early as 1999. Today, it touts the largest non-fossil energy installed capacity in India—an outcome not without social and ecological costs.

What makes Rajasthan particularly suited to solar energy expansion—especially post–National Solar Mission—is the availability of land officially classified as barren and uncultivable (informally "wasteland") and some of the highest solar irradiance levels in India. Its proximity to the Pakistan border has also ensured robust infrastructure—roads, transmission lines, and military build-up—supporting the installation of dozens of gigawatts of solar capacity within a decade. This rapid buildout, however, has severely impacted customary agropastoral land users, who for millennia have practiced lifeways dependent on the commons, ancient water harvesting, pastoralism, and subsistence crop cultivation. These users now face unprecedented pressures on lands critical for their social reproduction.

The loss of pasture and commons—whether due to land misclassifications, corruption, or disregard for local populations—has disproportionately affected people along lines of class, caste, and gender. As one informant (April 2024) from Jaisalmer explained: "Rajputs used to be the sole landowners before independence; now they are the largest landowners." In practice, this means they benefit from leasing private land to solar developers while retaining other properties for agropastoral activities. In contrast, Scheduled Castes and Tribes such as the Meghwal and Bhil often lack private property and are restricted from accessing enclosed commons.

These benefits also extend to local power brokers, such as land aggregators, who secure lucrative contracts by what one of them (November 2023) described as "playing psychological tricks" on landowners—through deception, lies, and even threats of violence. The convergence of state-capital alliances to enable rapid solar expansion thus comes at great cost to land-dependent practices and communities, often justified in the name of development, climate action, or national security.

Yet, these frontier zones are also sites of resistance and alternative visions. Resistance takes the form of protests, sit-ins, public interest litigations, and media coverage. Additionally, new organisations have emerged to monitor impacts on wildlife (like the critically endangered Indian Bustard), promote ancient farming methods, and document sacred forests (orans). Collaborations across institutions—district, state, national, and even international—have supported efforts to protect habitat and develop grasslands. In other words, fears over the loss of ancient lifeworlds and intangible heritage have mobilised counter-efforts to resist the state–capital–broker assemblage and empower place-based practices.

3.3   Gujarat

Situated in Western India, Gujarat has been working as a laboratory for liberalisation, energy development, and Hindu nationalism: the state liberalised land acquisition laws in the early 1990s and transferred massive amounts of common pasture and "wastelands" to industrialists and private infrastructural projects.

This new state-capital alliance found avenues for expansion and consolidation in the aftermath of the 2001 Gujarat earthquake that hit the district of Kutch, further West, closer to the border with Pakistan. Rehabilitation and resettlement programs led to a critical shift: the central government declared a five-year excise duty exemption, and Modi's state government announced a package of tax incentives for industries and private power plants. It is no coincidence that the first private Special Economic Zone (SEZ) port in India was created in Mundra (Kutch) and entrusted to Adani Power.

The district was subsequently identified as a rich windy region by the National Institute of Wind Energy (NIWE) and therefore framed as a major wind corridor, where thousands of wind turbines have been installed by established companies like Adani Power or Suzlon for well over a decade. The land dispossession that enfolded with the arrival of these wind turbines in mainland and borderland Kutch has been aggregating unique material characteristics and relied on a strong alliance with local land brokers, fixers, and middlemen who were tasked to find suitable land: apart from a minority of private lands, wind power projects encompassed all types of state lands, mostly so-called government "wastelands" while common pastures were increasingly portrayed as waste (Baka, 2017).

The proximity of the border with Pakistan also gave another incentive to develop wind infrastructural projects in the name of national security. Indeed, the historical presence of pastoral Muslim populations on both sides of the Kutch-Sindh border had generated rumours and whispers questioning the loyalty of Indian Muslims since Independence and justifying later on intense state surveillance and securitisation in the region. In the early 2000s, the Border Area Development Program advised developing mining industries and tourism projects in the middle of the Rann, a semi-arid land separating India and Pakistan and populated mainly by Muslim and Dalit pastoral populations (Ibrahim, 2021).

The acquisition of government-owned pastures left the villagers of Kutch divided in resisting or negotiating dispossession: those who owned private lands had historically accumulated enough socio-economic capital and were therefore in capacity to negotiate directly with energy companies. They were the best positioned to benefit from the new labour and land processes produced by dispossession, at the expense of already marginalised Dalit and Muslim pastoral communities who lost access to commons. Other caste configurations and political scenarios nurtured the emergence of collective resistance movements and alternative futures. In the presence of enduring inter-caste relations around land, pastures and water, Samio village saw the only organised, collective, and victorious social movement against the arrival of wind turbines in central Kutch.

There, pastoralist and non-pastoralist groups united in an inter-caste alliance as they shared common interests in relation to land and common fears about the future. Villagers contested the core of the interrelated "wasteness" and "greenness" discourses mobilised by the state to justify the private takeover of revenue grasslands and imagined a future that re-energised traditional agrarian struggles around the defence of common lands and attached livelihoods practices.

4   Discussion and Conclusion

Together the three cases demonstrate how statist territories logics embedded in energy extraction imperatives come into daggers with place-based lifeways revolving around agropastoralism. Despite a tense border dynamic and a regular outburst of cross-border violence rooted in us vs. them (and similar internal dynamics across ethnic and religious line), the disregard for oral and local epistemologies and the imposition of state endorsed surveyors, traceable to the rule of the British Raj is evident in all cases. Across Sindh, Rajasthan, and Gujarat, infrastructures are used as a vector for organising life, reproduction, and vitality and controlling territory. These infrastructural developments are carried out through thermal and lower-carbon energy and cognate infrastructures, justified in the language of development, climate change, and national security, but resulting in enclosure and exclusion. These interventions, "green" or not, blended discourses of development, modernism, and technology with strategies to exert control and surveillance over margins of the state and "create new citizens of various kinds, with loyalties that lead them to look affectionately to the east [i.e., the Indian nation]" (Simpson, 2014, 75), and the Pakistani nation for those across the border (Khan, 2020).

In the absence of a sympathetic state, corporate, and intermediaries towards ensuring the continuity of place-based lifeways and livelihood practices, as opposed to modernist ones, mutual aid, solidarity, and grassroots movements can enact a critical role in resisting infrastructural harm and guiding socio-material change towards recognising and practicing plurality. We see these counter-majoritarian and anti-casteist efforts towards transformative change through prefigurative practices

(Sharma, 2023). These practices include inter-class, caste, and religious alliance building to confront capitalist logic of frontier expansion as well as strengthening socio-ecological harmony. Therefore, many of these struggles are framed in the framework of livelihood and ecological sustainability. They do not wait for the perfect time and moment sometime in the future for the appropriate decisions to be made but instead practice community building and enhancing social cohesion and ecological wellbeing in the here and now.

At the heart of these contested frontiers lies a struggle over meanings—of land, development, and citizenship—that transcends extractivism as a material process and reveals it as a symbolic and institutional force (Peluso & Lund, 2011; Shapiro & McNeish, 2021). What is at stake is not simply access to territory, but the redefinition of who belongs and who is rendered disposable (Khan, 2020; Lerche & Shah, 2018). By tracking extractive expansions along three regions of the Indo-Pak border, we show how energy infrastructures do more than generate power: they naturalise hierarchies, render violence legible as progress, and transform spatial peripheries into laboratories of nationalist modernity. Yet, the same margins become incubators for alternative imaginaries—where disempowered groups prefigure futures grounded in reciprocity, pluralism, and ecological embeddedness. These are not merely acts of resistance, but active world-making practices that reclaim space from the territorial logics of the nation-state and its alliances with capital.

Despite bleak possibilities of breaking away from national antagonistic cross border and multiethnic politics, initiatives such as Fearless collective and Thari movement show that solidarity across borders is possible in some ways. Whether these movements can challenge the dominance of capital-state-powerbroker assemblages and offer alternative pathways for infrastructural development that foster social cohesion and ecological well-being is yet to be known.

References

Baka, J. (2017). Making space for energy: Wasteland development, enclosures, and energy dispossessions. *Antipode, 49*(4), 977–996. https://doi.org/10.1111/anti.12219

Bluwstein, J., & Cavanagh, C. (2023). Rescaling the land rush? Global political ecologies of land use and cover change in key scenario archetypes for achieving the 1.5 °C Paris agreement target. *The Journal of Peasant Studies, 50*(1), 262–294. https://doi.org/10.1080/03066150.2022.2125386

Bluwstein, J., & De Rosa, S. P. (2024). Political ecologies of the future: Introduction to the special issue. *Geoforum, 153,* 104023. https://doi.org/10.1016/j.geoforum.2024.104023

Borras, S. M., Jr., & Franco, J. C. (2013). Global land grabbing and political reactions 'from below.' *Third World Quarterly, 34*(9), 1723–1747. https://doi.org/10.1080/01436597.2013.843845

Brisbois, M. C., & Cantoni, R. (2025). Coping with decarbonisation: An inventory of strategies from resistance to transformation. *Global Environmental Change, 90,* Article 102968. https://doi.org/10.1016/j.gloenvcha.2025.102968

Chagnon, C. W., Durante, F., Gills, B. K., Hagolani-Albov, S. E., Hokkanen, S., Kangasluoma, S. M., & Vuola, M. P. (2022). From extractivism to global extractivism: The evolution of an organizing concept. *The Journal of Peasant Studies, 49*(4), 760–792. https://doi.org/10.1080/03066150.2022.2069015

Elden, S. (2010). Land, terrain, territory. *Progress in Human Geography, 34*(6), 799–817. https://doi.org/10.1177/0309132510362603

Ibrahim, F. (2021). *From family to police force: Security and belonging on a south asian border.* Cornell University Press.

Jamali, H. A. (2013). *The anxiety of development: Megaprojects and the politics of place in Gwadar, Pakistan* (Crossroads Asia Working Paper Series, No. 6).

Jeffrey, C., & Dyson, J. (2021). Geographies of the future: Prefigurative politics. *Progress in Human Geography, 45*(4), 641–658. https://doi.org/10.1177/0309132520926569

Khan, M. A. (2020). *Making them look the other way! The (ir)rationality of road building in the Sindh borderlands of Pakistan* (Doctoral dissertation). SOAS University of London. https://doi.org/10.25501/SOAS.00040030

Kröger, M. (2021). *Iron will: Global extractivism and mining resistance in Brazil and India.* University of Michigan Press. https://doi.org/10.3998/mpub.11533186

Lerche, J., & Shah, A. (2018). Conjugated oppression within contemporary capitalism: Class, caste, tribe and agrarian change in India. *The Journal of Peasant Studies, 45*(5–6), 927–949. https://doi.org/10.1080/03066150.2018.1463217

Levien, M. (2018). *Dispossession without development: Land grabs in neoliberal India.* Oxford University Press.

Li, T. M. (2014). *Land's end: Capitalist relations on an indigenous frontier.* Duke University Press. https://ebookcentral.proquest.com/lib/uea/detail.action?docID=1757712

Naess, L. O., Wangari-Muneri, E., Nightingale, A. J., & Mehta, L. (2025). Climate change and the operation of power: Intersectionality, dispossession, and knowledge politics in pastoral communities. *The Journal of Peasant Studies 52*(6), 1323–48. https://doi.org/10.1080/03066150.2025.2451288

Nygren, A., Kröger, M., & Gills, B. (2022). Global extractivisms and transformative alternatives. *The Journal of Peasant Studies, 49*(4), 734–759. https://doi.org/10.1080/03066150.2022.2069495

Paprocki, K., & McCarthy, J. (2024). The agrarian question of climate change. *Progress in Human Geography, 48*(6), 691–715. https://doi.org/10.1177/03091325241269701

Peluso, N. L., & Lund, C. (2011). New frontiers of land control: Introduction. *The Journal of Peasant Studies, 38*(4), 667–681. https://doi.org/10.1080/03066150.2011.607692

Rasmussen, M. B., & Lund, C. (2018). Reconfiguring frontier spaces: The territorialization of resource control. *World Development, 101*, 388–399.

Sareen, S., Haarstad, H., Gong, H., Aiken, G., Skjølsvold, T. M., Silvester, B. R., Popovic-Neuber, J., Stopa, M., Lindkvist, M., Pezzotta, M., Sasse, L., Shokrgozar, S., Haugland, B. T., Langhelle, O., & Inderberg, T. H. J. (2024). Watt sense of community? A human geography agenda on energy communities. *Progress in Environmental Geography*, 27539687241287795. https://doi.org/10.1177/27539687241287795

Sareen, S., & Shokrgozar, S. (2022). Desert geographies: Solar energy governance for just transitions. *Globalizations*, 1–17. https://doi.org/10.1080/14747731.2022.2095116

Sareen, S., Shokrgozar, S., Neven-Scharnigg, R., Girard, B., Martin, A., & Wolf, S. A. (2023). Accountable solar energy transitions in financially constrained contexts. In B. Edmondson (Ed.), *Sustainability transformations, social transitions and environmental accountabilities* (pp. 141–166). Springer International Publishing. https://doi.org/10.1007/978-3-031-18268-6_6

Shapiro, J., & McNeish, J. A. (2021). Our extractive age: Expressions of violence and resistance. *Taylor & Francis*. https://doi.org/10.4324/978100 3127611

Sharma, M. (2023). Homeland, cows and climate change. In B. Forchtner (Ed.), *Visualising far-right environments* (pp. 206–228). Manchester University Press. https://www.manchesterhive.com/display/9781526165398/978 1526165398.00017.xml

Simpson, E. (2014). *The political biography of an earthquake: Aftermath and Amnesia in Gujarat*. Oxford University Press.

Stoddard, I., Anderson, K., Capstick, S., Carton, W., Depledge, J., Facer, K., Gough, C., Hache, F., Hoolohan, C., Hultman, M., Hällström, N., Kartha, S., Klinsky, S., Kuchler, M., Lövbrand, E., Nasiritousi, N., Newell, P., Peters, G. P., Sokona, Y., & Williams, M. (2021). Three decades of climate mitigation: Why haven't we bent the global emissions curve? *Annual Review of Environment and Resources, 46*(1), 653–689. https://doi.org/10.1146/ann urev-environ-012220-011104

Open Access This chapter is licensed under the terms of the Creative Commons Attribution-NonCommercial-NoDerivatives 4.0 International License (http://creativecommons.org/licenses/by-nc-nd/4.0/), which permits any noncommercial use, sharing, distribution and reproduction in any medium or format, as long as you give appropriate credit to the original author(s) and the source, provide a link to the Creative Commons license and indicate if you modified the licensed material. You do not have permission under this license to share adapted material derived from this chapter or parts of it.

The images or other third party material in this chapter are included in the chapter's Creative Commons license, unless indicated otherwise in a credit line to the material. If material is not included in the chapter's Creative Commons license and your intended use is not permitted by statutory regulation or exceeds the permitted use, you will need to obtain permission directly from the copyright holder.

23

Towards a Transdisciplinary Approach to Prefiguration Experiments for Food System Transformation

Dagmar Diesner and Christian Scholl

1 Introduction: The Missing Link between Prefiguration and Transformation

Social movements do not just protest. They also build up bottom-up infrastructures and service provision for catering societal needs beyond capitalist markets. The social movement literature looks at these attempts as a form of "prefiguration" or "prefigurative politics" (Breines, 1982). This concept of prefiguration has been used both descriptively, to analyse organisational practices of social movements (Maeckelbergh, 2009) and normatively, to emphasise its theoretical significance (Yates, 2015). Scholl et al. (2017) identifies two dominant interpretations of prefiguration: as an ethical stance of "leading by example" and as a pragmatic, instrumental effort to build counter-structures that

D. Diesner (✉) · C. Scholl
Maastricht Sustainability Institute, Maastricht University, Maastricht, The Netherlands
e-mail: dagmar.diesner@maastrichtuniversity.nl

© The Author(s) 2026
S. Sareen and S. Juhola (eds.), *Societal Transitions to Sustainability*, Palgrave Studies in Environmental Transformation, Transition and Accountability,
https://doi.org/10.1007/978-3-032-07395-2_23

support transformation. He argues for a reconciliation between these approaches. In this chapter, we build on this and propose to connect the concept of prefiguration with the research of transdisciplinary action research (TAR), particularly emphasising cycles of experimentation and learning for transformative change. What remains underexplored in much of the literature is how prefiguration contributes—or is expected to contribute—to broader, long-term societal transformation beyond the immediate settings in which alternative practices are enacted. The underlying theory of change seems to expect a spread and take-up of the good example, but the exact mechanisms are not spelled out. Therefore, we consider a more sophisticated theoretical and empirical framework necessary.

This chapter takes a first step towards addressing this gap by combining the practice of prefiguration with the practice of transdisciplinary action research. Our main argument here is that prefigurative experiments must be connected to joint learning processes and the active dissemination of lessons learnt to effectively inform ongoing processes of transformations. The aim of this paper is to develop a more systematic and structured understanding of prefiguration as a strategy within transformation-oriented transdisciplinary action research, an approach that actively links experimentation to collective learning and diffusion/ dissemination.

Building on a review of the literature on prefiguration, we identified two key processes that require greater attention: (1) learning processes that emerge dynamically within, across and beyond prefigurative initiatives/episodes, and (2) the circulation and diffusion of these learnings into wider society, more recently discussed as "scaling", a process increasingly discussed on the multiple-scalar-level: scaling out, scaling across, scaling deep, and scaling up.

To address these dimensions, the paper draws on recent research on experimental learning in urban living labs (Scholl & De Kraker, 2021; van Oers et al., 2024) and conceptualises prefigurative practices as forms of "constituent experimentation" where new forms of collective organisation are capable of catalysing transformative change. Thereby the paper proceeds as follows: First, we connect insights from social movement studies, sustainability transition studies, and social learning to demonstrate that learning is the missing link between prefiguration and transformation. Next, we introduce the framework of experimentation and learning in transdisciplinary action research, showing how this can be used as a form of prefiguration. Finally, we illustrate this approach, termed as experimental prefiguration, through our ongoing research with Urban Food Labs and their role in transforming local food systems.

2   Collective Learning and Diffusion as the Missing Link between Prefiguration and Transformation

Learning takes place not only at the individual level but also collectively—within groups, communities, and in networks. Collective learning can be broadly described: "a process of acquiring and generating new knowledge and insights, and of meaning-making of experiences in communicative interaction, in a reciprocal relationship with the social, (bio-)physical and institutional context" (van Mierlo & Beers, 2020, 3).

Recent scholarship distinguishes between different types of collective learning processes. One of the most extensively studied is social learning (Reed et al., 2010), originated from fields such as natural resources management and complex system thinking. Scholars conceptualised learning as a social process in which stakeholders align, change, and transform their perspectives on problems and potential solutions (Macintyre et al., 2018). Given to the complexity of learning processes, it can be categorised into three dimensions: collaborative, interactive, organisational.

Collaborative learning occurs in situations where a group of individuals address a shared problem, critically exchanging, and reflecting on their knowledge and experiences (Laal & Ghodsi, 2012). Interactive learning, rooted in institutional economics, focuses on the integration of codified and tacit knowledge to foster deeper understanding and innovation (Nelson & Winter, 1985). Organisational learning examines how knowledge is created and disseminated within and between organisations, focusing on adaptive and generative learning processes (Antonacopoulou & Chiva, 2007; Crossan et al., 1999).

Although these learning modes are mainly studied separately, they are not mutually exclusive. Nevertheless, potential overlaps, interdependencies and interaction between them do exist, contributing to more holistic understandings and processes of change.

Collective learning can also be analysed in terms of its outcomes, which vary by type and depth. Outcomes may involve skills, competencies, or knowledge. The depth of learning is often conceptualised using the framework of multiple learning loops (Pahl-Wostl, 2009). For studying collective learning in prefigurative politics van Poeck et al. (2020)'s distinction between practical learning (i.e., novel practices, habits, routines, and technologies), relational learning (i.e., new relationships, new networks and governance arrangements, new business models), and conceptual learning (e.g., understanding how problems are framed and defined, engaging in critical thinking, and reframing, such as considering energy justice concerns) is relevant.

The last type of learning aligns with what is often termed as transformative learning, which can be considered as a process involving a radical shift of one's worldview through critical reflection, leading to new perspectives and deeper understandings (Illeris, 2014; Macintyre et al., 2018).

If transformative learning is indeed possible, it is highly promising for advancing prefigurative politics. For seizing its potential, however, prefiguration may need to be re-conceptualised. For this purpose, we frame prefiguration within ULLs, a physical space for collaboratively design experimentations.

3   Transdisciplinary Action Research: Experimentation and Learning in Urban Living Labs

Over the past decade, Urban Living Labs (ULLs) have become an increasingly prominent framework for addressing complex urban challenges. These are real-world experimental settings that bring together diverse stakeholders to co-create innovative solutions to urban problems (Kemp & Scholl, 2016; Voytenko et al., 2016). Within ULLs, real-life experimentation is paired with collaborative learning processes. But while experimentation has gained significant attention, the dimension of learning remains still underexplored (Scholl & De Kraker, 2021).

Recent literature on ULLs shows that the strategic combination of experimentation and learning is essential, if ULLs are to contribute meaningfully to transformative change (Voytenko et al., 2016). Differently put, facilitating collaborative learning processes in ULLs are a condition for making experimentation in real-life-situations more effective. This insight suggests that the better experimentations and learning processes are coupled in prefigurative settings, the impact of intentionally implemented experiments onto the different scaling measures in different social dimensions, its impact is far greater. For achieving a successful outcome, however, the methodological framework needs to be robust and flexible at the same time, it needs to respond effectively to the fluid dynamics in the experimentation processes within and outside of ULLs. As Yates (2015) argues, methodological rigour will allow strategies to evolve, strengthen, and adapt over time. This is particularly important for navigating the challenges involved in scaling prefigurative practices/initiatives from local controlled social settings to broader societal contexts/trajectories. For prefiguration to become a viable concept capable of fostering radical societal shifts, as such, continuous communication with and among stakeholders becomes a necessity during the preparation and experimentation phase.

However, preparing the experimentation phase in ULLs and in the wider society, it needs specific theoretical frameworks (Riddell & Moore, 2015) to address the learning agendas in the social, political, and economic sphere. For integrating experimentation and learning in a more systematic way, the Transdisciplinary Action Research provides a useful framework for its structured methodology in the context of sustainability transitions. It addresses real-life societal challenges by convening different academic disciplines and societal actors in collaborative knowledge production (Hadorn et al., 2008). Knowledge production is structured around specific processes for understanding the complexity of a real-life problem (system), the needs and reasons of change (target), and what kind of political, technological changes should occur to achieve the desired transformation (transformation). These three clusters of knowledge production (system, target, transformation), as seen in Fig. 1, occur not in isolation to each other but are developed in a continuous interplay.

TAR is inherently participatory with researchers, as it facilitates collective reflection among all participants (Kemp & Scholl, 2016). It supports the implementation of structured experimentations and the monitoring and evaluation of these structured experimentations navigating societal change processes. The structure and implementation of experiments follows four steps, as illustrated in Fig. 2.

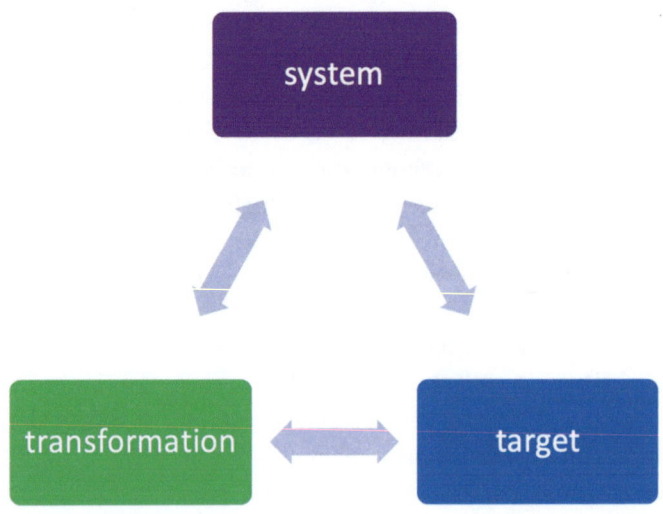

Fig. 1 Collaborative knowledge production (*Source* Authors)

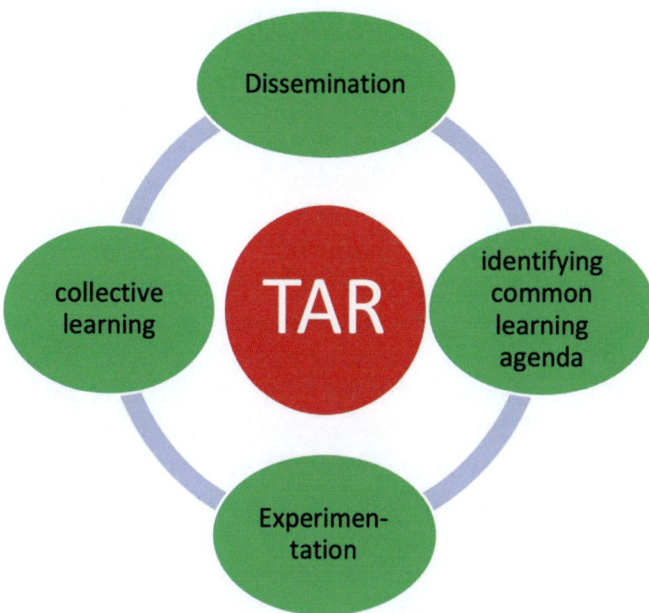

Fig. 2 Transactional action research (*Source* Authors)

- **Setting explicit learning goals** through a "learning agenda" to guide and focus on strategic learning
- **Co-designing the experiment** to achieve a high level of involvement from a diverse range of actors
- **Regular reflection and evaluation** of learning to capture the gained insights
- **Dissemination and embedding** gained insights to integrate them in local governance structures.

4   Prefiguration and Multiple Scaling Processes

The challenge on completing the experimentations in specific settings is to capture and disseminate the learning agendas, while developing a methodological framework that assesses the different scaling processes on their strategic prefigurative potential.

To understand better the processes in prefiguration and its contribution to systemic change, we use the concepts of scale applied in the literature on transformations in socio-ecological systems (Westley et al., 2014). Scaling is widely used in the transition literature using a multi-level perspective (Riddell & Moore, 2015), since it acknowledges the necessity to make simultaneously changes on different levels rather than only on one level. To quote Hermans et al.: "in order to achieve certain goals on one level it may be necessary to also make changes at another level because cause and effects are typically linked at different scales in complex systems" (2016, 286).

Therefore, it is necessary to consider alongside the measures of scaling also on what level (horizontal, administrative) scaling occurs. In our paper, we use four different scaling measures (up, out, deep, through) intersecting with two levels: the horizontal level, that is other stakeholder's initiatives and civil society, and the administrative level, local municipality, and authorities, see Table 1.

The multi-scale of grassroots innovation focuses here on two dimensions, closed and open. The former refers to the collaborative knowledge production within, across, beyond prefigurative initiatives identifying the learning agendas in a prefigurative context (ULLs), while the latter is translated to how these identified learning agendas are adopted and disseminated into the wider society, and what impact these learning agendas have in transforming the food system. Upscaling is concerned with identifying opportunities and barriers within institutional structures to embed prefigured initiatives' innovations and actions, such as generating specific projects; liaising with authorities to secure vital resources, such as public land or funding (Farla et al., 2012; Hermans et al., 2016). Outscaling is defined as efforts to replicate and disseminate programs, products, ideas, or innovative approaches (Riddell &

Table 1 Multi-scale process. *Source* Authors

Different levels of scaling	Practice
Scaling up	It is associated with vertical expansion, aiming to increase market share. The goal is to broaden the base of food citizens, enhancing the economic sustainability of the network and potentially enabling broader expansion
Scaling out	It involves the horizontal expansion of Sustainable Food Networks (SFNs) by replicating or adapting practices, models, or organisational forms across different settings or networks. It relies on spreading initiatives throughout multiple sites while preserving local roots and maintaining flexible interconnections
Scaling deep	It focuses on reinforcing the core values, practices and collective identities that sustain SFNs. It aims to embed sustainability, equity, and community engagement more deeply in everyday practices, promoting lasting cultural and behavioural shifts
Scaling through	It emphasises the political and democratic aspects of transformation. It involves co-creating counter structures within participatory governance structures, such as food policy councils, participatory-guarantee-system, citizen's assemblies or partnerships with public institutions to enhance their legitimacy and influence, aiming to reshape the rules that govern food systems

Moore, 2015) "in order to affect more people or to cover a larger geographical area" (Hermans et al., 2016, 287). This is associated with disseminating to part of society that is yet to be part of the transformation process. This outreach process is linked to deep scaling. Deep scaling is attributed to overriding habits and experiences and to adapt to changing circumstances (Riddell & Moore, 2015; Ohlssen, 2011). The challenges here is to make radical shifts in citizens' behaviour towards consumer citizenship through intersecting with the passive consumer behaviour (Gliessman, 2011; Renting et al., 2012). Food citizenship transforms consumers into citizens by raising awareness of their potential of power as citizens (Renting et al., 2012). Finally, while scaling through is often referred to as a collaborative effort to invoke scaling of other types of scaling, we define scaling through as changing the institutional context in order to achieve the desired systemic transformative change in food system. Prefigurative politics is about igniting radical perspectives, that runs contrary to the current institutional food policy order. Alone the European Common Agricultural Policy is the vehicle for maintaining the neo-liberal status quo in the food system and exports its political agenda through the so-called Free Trade Agreements. The overdue radical transformative shifts in the food system requires new policies structures (Gonzales De Molina, 2013) whereby prefigurative politics can play a significant role. Through the lens of prefiguration counter structures, such as agroecology, a socio-political, and scientific concept, could become a vital tool for alternative food systems.

5   Applying Experimental Prefiguration to the Case of Urban Food Transformations

The ULLs play a central role in the methodological framework of the three-years transnational research project SURFIT—Scaling Urban Regenerative Food Integrated Transition. This project aims to support the scaling of existing food initiatives in Malmo, Krakow, and Trento through convening local actors at ULLs. In each city, six ULL sessions are held over an 18-month period to facilitate collective learning, strategic experimentation for catalysing a systemic food transformation. The challenge in employing a multi-scale strategy raises both conceptual and practical questions: How can fragmented efforts be aligned into a broader collective force? What new strategies might be required? How can existing initiatives be strengthened? What support system is needed for new initiatives to develop?

Most of the participants are individual entrepreneurs, such as flower, soap, or agricultural producers, others are engaged in educational nature programmes or managing an organic kitchen or are cooks, while others are food waste collectors and redistributors, and some are citizens engaged in Community-Support-Agriculture (CSA) projects, in self-organised consumer shops, or in self-organised consumer buying collectives. Participation from municipality varies significantly by city. For example, the municipality of Trento is embedded within the project, in contrast to Maastricht where the local government is more careful at participating at the project.

Despite contextual differences, all four cities share a common aim, namely, to catalyse alternative food systems and overcome substantial structural barriers in the market, finance, policy, and society. Building horizontal network across autonomous food production systems proofs challenging in every city. These networks must navigate and resist a dominant export-oriented model, that shapes consumption habits, production methods, distribution infrastructures, and policy frameworks—while treating food as a speculative object (Ferrando, 2020; Renting et al., 2012).

This tension was clearly articulated during a ULL session in Maastricht, when a local producer mentioned: "I am the only organic food producer in Riemst, Belgium (3km distanced to Maastricht), supplying the urban centre in Maastricht. This wasn't always like this. In the 1950s and 60s we had lots of people that worked in the fields doing horticulture. These people are gone, and with them, the farming knowledge."

The producer's statement captured a widely shared concern across ULL discussion, that also interestingly emerged in the food labs in Trento and Krakow. Farming as a profession has been continuously declining with the neo-liberalisation of agriculture, making it very difficult to commence building a farmer's network or a farmer's market, or even re-establishing farmer's knowledge. The decline of farmers is associated with knowledge loss in farming (Šūmane et al., 2018), and has subsequently effected consumers' knowledge on food, such as on seasonality or farmer's conditions, that has influenced consumption behaviour. Modern food habits revolve around take-aways, dining-out or pre-packaged meals, re-enforcing the model where citizens are primarily treated as consumers (Renting et al., 2012). This is a difficult premise for 'disruptors' to enter into the tightly-controlled relationship between consumers and producers, and build a food system from the bottom-up.

One proposed response to alter consumer behaviour during the Maastricht ULL was to approach local retailers and negotiate for dedicated spaces where regional producers could get visibility and customers receive an alternative for purchasing local products. However, the challenge here is, that the retail industry is electronically connected directly to its suppliers and navigate the demands efficiently, which local and regional small-scale producers cannot meet. Besides, embedding sustainable food production systems into the retail industry poses politically risks. Already the previous organic movement (1965–2000) used the retail industry to increase their market share, which led to the bifurcation of the organic food movement (Fonte & Cucco, 2015), and to the mainstreaming and dilution of organic values, undermining the transformative potential of the movement.

This highlights the importance of collective learning in the ULL, where participants actively interrogate not just practical barriers, but also the conceptualisation of the modern food system. However, one pressing question remains, namely, how can sustainable food initiatives reach sections of society without losing their ecological, economic values, and ethical integrity?

Participants of the Maastricht food lab proposed to enlarge the production capacity on the Community-Support-Agriculture (CSA) fields by establishing an on-site food processing for take-aways. This experiment speaks directly to the needs of busy family lives and could have the ambitious outcome to make an impact on all four scaling levels. CSA's are typically located within a neighbourhood and thus have a high potential for attracting citizens that may not necessarily consider organic, seasonal food as a priority, or is a choice for those citizens who do not have the time to prepare local healthy food and rely on pre-cooked supermarket dinners. However, this strategy is complex and long-term, as one producer noted: "There is no current regulatory framework prohibits to undertake experimentations even though there is no permit system for this process."

This intervention by the producer highlights the policy loopholes in the regulatory framework. It diminishes the capacity of experimenting with possibilities to generate a food system for the citizens, producers and for the climate.

6   Discussion and Conclusion

The two examples discussed demonstrated the complexity of co-designing experiments in ULLs, particularly because they engage multiple levels of scaling. It might be useful to think of these two examples as strategic experiments, since they aim is to transform the food system long-term. Historically, prefigurative politics has often been framed as a short-term disruption, offering "an experiment that breaks with that society, an experiment that will not last, but which allows glimpse of possibility" (Cohn-Benedit in Breinier, 1982, 30). In this framing, experimentation becomes a symbolic act, such as organising a food festival, instead of sustaining the effort and commitment toward systemic change.

This paper has argued that prefiguration when embedded in trans-disciplinary action research, can serve as a strategic framework for transformative change. While prefiguration has been widely used in social movements and in grassroots initiatives, their potential to influence systemic transformation has remained limited because of the absence of a collective structured learning process. In this chapter, we proposed a re-conceptualisation of prefiguration as structured experimentation linked to joint learning agendas. We illustrated how this approach can be operationalised through urban food labs, which function as both, experimental and prefigurative spaces.

However, our preliminary findings from the Maastricht urban food lab highlight the potential and the limitations of prefigurative experimentation. While such spaces act as catalysts for change, their success depends on sustained commitment of citizens and communities, the ability to maintain structured, reflexive, and collaborative learning processes, and the capacity to navigate institutional dynamics without compromising core values. Here, the local municipality can take a more proactive role in supporting and co-developing initiatives with local food actors and rural communities regardless of political limitations.

Competing Interests   The authors have no conflicts of interest to declare that are relevant to the content of this chapter.

References

Antonacopoulou E, & Chiva, R. (2007). The social complexity of organizational learning: The dynamics of learning and organizing. *Management learning, 38*(3), 277–295.

Breines, W. (1982). *Community and organization in the new left, 1962–1968: The great refusal*. Rutgers University Press.

Crossan, M. M., Lane, H. W., & White, R. E. (1999). An organizational learning framework: From intuition to institution. *Academy of Management Review, 24*(3), 522–537.

Farla, J., Markard, J., Raven, R., & Coenen, L. (2012). Sustainability transitions in the making: A closer look at actors, strategies and resources. *Technological Forecasting and Social Change, 79*(6), 991–998.

Ferrando, T. (2020). Financialisation of the transnational food chain: From threat to leverage point? *Transnational food security* (pp. 142–168). Routledge.

Fonte, M., & Cucco, I. (2015). The political economy of alternative agriculture in Italy. *Handbook of the international political economy of agriculture and food* (pp. 264–294). Edward Elgar Publishing.

Gliessman, S. (2011). Transforming food systems to sustainability with agroecology. *Journal of Sustainable Agriculture, 35*(8), 823–825.

Gonzalez de Molina, M. (2013). Agroecology and politics. How to get sustainability? About the necessity for a political agroecology. *Agroecology and Sustainable Food Systems, 37*(1), 45–59.

Hadorn, G. H., Hoffmann-Riem, H., Biber-Klemm, S., Grossenbacher-Mansuy, W., Joye, D., Pohl, C., Wiemann, U., & Zemp, E. (Eds.). (2008). *Handbook of transdisciplinary research* (10: 978-1). Springer.

Hermans, F., Roep, D., & Klerkx, L. (2016). Scale dynamics of grassroots innovations through parallel pathways of transformative change. *Ecological Economics, 130*, 285–295.

Hoeflehner, T., de Kraker, J., Rijkens-Klomp, N., Seravalli, A., Wachtmeister, A., & Illeris, K. (2014). Transformative learning and identity. *Journal of Transformative Education, 12*(2), 148–163.

Kemp, R., & Scholl, C. (2016). City labs as vehicles for innovation in urban planning processes. *Urban Planning, 1*(4), 89–102.

Laal, M., & Ghodsi, S. M. (2012). Benefits of collaborative learning. *Procedia-Social and Behavioral Sciences, 31*, 486–490.

Macintyre, T., Lotz-Sisitka, H., Wals, A., Vogel, C., & Tassone, V. (2018). Towards transformative social learning on the path to 1.5 degrees. *Current Opinion in Environmental Sustainability, 31*, 80–87.

Maeckelbergh, M. (2009). *The will of the many: How the alterglobalisation movement is changing the face of democracy.* Pluto Press.

Nelson, R. R., & Winter, S. G. (1985). *An evolutionary theory of economic change.* Harvard University Press.

Ohlsson, S. (2011). *Deep learning: How the mind overrides experience.* Cambridge University Press.

Pahl-Wostl, C. (2009). A conceptual framework for analysing adaptive capacity and multi-level learning processes in resource governance regimes. *Global Environmental Change, 19*(3), 354–365.

Reed, M. S., Evely, A. C., Cundill, G., Fazey, I., Glass, J., Laing A., Newig, J., Renting, H., Schermer, M., & Rossi, A. (2012). Building food democracy: Exploring civic food networks and newly emerging forms of food citizenship. *The International Journal of Sociology of Agriculture and Food, 19*(3), 289–307.

Riddell, D., & Moore, M. L. (2015). *Scaling out, scaling up, scaling deep.* McConnell Foundation. JW McConnell Family Foundation & Tamarack Institute, Waterloo, Ontario.

Scholl, C., Agger Eriksen, M., Baerten, N., Clark, E., Drage, T., Essebo, M., Scholl, C., De Kraker, J. D. (2021). The practice of urban experimentation in Dutch city labs. *Urban Planning, 6*(1), 161–170.

Scholl, C., & De Kraker, J. D. (2021b). Urban planning by experiment: Practices, outcomes, and impacts. *Urban Planning, 6*(1), 156–160.

Šūmane, S., Kunda, I., Knickel, K., Strauss, A., Tisenkopfs, T., des Ios Rios, I., Rivera, M., Chebach, T., & Ashkenazy, A. (2018). Local and farmers' knowledge matters! How integrating informal and formal knowledge enhances sustainable and resilient agriculture. *Journal of Rural Studies, 59*, 232–241.

Van Mierlo, B., & Beers, P. J. (2020). Understanding and governing learning in sustainability transitions: A review. *Environmental Innovation and Societal Transitions, 34*, 255–269.

van Oers, L., Smessaert, J., & Feola, G. (2024). Initiating transformation within a Dutch grassroots agri-food initiative: An analysis of social processes. *Sociologia Ruralis, 64*(4), 571–591.

Van Poeck, K., Östman, L., & Block, T. (2020). Opening up the black box of learning-by-doing in sustainability transitions. *Environmental Innovation and Societal Transitions, 34*, 298–310.

Voytenko, Y., McCormick, K., Evans, J., & Schliwa, G. (2016). Urban living labs for sustainability and low carbon cities in Europe: Towards a research agenda. *Journal of Cleaner Production, 123*, 45–54.

Westley, F., Antadze, N., Riddell, D. J., Robinson, K., & Geobey, S. (2014). Five configurations for scaling up social innovation: Case examples of nonprofit organizations from Canada. *The Journal of Applied Behavioral Science, 50*(3), 234–260.

Wlasak, P. (2017). *Guidelines for urban labs.* https://adk.elsevierpure.com/ws/files/61301561/Scholl_et_al_2017_GUIDELINESforURBAN_LABS_U RBExp_FINAL.pdf. Accessed May 10, 2025.

Yates, L. (2015). Rethinking prefiguration: Alternatives, micropolitics and goals in social movements. *Social Movement Studies, 14*(1), 1–21.

Open Access This chapter is licensed under the terms of the Creative Commons Attribution-NonCommercial-NoDerivatives 4.0 International License (http://creativecommons.org/licenses/by-nc-nd/4.0/), which permits any noncommercial use, sharing, distribution and reproduction in any medium or format, as long as you give appropriate credit to the original author(s) and the source, provide a link to the Creative Commons license and indicate if you modified the licensed material. You do not have permission under this license to share adapted material derived from this chapter or parts of it.

The images or other third party material in this chapter are included in the chapter's Creative Commons license, unless indicated otherwise in a credit line to the material. If material is not included in the chapter's Creative Commons license and your intended use is not permitted by statutory regulation or exceeds the permitted use, you will need to obtain permission directly from the copyright holder.

24

"Mercato Brado" and Future-Making in the Time of Toxics

Bianca Griffani

1   Introduction

Building on long-term fieldwork in a steel-town in central Italy, in this contribution I try to think prefigurative politics through the latent presence of toxicants in the soil, air, and water of places wasted by industrial production. In dialogue with Ernst Bloch's philosophy of Hope, as recently reconstructed by Ana Cecilia Dinerstein (2017, 2022) in relation to prefiguration, and with Kim Fortun's notion of "late industrial" (Fortun, 2012, 2014, 2020) as the longue durée in which the socio-ecological injuries of high-industrialism become manifest, I wish to reflect particularly on the implications that the unassumed temporal work of toxicants has for political projects that aim to embody in the present the future socio-ecological relations they—we—desire.

B. Griffani (✉)
Department of Anthropology, Goldsmiths College, University of London, London, UK
e-mail: bgrif004@gold.ac.uk

© The Author(s) 2026

S. Sareen and S. Juhola (eds.), *Societal Transitions to Sustainability*, Palgrave Studies in Environmental Transformation, Transition and Accountability,
https://doi.org/10.1007/978-3-032-07395-2_24

I explore this question through the experience *Mercato Brado* [en: Feral Market], an agroecological network active between 2011 and 2020 in Terni, where I grew up, and where I also conducted my doctoral fieldwork between 2021 and 2023. I focus on a specific prefigurative device, not invented but perfected by my interlocutors: the "Participated Guarantee System" [it: *Garanzia Partecipata*, heretofore 'PGS']. I argue that, through the PGS, my interlocutors negotiated and mediated exposure to contamination through food consumption, as an intentional and reparative act, fundamentally different from the forced subjection to harm from pollutants which characterises the everyday experience of late industrialism in Terni. Thus, through exposure, new relationships and common meanings were bodied-forth and made present.

2   Future-Making in the Time of Toxics

As this edited collection demonstrates, prefiguration, as a mode of organisation and practice that seeks to actualise desired futures, has multiple genealogies and present uses. In this contribution, I draw on the paradigm of prefigurative politics that emerges from Marxist, decolonial, and abolitionist movements, where prefigurative praxis is positioned as antagonistic to the "rule of myth" of the present in which it unfolds (Benjamin, 2004, p. 252)—a collective act of "venturing beyond" hetero-sexist, colonial capitalism (Dinerstein, 2022, p. 51). Ana Cecilia Dinerstein argues that such prefigurative politics are necessarily premised on an "ontology of the not-yet-being" that recognises us and the world as being in a process of becoming shorn of teleological commitments. Prefiguration is thus anticipatory of a possible which *must be* unexhausted and inexhaustible, in the understanding that, as the Marxist philosopher of Hope Ernst Bloch wrote, "there is no true realism without the true dimension of the openness of the world" (Bloch in Dinerstein, 2022, p. 50).

For Dinerstein, Bloch's non-linear, dialectical interpretation of time, where unrealised future possibilities inhabit latently both past(s) and present(s), aligns closely with the tempor-ontological pluralism underpinning decolonial "concrete utopias"—as Bloch might have called them—like the Zapatista self-governed communities of Chapas, insurrections against the homogenising abstractions of capitalist contemporaneity (Dinerstein, 2022, pp. 57–58). Latency, however, is not only the quality of an unmeasured *not-yet*. Ethnographers of wasted ecologies like Pine (2019) and Tironi (2023) have documented how, in the ruins of colonial industrial modernity, prefigurative and/or reparative projects are necessarily undertaken in dialogue with the ghostly presence of capitalist systems of production, materialised in polluted waters, soil, and air, as well as in social relations shaped by proletarianisation, exploitation, dispossession, and difference.

These "toxic" externalities often outlive the processes of value-production which they are integral to but artificially separated from—an "away to keep the centre clean" (Liboiron & Lepawsky, 2022, p. 21). They linger on, not merely as fossil records of the unstable materiality of industrial life, populated by noxious substances that leak, migrate, and trespass fencelines and flesh, relatively undisciplined by states and supranational entities. Rather, as the anthropologist Kim Fortun suggests, they exist as late-industrial "subalterns" (Fortun, 2014, p. 315), constantly engaged in often-illegible labours of future-making of their own—what Manuel Tironi, from his vantage-point on the Tubul-Raqui wetlands, describes as the "accretional and unspectacular damage" of toxicants, "accumulating slowly and below the threshold of technocratic perceptibility" (Tironi, 2023, p. 41).

Fortun theorised late industrialism thinking through her experience of advocacy in the aftermath of the 1984 Bhopal disaster, a mass-casualty industrial incident in Madhya Pradesh, India, where a leak in a pesticide factory—majority-owned by the American conglomerate *Union Carbide Corporation*—exposed over half a million residents to the poisonous gas methyl isocyanate. Her explicit aim was to "[weave] the problem of toxicity into sociocultural theory" (Fortun, 2016) in a way that is attuned to the particular temporalities of toxicants, which in inhabiting the body lay the ground for a future not yet manifest as disease. For Fortun, understanding toxicity thus requires a concept of time aligned to the "future anterior" theorised by Jacques Derrida after Emmanuel Levinas, where "the past is folded into the way reality presents itself, setting up both the structures and the obligations of the future" (Derrida in Fortun, 2012,

p. 450). I offer this anteriorised future as a complement to Bloch's dialectical, non-synchronous temporality proposed by Dinerstein as a shield and a weapon against the synchronising violence of globalised capitalism and its ontological fictions. Both temporal framings are multilayered, and express a speculative, non-teleological materialism. However, I suggest that Fortun's bid for a politically useful conception of time and change might afford us more precise tools to engage the concreteness of prefiguration in eco-systems "always weighted (and often soiled) by history" (Fortun, 2014, p. 323). Note that I use the term "concrete" in the meaning articulated by Marx in the *Gundrisse*, as the "wholeness" of an organic system of mutually conditioning phenomena, in contradiction to the "abstract" understood as a partial, incomplete reflection of concrete phenomena by way of thought.

Might we not combine ontological openness with a materialist ethics of historical responsibility, and give up nothing? Afterall, concrete utopias—"concrete action toward the anticipation of the *not-yet*" (Dinerstein, 2017)—are enformed by a commitment to radical openness, as by the historically-produced structures and systems they critique (themselves mattered through by a myriad agencies, human, and more-than-human, c/f Hugøy 2025). To illustrate this point, I turn to Terni, one of those once-thriving industrial centres which have become peripherical to the circuits of productive capital and thus come to be understood as wasted in themselves. Here, for a good decade, a motley crew of "outlaw" farmers imagined and practiced other ways of being in the world and of it, other forms of labour and cooperation, and other means of attending to the interdependencies involved in living on and from the land. All of this, they did in and against the toxic inheritances of one-hundred-and-forty years of steelmaking. While studying local mutual-aid projects in the covid-19 pandemic, I got to know them, and some of the comrades and fellow travels they once called "co-producers" [it: *coproduttori*]. This essay draws on our shared practice of "deep hanging out" (Rosaldo in Clifford, 1996), and on a series of unstructured interviews carried out over 2021 and 2022.

3   Specters of Value in Terni

Imagine a community which is known, if it is known at all, and most importantly has come to *know itself*, by its factory. From the 1880s until the 1970s steel crisis, when the workforce started to shrink and town-life began to lose its centre, *Terni Special Steel,* one of the first and largest steel mills in the country, controlled everyday life and most of its institutions in Terni. Politics, public life, work, and leisure, even family life, were organised by the time and relationality of industrial labour. So long as work paid off and the future looked bright, imagination too submitted to this "foreign potentate", so that "for years, we lived exclusively for [the factory]" (Portelli, 2017, p. 141). Yet the processes which made a rural market town of a few thousands inhabitants into a one-hundred-thousand-strong "city of steel" [it: *città dell'acciaio*] in little more than fifty years were already producing Terni as a "sacrifice zone": a place whose residents must pay the price, in both health and economic terms, for the imperatives of national development (Lerner, 2010).

Our oak-trees hold the traces of siderurgical production, archiving high concentrations of mercury, lead, thallium, chromium, and nickel in the soil (Perone et al., 2018). Even now, the steel-mill admits to releasing between 1 and 1.5 tonnes of carcinogenic substances in the air per year—mostly hexavalent chromium and nickel, but also dioxins and polychlorinated biphenyls (among the highest volumes in Europe by an industrial group), arsenic, and lead (Romagnoli, 2019). In the 1990s, amid one of many crises of production, rapidly depreciating land was designated for waste disposal and put to profit. The construction of a special and hazardous waste landfill and of two incinerators in public–private-partnerships benefited waste capitalists, while exposing residents to even higher levels of dioxins. Today, with only one incinerator in activity, about thirty-thousand people in the suburbs breathe air deemed "harmful" by the same public authorities that are currently brokering deals for new (dioxin-producing) waste-to-energy plants (see Comune di Terni, 2019).

Local environmental activists and advocates reunited around the *Comitato NO INC* [en: Committee Against the Incinerators]—many of whom were also involved in *Mercato Brado*—have denounced how regulatory governance protects the production of value (and risk) above all, so that industrial polluters like the steel-mill are allowed to introduce into air, rivers, soil, subsoil, and groundwater, significant quantities of substances known for their carcinogenic, mutagenic, epigenotoxic, and toxic qualities, without accounting for the accretion and interaction of their impacts on the human and non-human lives facing exposure (Romagnoli et al., 2020). The spectral presence of toxicants and of their compound effects, unseen by regulatory apparatuses, has other means of making itself known. The latest *National Epidemiological Study of Territories and Settlements Exposed to Pollution Risk* (SENTIERI 6) signals excess incidence of diseases associated with environmental exposure in the Terni valley, including all malignant tumours, lung cancers, bladder cancer, and breast cancer, as well as kidney diseases and kidney failures in women, with excess mortality estimated between 40% and 68% for some types of carcinomas and sarcomas (Zona et al., 2023, pp. 173–175).

Socialised pollution and its injuries are not the only negative "externalities" which people have come to accept as a function of late industrialism in Terni. With the decline of local steel and chemical production, the city has experienced significant capital flight, diminishing returns from waged work, and a rise in household indebtedness, as industrial labour—itself increasingly de-skilled and flexibilised—is partly replaced with employment in social care and service work, where pay is low and contractual precarity is commonplace. Turistification is routinely offered as a remedy for industrial displacement, but its promises have never come to fulfilment—prompting local politicians to double down on defending *Terni Special Steel* as the city's permanent centre of gravity. As the youth emigrates, and the local population ages and dwindles, the future appears limited.

4 *Mercado Brado*'s Prefigurative Agroecology

Still, some do try to summon a sense of futurity, something like hope (Pine, 2019, p. 12). *Mercato Brado* [en: Feral Market] was a network of food growers and consumers born in 2011 in Terni from the meeting of a heterogeneous group of people interested in food sovereignty, agroecology, non-capitalist economies, re-ruralisation, and self-sufficient living. Most of its founding members were coming of age in the late 1990s and early 2000s, the years of the *"Pacchetto Treu"* and *"Legge Biagi"* reforms that ushered in labour deregulation and flexibilisation in the country—a radicalising experience. Coming from working-class and lower-middle-class households, many completed a university degree but found little waiting for them in the jobs market. Disillusioned with the work society, and looking for alternative livelihood strategies, like Dinerstein's Zapatistas my interlocutors turned to the pre-industrial past, not as a way of looking "backwards", but rather as a reference point from which to imagine different possible modes of dwelling in a territory to which industrial modernity had been so injurious.

Before founding *Mercato*, most of my interlocutors had encountered, either by chance or by design, *Genuino Clandestino* [en: Genuine Clandestine]. This is an alliance of small farmers created in 2010 to fight against new legislation which de-facto illegalised low-tech 'peasant' farming and food-production by subjecting their products to the same regulatory requirements as industrially processed foods. The network was composed of territorial nodes supporting small producers and organising clandestine markets for their now illicit goods—habitually held in one of the many squatted community centres still to be found around Italy. As my informants put it to me, *Genuino Clandestino* fought against a system that had made living from the land into "a rich people's business (…) so that if you don't have family money or access to lines of credit, you should just give it up" (B). The point "was not to romanticise our 'outlaw' status, but to highlight that clandestinity is a condition which the state had imposed on us" by illegalising "small artisanal food production practices which do not require start-up capital investment to the tune of one-hundred-thousand-euros" (E).

Started as a national campaign to reclaim an alienated right, in the years since, the network has become a focal actor in domestic and European struggles to transform the economic and normative contexts of organic food production "from below", as well as a vital infrastructure for the circulation of knowledges and resources that sustain experiments in commoning and reparative forms of rural living. *Mercato Brado* was thus born to bring home "on soiled grounds" (Fortun, 2020, p. 113) the practices its founding members had scouted in the squatted communal farm of *Mondeggi* near Florence (Ghelfi, 2023) or in the self-managed market *Mercati Aperti* [en: Open Markets] in Bologna (Diesner, 2020), but also to preserve and re-activate the local peasant biocultural heritage. Terni's "outlaw farmers" first met around the Occupied Autonomous Social Centre *Germinal Cimarelli*, less than a mile from the *Terni Special Steel* plant. This is the city's only grassroot community centre, a collective space created in 2009 by anarchist and left-communist activists and militant football fans out of the paraffin and tar-saturated ruins of the Gruber industrial wool mill. The *Cimarelli* is also where I met them in the Summer of 2021, a couple of years after *Mercato Brado* had ceased to exist as a network of co-production and exchange, yet another casualty of the Covid-19 pandemic.

Mercato Brado was meant to achieve several things: connecting local "peasant" farmers and artisanal producers to comrades across the country, and to the city's residents; in so doing, creating a network of exchange that might sustain the illegalised but reparative practices of *Mercato Brado*'s producers; giving access, to those who chose re-ruralisation having only ever lived in proletarianised communities, to the peasant knowledges they wished to body forth, through courses, workshops, and mentorship schemes. All of this depended on acts of companionship, especially when it came to nurturing a self-organised and stable food community that could provide farmers with a market for their goods and thus with the income needed to continue working the land. A typical market day, held in the garden outside of the social centre, or in the public park adjoining a nearby social housing estate, included diffuse moments of sociality—for chit-chatting and day-drinking—and always culminated in a social meal that was cooked collaboratively off the markets' left-over produce, and eaten together while sitting down at a long table. This practice radically diverges from the mundane, alienated and alienating experience of purchasing food-as-commodity—that is, food abstracted from the social relations of its production—in a store, be it a discount market or bespoke organic shop.

5   Facing Toxicity with Care

Collective moments of sociality and exchange like the one I have just described were also necessary for the production of new common meanings—that shared sense of our social life we can articulate in speech, which both reflects and shapes the grounds of political possibility (Riofrancos, 2020, p. 191). My interlocutors explicitly contrasted "*vivere la campagna* [en: to live the land]" to "*vivere in campagna* [en: to live in the countryside]"—the latter being what "plenty of people have always done" in a rural-industrial county like Terni, a modern way of being in the rural without paying mind to what is in the air and the soil, to what the land gives spontaneously, or to what it might need and give in return. Involving urbanite consumers into "living the land" as a mode of dwelling meant negotiating a shared understanding of what food is really *worth* when it is re-conceived as an agent of socio-ecological transformation. A key instrument of this process of resignification, at the centre of *Mercato Brado*'s self-governance structure, was the Participated Guarantee System (PGS) [*it: Sistema di Garanzia Partecipata*].

The PGS was, in short, a collaborative methodology to determine the "provenance" [it: *provenienza*] of the produce sold in the market—where provenance indicates the sum-total of the processes and agencies which accumulate in a bottle of oil, in a bag of dried chickpeas, or in a jar of honey sold in the market. In *Mercato Brado*, "provenance" determines the social value of food. It ought to index, in as much as possible, the resources, agencies, techniques, and forms of labour which have participated in the becoming of a specific agricultural product form its beginning. Thus, price, as a democratic valuation of provenance, must account for "workers's rights and [for] the actual value of their labour (…) [for] the organisation of production in the farm" but also "[for] an ethics of land repair" (F).

Every time a new producer joined the network, a delegation comprising of both those who make and buy "feral" goods would carry out a comradely inspection. Together, host(s) and guests would visit the site and do some work—herding animals, pruning vines, or making cheese. In labouring together, there would be time to talk about the soil, water sources, the beasts, or the techniques employed. These conversations would often carry over into the time of rest, when people would sit around a collective lunch or snack [*it: merenda*]. This was a moment too for "getting to know each other as people" (F). Moving through the Participated Guarantee System thus transformed consumers into "*co-producers* (…) people who do not only come to buy something from you, but who inform the choices you make as a producer (…) and enable them" (F).

On soiled grounds like ours, the list of commensals that participate in every act of growing necessarily includes toxicants, and these too were credited for their role as co-actors through the PGS. However, proximity to or compresence with toxicants would not necessarily determine the price of a product to be "low", for it does not take away from its social value. In fact, food that is grown on toxic ground might be very valuable, where it is the outcome of regenerative or restorative agricultural practices. Armed with all the knowledge conveyed in the PGS, consumers-*cum*-co-producers could participate in the realisation of alternative economic and ecological relations through intentional acts of exposure—for example, by purchasing produce grown in the steelworks' deposition zone to support a farmer's phytoremediation project. These acts of exposure stand in contrast with the imposed—and often unacknowledged—subjection to harm from pollutants that characterises life in late industrial contexts. In this sense, co-producers' practices are also radically different from those of the ethical "citizen-consumer who carries the responsibility, as well as the risk, of making wise spending choices as an independent individual" (Kosnik, 2018, p. 127)—because co-producers' sense of what is wise, and what one ought to be responsible for, is attained through collective and intentional processes of meaning-making.

The ethics of repair of *Mercato Brado*, resting on collective responsibility rather than individual choice, was not only oriented towards caring for the environment as a common. Indeed, *Mercato Brado* was one of the few territorial nodes in *Genuino Clandestino* to include "economic need" in the list of indexes determining the price of a product, in recognition of the harsh material conditions against which people in Terni were trying to make and keep life (F). Higher economic need—say, the fact of having lost one's job and/or being responsible for dependants—might justify a higher price. In doing so, my interlocutors confronted together two phenomena which are presented as inarticulate in emic theories of deindustrialisation, namely economic precarity and environmental injury, habitually cast as opposing terms in the false dichotomy between economic growth and ecosystemic health.

6   Conclusion

In this contribution, thinking with the work of Ana Cecilia Dinerstein and Kim Fortun, I have sought to theorise the prefigurative agroecological practices of *Mercato Brado* as a hopeful act of venturing beyond the present, well attuned to the concrete realities of toxicity in a late industrial setting like Terni. I have argued that my interlocutors' praxis invented and maintained new relationships of production and exchange, as well as new common meanings—about food, and about industrial production and its eco-social legacies. Their efforts suggest that prefiguration is about conjuring and presencing the not-yet-there, as much as it is about attending to what is there but will not announce itself—those bad omens that are still so hard to look at, and even harder to name. In this sense, Terni's "feral" farmers offer a concrete utopia, powerful in its insistence that even on soiled ground, new worlds can take root.

Acknowledgements The author would like to thank Alexandra Dragin, João Rocha Gomes, Tyler Schuenemann, and the editors of this volume for their generous feedback on an earlier draft of this chapter. Deepest gratitude goes to the members of Mercato Brado for generously sharing their food, time, recollections, and reflections with me.

Competing Interests The author declares no conflict of interest in relation to the research, authorship, or publication of this chapter.

References

Benjamin, W. (2004). *Walter Benjamin: Selected writings* (M. P. Bullock & M. W. Jennings, Eds.). Belknap Press of Harvard University Press.

Clifford, J. (1996). Anthropology and/as travel. *Etnofoor, 9*(2), 5–15.

Comune di Terni. (2019). *Relazione sull'attività di indagine della IV Commissione Consiliare Garanzia e Controllo* (No. 1). Comune di Terni.

Diesner, D. (2020). A revolution under our feet: Food sovereignty and the commons in the case of Campi Aperti. In *A revolution under our feet: Food sovereignty and the commons in the case of CampiAperti, Italy*. https://doi.org/10.4324/9780429021886-6

Dinerstein, A. C. (2017). Concrete utopia. *Public Seminar*. https://publicseminar.org/2017/12/concrete-utopia/

Dinerstein, A. C. (2022). Decolonizing prefiguration: Ernst Bloch's philosophy of hope and the multiversum. In L. Monticelli (Ed.), *The future is now: An introduction to prefigurative politics* (Alternatives to Capitalism in the 21st Century). Bristol University Press. https://doi.org/10.56687/9781529215687-009

Fortun, K. (2012). Ethnography in late industrialism. *Cultural Anthropology, 27*(3), 446–464. https://doi.org/10.1111/j.1548-1360.2012.01153.x

Fortun, K. (2014). From Latour to late industrialism. *HAU: Journal of Ethnographic Theory, 4*(1), 309–329. https://doi.org/10.14318/hau4.1.017

Fortun, K. (2016). Essentially late industrial. *Toxic: A Symposium on Exposure, Entanglement, and Endurance*. http://www.toxicsymposium.org/conversations-1/2016/3/2/essentially-late-industrial

Fortun, K. (2020). To fieldwork, to write. In C. McGranahan (Ed.), *Writing anthropology: Essays on craft and commitment*. Duke University Press. https://doi.org/10.1215/9781478009160

Ghelfi, A. (2023). New peasantries in Italy: Eco-commons, agroecology and food communities. In D. Papadopoulos, M. Puig de la Bellacasa, & M. Tacchetti (Eds.), *Ecological reparation: Repair, remediation and resurgence in social and environmental conflict*. Policy Press. https://doi.org/10.1332/policypress/9781529239546.003.0014

Kosnik, E. (2018). Production for consumption: Prosumer, citizen-consumer, and ethical consumption in a postgrowth context. *Economic Anthropology, 5*(1), 123–134. https://doi.org/10.1002/sea2.12107

Lerner, S. (2010). Sacrifice zones: The front lines of toxic chemical exposure in the United States. *MIT Press*. https://doi.org/10.7551/mitpress/8157.001.0001

Liboiron, M., & Lepawsky, J. (2022). Discard studies: Wasting, systems, and power. *MIT Press*. https://doi.org/10.7551/mitpress/12442.001.0001

Perone, A., et al. (2018). Oak tree-rings record spatial-temporal pollution trends from different sources in Terni (Central Italy). *Environmental Pollution, 233*, 278–289. https://doi.org/10.1016/j.envpol.2017.10.062

Pine, J. (2019). The alchemy of meth: A decomposition. *University of Minnesota Press*. https://doi.org/10.5749/j.ctvthhd6p

Portelli, A. (2017). *Biography of an industrial town: Terni, Italy, 1831–2014* (Palgrave Studies in Oral History). Springer International Publishing. https://doi.org/10.1007/978-3-319-50898-6

Riofrancos, T. (2020). Resource radicals: From petro-nationalism to post-extractivism in Ecuador. *Duke University Press*. https://doi.org/10.1215/9781478012122

Romagnoli, C. (2019). Report to the Parliamentary Commission of inquiry into illegal activities related to the waste cycle and related environmental crimes, Terni Mission. ISDE.

Romagnoli, C., Neri, F., & Guarducci, A. R. (2020). Condivisione e sviluppo di un modello per la prevenzione primaria territoriale: l'ecodistretto. *SISTEMA SALUTE, 64*(3).

Tironi, M. (2023). Hesitant: Three theses on ecological reparation (otherwise). In D. Papadopoulos, M. Puig de la Bellacasa, & M. Tacchetti (Eds.), *Ecological reparation: Repair, remediation and resurgence in social and environmental conflict*. Policy Press. https://doi.org/10.1332/policypress/9781529239546.003.0003

Zona, A., et al. (2023). SENTIERI 6. https://epiprev.it/pubblicazioni/sentieri-studio-epidemiologico-nazionale-dei-territori-e-degli-insediamenti-esposti-a-rischio-da-inquinamento-sesto-rapporto

Open Access This chapter is licensed under the terms of the Creative Commons Attribution-NonCommercial-NoDerivatives 4.0 International License (http://creativecommons.org/licenses/by-nc-nd/4.0/), which permits any noncommercial use, sharing, distribution and reproduction in any medium or format, as long as you give appropriate credit to the original author(s) and the source, provide a link to the Creative Commons license and indicate if you modified the licensed material. You do not have permission under this license to share adapted material derived from this chapter or parts of it.

The images or other third party material in this chapter are included in the chapter's Creative Commons license, unless indicated otherwise in a credit line to the material. If material is not included in the chapter's Creative Commons license and your intended use is not permitted by statutory regulation or exceeds the permitted use, you will need to obtain permission directly from the copyright holder.

25

The Role of Art Activism in the Prefigurative Politics of Food System Transformation

Sarah Milliken

1 Introduction

In 1982 artist Agnes Denes created *Wheatfield—A Confrontation* on a two-acre site on the Battery Park landfill in Lower Manhattan. Together with a group of volunteers she prepared the soil, planted the seeds, fertilised, weeded and irrigated the crop, harvested it four months later, and sent the seeds to art galleries around the world as part of The International Art Show for the End of World Hunger (1987–1990). Visitors to the exhibits could take the seeds away and plant them, thereby linking the public act of cultivation to a global community of land stewards. Contrasting sharply with the surrounding Wall Street financial district and its associations with inequality and exploitation, *Wheatfield* was the embodiment of alternative values, 'a symbol, a universal concept; it represented food, energy, commerce, world trade, and economics. It referred

S. Milliken (✉)
School of Design and Creative Industries, University of Greenwich, London, UK
e-mail: S.Milliken@greenwich.ac.uk

© The Author(s) 2026
S. Sareen and S. Juhola (eds.), *Societal Transitions to Sustainability*, Palgrave Studies in Environmental Transformation, Transition and Accountability,
https://doi.org/10.1007/978-3-032-07395-2_25

to mismanagement, waste, world hunger and ecological concerns. It called attention to our misplaced priorities' (Denes, 1982). *Wheatfield* therefore served as both a critique of urban capitalist priorities and a vision of an alternative future where agriculture, nature, and human values coexist harmoniously within cities. Despite being a temporary project, *Wheatfield* has entered the canon of art history and has inspired numerous artists to use their work to advance a critique of our food system and actively engage in the search for solutions.

Adopting the definition of prefigurative politics as 'embodied strategies to render desirable futures with immediacy' (Sareen and Juhola, this volume), I explore the role of art activism in the prefigurative politics of food system transformation. Art activism refers to creative practice that intentionally addresses political or social issues, aiming to challenge existing power structures and bring about change. The relationship between art and the politics of food is multifaceted and deeply rooted in how both food and art serve as powerful media for social commentary, resistance, and community engagement. While various authors have started to explore the role of art in prefigurative politics, or the role of prefigurative politics in food system transformation, the disciplinary spaces in which they are discussed and theorised are largely separate, thereby creating a boundary that may hinder productive dialogue and mutual learning. This chapter therefore offers an original contribution by combining the two strands of enquiry.

First, I briefly summarise the nature of the problem with our food system and present Meadows' (1999) leverage points framework which provides a structured way to think about where to intervene in a system to achieve transformative change. I then outline how prefigurative politics manifests in food systems and how these initiatives can be mapped onto the leverage points framework. The next section examines how creative practice has unique potential to engage with the most transformative leverage points by focusing on fundamental myths, paradigms, and systems of meaning-making. This is followed by a brief overview of the literature on creative practice and prefigurative politics. In the final section I bring these various elements together to analyse *Wheatfield—A Confrontation* through the lens of the prefigurative politics of food system transformation.

2   Leverage Points in the Food System

It is widely acknowledged that the current food system generates problematic, undesired, and often unplanned consequences—such as environmental degradation, economic exploitation, malnutrition, food insecurity, increased inequalities, and poverty—and that a transformation towards sustainability is needed. This will be a complex process that must involve changes in many different areas—including consumption, production, trade, and governance—at multiple levels (local, subregional, national, and international) and for a wide variety of stakeholders (producers, traders, processors, retailers, consumers, and policy makers).

While the term 'food system transformation' is widely used, its meaning often becomes vague or co-opted, and it is used to refer to various types of change, ranging from minor adaptations to radical overhauls (Ingram & Thornton, 2022). Answers to questions about how to change (strategies) or what to change (goals and desired outcomes), and by whom (agents of change), are not always made explicit. While there is a growing recognition that addressing food systems demands systems-based approaches which integrate socio-technical transitions or social-ecological transformation theories, not all studies adopt approaches sensitive to systemic features like complexity and feedback loops (Juri et al., 2024).

In her hierarchy of intervention points for leveraging change, Meadows (1999) argued that the transformational capacity of a given intervention would depend on the characteristics of the system properties that a given intervention acts upon, with some interventions likely to cause transformational change while others will only induce minor changes in outcomes. The twelve intervention points are ranked, progressing from shallow interventions at the top of the hierarchy (easy but with limited impact) to deep interventions at the bottom (challenging but potentially transformative). The least impactful interventions are those that address parameters—the relatively mechanistic characteristics typically targeted by policy makers. These are followed by interventions that address feedbacks—the interactions between elements within a system that drive internal dynamics—and design—the social structures and institutions that manage feedbacks and parameters. The most impactful interventions are those that address intent—the underpinning values, goals, and worldviews of actors that shape the emergent direction to which a system is oriented (Abson et al., 2017).

The leverage points framework is a meta-perspective that integrates two frequently conflicting perspectives—causality at the top of the hierarchy, and teleology at the bottom—and provides a conceptual space where quite fundamentally different modes of thinking can meet. By recognising the joint importance of both teleology and causality as mechanisms of change it provides a practical way of holistic thinking and intervening in systems by paying attention to what is important, not just what is quantifiable (Fischer & Riechers, 2019).

An intervention is rarely a single intervention at a single leverage point, within a single system (Leventon et al., 2021). Shallow and deep interventions may interact through 'chains of leverage', with one type of change in a system precipitating another across different depths of leverage. If such chains do extend to deep leverage points, then a given chain of leverage has the potential to bring about transformative change. In contrast, a chain that only involves shallow leverage points is unlikely to effect transformation (Fischer & Riechers, 2019).

Abson et al. (2017) highlighted the importance of three realms of deep leverage for sustainability transformations—reconnecting people to nature, restructuring institutions, and rethinking how knowledge is created and used in pursuit of sustainability—because of the strong interactions within and between them. For example, changing how knowledge is produced might influence mid-level (feedback, design) and deep (intent) leverage points.

Davelaar (2021) argued that defining broad realms of deep leverage does not bring into sharp focus the most influential and deepest leverage points—paradigm and the power to transcend it—which are the final (teleological) root cause of unsustainability. When the goal of transformation is sustainable practice, it is in the understanding of sustainability that a paradigmatic shift is needed. This requires a radical shift in our worldview towards one which reconceives the human–world bond through the systemic lens of dynamic inclusion, aliveness, purpose and value, and a corresponding change in how this experience is translated into ways of thinking and doing.

Many sustainability interventions target highly tangible but essentially ineffective leverage points, thus there is an urgent need to focus on less obvious but potentially far more powerful ones (Abson et al., 2017). The promotion of 'technical fixes' to address what are often complex multi-dimensional problems implies that sustainability problems can be resolved without consideration of the underpinning structures, values and goals. Policy interventions, for example, typically target shallow leverage points. This is apparent in the many policy instruments that focus on simply adjusting parameters, for example by setting targets or providing financial incentives within existing structures. While such interventions are important and can generate beneficial outcomes, on their own they are unlikely to lead to transformational change. For maximum impact, the transformation strategies should combine deep interventions (shifting goals/rules) with shallow ones (technical fixes) to create synergistic change (Abson et al., 2017).

A balanced approach that combines shallow, mid-level, and deep interventions is therefore necessary for meaningful food system trans-formation. However, scholarly works (see Juri et al., 2024 for a recent review) reveal a predominant focus on mid-to-shallow leverage points, with emerging but limited engagement with deeper systemic interven-tions. Shallow leverage points include providing financial incentives for organic farming practices, improving food supply chain logistics, and promoting precision agriculture technologies. Mid-level interventions include enforcing carbon pricing in agriculture, creating participatory governance structure in food systems, and strengthening local food networks. While these can create significant systemic shifts, they often stop short of addressing root causes. Deep leverage points such as fostering shared values about food sovereignty and justice, for example by educating communities about the importance of equitable access to nutritious food and empowering them to demand systemic change, which in turn has the potential to shift societal mindsets and paradigms towards equity and sustainability, are rarely addressed. Dominant scien-tific discourses and policy interventions tend to mutually reinforce one another, such that shallower interventions are also favoured in policy at the expense of deep leverage points (Dorninger et al., 2020).

3   The Prefigurative Politics of Food System Transformation

In the context of food system transformation, prefigurative politics refers to actions and practices that embody the principles, values, and structures of a desired future food system within present-day activities. Prefigurative politics in food systems encourages the creation of 'islands' of alternative practices—such as ethical consumption, local food networks, and community gardens—that model the principles of justice, sustainability, and participation. This approach is visible in community-supported agriculture, food cooperatives, urban agroecology projects, and food policy councils, where participants create new forms of economic and social organisation that challenge the dominant industrial agri-food system. These efforts are not just about practical alternatives but also about contesting existing power relations, ownership structures, and governance models in food systems, aiming for more distributed and participatory forms of control (Forno & Wahlen, 2024, Koensler, 2020, Tornaghi & Dehaene, 2020).

Prefigurative interventions in food system transformation can be directly mapped onto the leverage points framework. By experimenting with new forms of governance (like participatory food councils) and ownership (co-ops, community land trusts), these initiatives intervene at the level of system design by reshaping rules, information flows, and power dynamics, while those which seek to shift the underlying intent of the food system—from profit and efficiency towards justice, sustainability, and democracy—by creating local food networks or food sovereignty projects align with changing the goals (intent) and paradigms (mindsets) that underpin the system.

A central challenge for prefigurative interventions in food system transformation is scaling up—moving from small, often isolated projects to broader institutional change. This requires building coalitions and articulating 'chains of equivalence' across diverse struggles to universalise these alternatives and challenge the status quo. There is a risk that prefigurative efforts remain fragmented or symbolic if they do not engage with larger structures of power and address the contradictions and cleavages within both alternative and mainstream systems. Prefigurative food politics must also contend with resistance from entrenched interests and dominant institutions that benefit from the current system, making political strategy and coalition-building essential for transformation (Leitheiser & Vezzoni, 2024).

4   Creative Practice and Sustainability Transformation

Sustainability challenges are simultaneously scientific and cultural in nature. While scientific discourses have dominated the sustainability literature to date, as drivers of culture the arts have a critical role to play in societal transformation (Heras et al., 2021, Trott et al., 2020). The arts can engage people with new perspectives on sustainability issues by offering opportunities for critical reflection and providing spaces for creative imagination and experimentation. Such processes may be important for contributing to the changes needed to realise transformations to sustainability (Bentz et al., 2022).

Creative practice has unique potential to engage with deep leverage points in sustainability transformations by focusing on fundamental myths, paradigms, and systems of meaning-making (Vervoort et al., 2024). The 9 Dimensions tool was created to support reflective and evaluative dialogues about links between creative practice and sustainability transformations, by bringing together disciplinary perspectives on societal change from sociology, anthropology, psychology, and more. The framework consists of three categories of change, each covering three dimensions. Each dimension, briefly summarised below, provides a unique lens for understanding how creative practice contributes to transformations.

4.1   Changing Meanings (Embodying, Learning, Imagining)

Embodying allows people individually and together to understand different realities first-hand by tapping into the full intelligence of the body, the senses, and experience. Learning that leads to changing basic assumptions, worldviews and knowledge about the world is part of one of the deepest leverage points for sustainability transformations. When people see or experience new ways of being or doing in action that they did not consider possible before, this in turn opens new imaginative possibilities (Vervoort et al., 2024).

4.2   Changing Connections (Caring, Organising, Inspiring)

Care can shape individual and collective imaginations and open new possibilities by fostering relational responsibility and facilitating the trust needed for sustainability transformations to occur. Creative practice can generate reflection, dialogue, the exploration of alternatives, and the clarification of values needed as a basis for organisational change that supports sustainability transformations. It also has the potential to reach, inspire, and activate people and organisations beyond those involved directly in its practice. This can happen both through people directly engaging with the creative practice itself, or through secondary representations and communications (Vervoort et al., 2024).

4.3   Changing Power (Co-Creating, Empowering, Subverting)

Co-creation enables the perspectives and concerns of diverse and perhaps underrepresented groups to be included, and pre-existing notions about the world to be challenged, integrated, and altered. Creative practice has the potential to help empower individuals and groups who are in marginalised positions and/or who champion radical and novel perspectives. The subversion, disruption, and unmaking of current unsustainable societal structures, regimes, and institutions is a crucial component of transformations towards more sustainable futures. Playful subversive creative practice that challenges and inverts present realities and exposes their absurdity can provide the individual and collective emotional energy and hope needed to develop a shared critical consciousness and to engage with actual subversion (Vervoort et al., 2024).

Creative practice with the most transformative potential combines learning and imagining based in deep, situated embodiment with many possibilities for networked growth, adaptation, and mutation of the practice by others. Care is a powerful dimension that theory and practice show as having strong transformative potential, while co-creative approaches almost always seem to offer more benefits than less co-creative ones, allowing participants to develop co-ownership, relationships and skills, and to share ideas (Vervoort et al., 2024).

5 Creative Practice and Prefigurative Politics

Arts-based interventions have played a key role in enriching and mobilising prefigurative movements all over the world. Creative practice has been used in a myriad of subversive and thought-provoking ways to create counter-narratives, galvanise social movements, strengthen group identities, and communicate with a wider audience (Cathcart Frödén, 2023). Indeed, artistic expression can be seen as part of the strategic toolkit used by these movements to facilitate engagement, education, and outreach (Sanz & Rodriguez-Labajos, 2021, Serafini, 2015).

The practices and values of participatory art projects in particular share similar features with prefigurative social movements. Both are loosely bordered contexts made up of shifting communities and networks that pursue goals of social transformation in diverse and localised ways, and both aspire to create countercultural spaces where the collective imagination can be nurtured, where relationships and mutual care can be prioritised, where voices from the margins can become more audible, and where inequalities can be addressed. Participatory arts and prefigurative social movements can thus be seen as adjacent and intersecting epistemic communities—spaces in which individuals with different backgrounds and forms of expertise can contribute to collective and interactive efforts to deal with shared problems (Cathcart Frödén, 2023).

McNally (2024) argues that participatory art projects can be understood as a practice of prefigurative politics which involves activist artists directing effort into performing now their vision of a 'better world' to come, rather than simply protesting against a dominant regime. These projects assemble publics around specific social justice issues, such as redevelopment and the displacement of marginalised communities, or the decolonisation of public space, to critically reimagine and enact desired futures that radically reconfigure existing oppressive social conditions (McNally, 2024).

6   Art Activism and the Prefigurative Politics of Food System Transformation

In this final section I return to my point of departure, Agnes Denes' *Wheatfield—A Confrontation*, and present it using the 9 Dimensions tool for evaluating how creative practice stimulates societal transformations. I then briefly assess the extent to which the artwork represents 'embodied strategies to render desirable futures with immediacy'.

Volunteers helped to prepare the soil and plant, tend and harvest the wheat in a collective effort (co-creating), giving them a sense of the physical effort involved in food production (embodying), and encouraging care for the land and awareness of ecological processes (caring). The project highlighted issues of land use, food security, and ecological cycles, prompting public reflection (learning). By transforming a neglected urban space into a productive field, people were invited to imagine alternative uses for city land and a more ecologically integrated urban future (imagining). The project involved coordination with city agencies, volunteers, and organisations, demonstrating new models of civic and artistic collaboration (organising). It has inspired countless public art projects and is widely cited as a pioneering example of land art with a sustainability message (inspiring). The project empowered citizens and artists to believe in the possibility of transforming urban spaces for ecological and social purposes (empowering) by critiquing urban priorities, capitalism, and the disconnect between nature and finance (subverting).

Wheatfield can be interpreted as embodying aspects of prefigurative politics, though not in a direct or overtly organisational sense as a sustained, collective effort to build new social relations. Firstly, there is the symbolic embodiment of alternative values. By growing wheat—a staple food crop—on some of the most valuable real estate in the world, Denes symbolically enacted the possibility of prioritising basic human needs (food, ecology, sustainability) over speculative finance and urban development. This act can be seen as a prefigurative gesture: it models a world where land is used for nourishment and public good rather than profit and speculation. Secondly it exemplifies activism through action. The work was not merely a critique but an active intervention which involved preparing the soil and planting, tending, and harvesting the wheat, demonstrating the viability, and value of such labour and land use, even in the heart of a financial district. This hands-on approach mirrors the ethos of prefigurative politics, where the means (the act of planting and harvesting) are as important as the ends (the message about priorities and values). After harvest, the seeds were replanted globally— an act of redistribution and sharing that prefigures a society focused on collective welfare and ecological responsibility rather than individual accumulation.

7   Conclusion

In this chapter I have presented a theoretical and methodological framework for examining the role of art activism in the prefigurative politics of food system transformation and, by way of illustration, applied it to an analysis of Agnes Denes' *Wheatfield—A Confrontation*. It is necessary to point out that *Wheatfield* has been criticised on the basis that the monocrop field, which was fertilised, treated with pesticides, and eventually harvested using heavy machinery, inadvertently echoed the commodification of farming practices by corporate industrial-chemical agribusiness to which the work's anti-capitalist spirit was otherwise opposed (Demos, 2016, Voelcker, 2024). Nevertheless, *Wheatfield* remains a touchstone in discussions of public art and environmental activism, and while the artwork itself is long gone, the photograph of Denes standing in a golden field with Manhattan's towers looming behind has become a defining symbol of art's power to confront and transform public consciousness.

Art activism has a role to play in the prefigurative politics of food system transformation, by creating living examples of sustainable alternatives, fostering experimentation, challenging dominant systems, cultivating hope and agency, and integrating social and ecological concerns. With its potential to alter worldviews, creative practice can engage with deep leverage points in sustainability transformations and potentially link to structural and institutional change through chains of leverage.

Competing Interests The author has no conflicts of interest to declare that are relevant to the content of this chapter.

References

Abson, D. J., Fischer, J., Leventon, J., Newig, J., Schomerus, T., Vilsmaier, U., von Wehrden, H., Abernethy, P., Ives, C. D., Jager, N. W., & Lang, D. J. (2017). Leverage points for sustainability transformation. *Ambio, 46*(1), 30–39. https://doi.org/10.1007/s13280-016-0800-y

Bentz, J., do Carmo, L., Schafenacker, N., Schirok, J., & Dal Corso, S. (2022). Creative, embodied practices, and the potentialities for sustainability transformations. *Sustainability Science, 17*(2), 687–699. https://doi.org/10.1007/s11625-021-01000-2

Cathcart Fródén, L. (2023). "Be like water": Participatory arts, prefigurative social movements and democratic renewal. In A. Bua & S. Bussu (Eds.), *Reclaiming participatory governance: Social movements and the reinvention of democratic innovation* (pp. 104–119). Routledge.

Davelaar, D. (2021). Transformation for sustainability: A deep leverage points approach. *Sustainability Science, 16*(3), 727–747. https://doi.org/10.1007/s11625-020-00872-0

Demos, T. J. (2016). *Decolonizing nature: Contemporary art and the politics of ecology*. Sternberg Press.

Denes, A. (1982). *Wheatfield—A confrontation* (1982). Commissioned by the public art fund. Republished in E. Enderby (Ed.) (2019). *Agnes Denes: Absolutes and intermediates* (pp. 256–257). The Shed.

Dorninger, C., Abson, D. J., Apetrei, C. A., Derwort, P., Ives, C. D., Klaniecki, K., Lam, D. P. M., Langsenlehner, M., Riechers, M., Spittler, N., & von Wehrden, H. (2020). Leverage points for sustainability transformation: A review on interventions in food and energy systems. *Ecological Economics, 171,* Article 106570. https://doi.org/10.1016/j.ecolecon.2019.106570

Fischer, J., & Riechers, M. (2019). A leverage points perspective on sustainability. *People and Nature, 1*(1), 115–120. https://doi.org/10.1002/pan 3.13

Forno, F., & Wahlen, S. (2024). Prefiguration in everyday practices: When the mundane becomes political. In L. Monticelli (Ed.), *The future is now: An introduction to prefigurative politics* (pp. 119–129). Bristol University Press. https://doi.org/10.51952/978152915687.ch008

Heras, M., Galafassi, D., Oteros-Rozas, E., Ravera, F., Berraquero-Díaz, L., & Ruiz-Mallén, I. (2021). Realising potentials for arts-based sustainability science. *Sustainability Science, 16*(6), 1875–1889. https://doi.org/10.1007/s11625-021-01002-0

Ingram, J., & Thornton, P. (2022). What does transforming food systems actually mean? *Nature Food, 3,* 881–882. https://doi.org/10.1038/s43016-022-00620-w

Juri, S., Terry, N., & Pereira, L. M. (2024). Demystifying food systems transformation: A review of the state of the field. *Ecology and Society, 29*(2), Article 5. doi:https://doi.org/10.5751/ES-14525-290205.

Koensler, A. (2020). Prefigurative politics in practice: Concrete utopias in Italy's food sovereignty activism. *Mobilization: An International Quarterly, 25*(1), 133–150. https://doi.org/10.17813/1086-671-25-1-133

Leitheiser, S., & Vezzoni, R. (2024). Joining the ideational and the material: Transforming food systems toward radical food democracy. *Frontiers in Sustainable Food Systems, 8,* Article 1307759. https://doi.org/10.3389/fsufs.2024.1307759

Leventon, J., Abson, D. J., & Lang, D. J. (2021). Leverage points for sustainability transformations: Nine guiding questions for sustainability science and practice. *Sustainability Science, 16*(3), 721–726. https://doi.org/10.1007/s11625-021-00961-8

McNally, D. (2024). Participatory art and geography: Politics, publics, and space. *Progress in Human Geography, 48*(5), 537–551. https://doi.org/10.1177/03091325231219698

Meadows, D. (1999). *Leverage points. Places to intervene in a system.* Sustainability Institute. https://www.donellameadows.org/wp-content/userfiles/Leverage_Points.pdf

Sanz, T., & Rodriguez-Labajos, B. (2021). Does artistic activism change anything? Strategic and transformative effects of arts in anti-coal struggles in Oakland, CA. *Geoforum, 122*, 41–54. https://doi.org/10.1016/j.geoforum.2021.03.010

Serafini, P. (2015). Prefiguring performance: Participation and transgression in environmentalist activism. *Third Text, 29*(3), 195–206. https://doi.org/10.1080/09528822.2015.1082789

Tornaghi, C., & Dehaene, M. (2020). The prefigurative power of urban political agroecology: Rethinking the urbanisms of agroecological transitions for food system transformation. *Agroecology and Sustainable Food Systems, 44*(5), 594–610. https://doi.org/10.1080/21683565.2019.1680593

Trott, C. D., Even, T. L., & Frame, S. M. (2020). Merging the arts and sciences for collaborative sustainability action: A methodological framework. *Sustainability Science, 15*(4), 1067–1085. https://doi.org/10.1007/s11625-020-00798-7

Vervoort, J. M., Smeenk, T., Zamuruieva, I., Reichelt, L. L., van Veldhoven, M., Rutting, L., Light, A., Houston, L., Wolstenholme, R., Dolejšová, M., Jain, A., Ardern, J., Catlow, R., Vaajakallio, K., Falay von Flittner, Z., Putrle-Srdić, J., Lohmann, J. C., Moossdorff, C., Mattelmäki, T., Ampatzidou, C., Choi, J. H., Botero, A., Thompson, K. A., Torrens, J., Lane, R., & Mangnus, A. C. (2024). 9 dimensions for evaluating how art and creative practice stimulate societal transformations. *Ecology and Society, 29(1),* Article 29. https://doi.org/10.5751/ES-14739-290129

Voelcker, B. (2024). Cultivating hinterland: What lies behind Agnes Denes' *Wheatfield*? In P. Gupta, S. Nuttall, E. Peeren & H. Stuit (Eds.), *Planetary hinterlands: Extraction, abandonment and care* (pp. 51–63). Palgrave Macmillan.

Open Access This chapter is licensed under the terms of the Creative Commons Attribution-NonCommercial-NoDerivatives 4.0 International License (http://creativecommons.org/licenses/by-nc-nd/4.0/), which permits any noncommercial use, sharing, distribution and reproduction in any medium or format, as long as you give appropriate credit to the original author(s) and the source, provide a link to the Creative Commons license and indicate if you modified the licensed material. You do not have permission under this license to share adapted material derived from this chapter or parts of it.

The images or other third party material in this chapter are included in the chapter's Creative Commons license, unless indicated otherwise in a credit line to the material. If material is not included in the chapter's Creative Commons license and your intended use is not permitted by statutory regulation or exceeds the permitted use, you will need to obtain permission directly from the copyright holder.

26

Recovery or Transformation? The Prefigurative Politics of Hurricane Response in the US

Tyler Schuenemann

1   Introduction

Governments are rarely the first responders to disaster. Instead, it is neighbours, volunteers, and community organisations that form ad hoc relief efforts, what scholars call, "emergent organisations" (Stallings & Quarantelli, 1985). Although lacking in official expertise and the financial and logistical resources of established relief agencies, emergent organisations have shown themselves to be extremely effective in providing humanitarian assistance after disaster. Their success is often attributed to their comparative flexibility and possession of local knowledge that allows them to act quicker in an emergency than a large bureaucratic agency arriving from outside of the affected communities (Imperiale & Vanclay, 2016).

But not all civilian disaster responders see themselves as mere humanitarians filling the gaps of official relief agencies. Some see themselves

T. Schuenemann (✉)
Keene State College, Keene, NH, USA
e-mail: Tyler.Schuenemann@keene.edu

© The Author(s) 2026

401

S. Sareen and S. Juhola (eds.), *Societal Transitions to Sustainability*, Palgrave Studies in Environmental Transformation, Transition and Accountability, https://doi.org/10.1007/978-3-032-07395-2_26

as agents of political transformation. For example, "mutual aid" organisations have emerged in the aftermath of several disasters, including Hurricane Katrina (2005), Hurricane Sandy (2012), Hurricane Maria (2017), and the COVID-19 pandemic (Luft, 2009; Landau, 2022; Rodriguez Soto, 2020). These organisations are sceptical that established institutions can address our vulnerabilities to environmental disaster. They are also sceptical that the conventional arenas of political advocacy are the best path for redressing this shortcoming. As an alternative space of action, they see disasters as opportunities for instigating lasting change by directly building alternatives and spreading them by recruiting a larger movement within the disaster zone. Rather than an effort to "recover" after disaster, they seek to transform.

This paper examines the case of "Occupy Sandy" to better understand disasters as potential sites of political transformation, and what opportunities and traps these events provide to activists. Drawing upon interviews, ethnographic observations, and archival research, I argue that disaster recovery is a generative venue for collective action because it offers meaningful experiences of agency when so-called normal times do not. These politicised emergent organisations demonstrate the productive potential of horizontal organising in the face of emergency, acting as a democratic alternative to prominent administrative models of disaster response. However, these bursts of activity are often fleeting. Without a larger movement to attach themselves to, these emergent organisations fade as the recovery process progresses, and life returns to normal.

2   Recovery or Transformation?

Hurricane Sandy struck several Caribbean islands, the US, and parts of Canada in October of 2012. Most of the damage occurred in New York and New Jersey. In that debris, both those directly impacted by the disaster and those living nearby did what most people do after a disaster. They collaborated to help each other. New volunteer groups emerged, and already existing organisations pivoted to disaster response. Faith-based institutions as well as established community groups like the YANA community centre in Rockaway, Queens, and the Red Hook

Initiative in Brooklyn shifted to relief work. Some of them also provided a base of organising for volunteers coming from unaffected areas, which included working with what would become "Occupy Sandy" (O'Sullivan & Schwartz, 2023).

Occupy Sandy initially emerged as an extension of an existing network of activists coming out of the Occupy Wall Street movement which had largely gone dormant a year prior. The Occupy movement had focused on seizing public space as a collective assertion of the rights of the people to the city and using those spaces to showcase and legitimise alternative social relationships and political action (Sitrin, 2012). This experience of occupying public spaces gave activists the hard-skills of feeding, sheltering, and organising strangers for a common project, which came in handy after the hurricane. What started as a small group of political activists eventually absorbed tens of thousands of new volunteers and became an organisation largely defined by the activities of disaster response. As a result, the organisation became an aggregation of those who came from established activist networks and people with no prior activism background.

The initial activities of Occupy Sandy would be difficult to distinguish from other emergent groups operating in the disaster zone. They travelled to distant neighbourhoods scattered across New York City and New Jersey to knock on doors and check on people, drop off supplies, host school programs, clear debris, and remove mould from houses. They also created clinics and organised home visits for medical services. These efforts were incredibly effective, partially thanks to their horizontal style organising and a direct-action, "do-it-yourself" approach to solving problems. This allowed Occupy Sandy to quickly absorb volunteers and put them to work after orientation and training, while larger relief bureaucracies were slow to get started in their work and turned away volunteers—many of whom found their way to Occupy Sandy.

But Occupy Sandy was also doing something unique to other organisations on the ground. They utilised their recovery efforts as a space for prefigurative politics, experimenting with new forms of organisation, care, and consultation, what they often called "mutual aid," "solidarity," or a "people's recovery." These practices were largely defined in opposition to what they saw as the undemocratic, infantilising, ineffective, and politically conservative practices of established relief institutions. They also took advantage of the publicity that they were receiving for their relief work and used it to popularise their narrative of "mutual aid, not charity" and draw others into their political activism.

This organising had three major themes, each of which proposed alternatives to established conventions of disaster management. The first theme was a negation of the top-down approach of rational administration in disaster management. In contrast, Occupy Sandy emphasised the value of non-hierarchical organisation and "empowering" those affected by disaster to direct and participate in their own recovery. This would involve sending volunteers to knock on doors, asking people what they need, and building a project to directly address those needs. They also worked with community leaders to escort outside volunteers into neighbourhoods to consult locals, ultimately seeking to recruit them into leading the projects themselves (O'Sullivan & Schwartz, 2023, p. 115; Bondesson, 2017, pp. 140–146). Occupy Sandy participants publicised these efforts in their own social media posts and in their interviews with established media outlets. They took pictures of hand-written signs for online publication. One photograph shows no uniforms, but people dressed in every-day clothing, donning a simple nametag made from masking tape and magic marker. In contrast to FEMA jackets and Red Cross uniforms, their presentation did not signify that they are expert or official workers, just normal-looking people working together spontaneously, without rank. Pictures of ad hoc solutions were also common in their social media posts, including churches packed to the brim with donated supplies and volunteers. Rather than experts solving the problems of victims, they portrayed a spontaneous, almost romantic outpouring of solidarity from every-day people empowered to solve their own problems. In the words of one participant, "…if we want it done right, we need to take on responsibility and power ourselves" (Rampell, 2013).

The second theme that Occupy Sandy promoted in its work was the political nature of the destruction itself. Scholars of disasters have long pointed out that these are not "natural" events, but the consequences of long, unfolding political, economic, and ecological processes that make communities vulnerable to hazards (Hewitt, 1983). But this politicised interpretation of disaster is often absent from the humanitarian drama in media coverage. Occupy Sandy participants put the political nature of the destruction at the centre of their practice and of their media campaign.

Key to this political narrative was a focus on the inequalities of destruction and aid. In their narrative, the storm's destruction was an injustice, highlighting the abandonment, exploitation, and inequality of life under capitalism and climate change. Participants regularly remarked upon how wealthier, whiter neighbourhoods were quickly getting help and their electricity back while other neighbourhoods were not being visited by relief agencies. Hence, "recovery" is a struggle over who gets what. This notion is illustrated in the populist language used in the organisation's communications, including "the people" and "the 99%." It is for this reason that they gave journalists and activists from New Orleans a forum at the summit held for participants to share their accounts of the racist and classist character of post-Hurricane Katrina recovery (2005), and to warn of things to come post-Sandy, unless they organise to stop it. Occupy Sandy participants drew from the Katrina example, as well as Naomi Klein's popular book about the "disaster capitalism," to legitimise a political reading of disasters:

We're well aware that disasters are often opportunities for the government to take valuable land or rebuild on land that they see as valuable in some way. That benefits people with the big bucks, but not the people living on the ground... Part of our work is building strong communities and networks with the knowledge that there might be a fight coming towards us... (Rampell, 2013).

For this reason, Occupy Sandy participants began training disaster victims in the tactics of holding sit-ins in anticipation of government agencies refusing to allow people to continue living in flood zones on the shore (Nir, 2013). They believed that recovery required more than humanitarian aid. It required organising to force authorities to legitimise their actions by being more transparent and inclusive in their decision-making.

The third theme that Occupy Sandy promoted was the need for democratic consultation in determining the goals of recovery. The received wisdom in the field of disaster management is that disasters are a break from the normal. Thus, institutions like FEMA and the Red Cross are tasked with returning to the status-quo (Hewitt, 1983). It is a judgement that the goals of recovery are to restore what was lost, not to transform it.

Occupy Sandy participants challenged this assumption, opening it up for debate. For example, in a group discussion circle wherein participants were reflecting upon their ongoing work several months after the hurricane, one person described their work as "recovery" but immediately rejected the term's implications. He corrected himself: "Occupy Sandy is about making things better than they were, not recovering back to the poverty that was there pre- [Hurricane] Sandy." Others in the same discussion echoed these sentiments in stating that the hurricane "exposed the wounds of poverty" that were there prior. In their eyes, only parts of the previous status-quo are deemed worthy of restoration, including housing, schools, and re-opening businesses. And even these are objects to be radically re-thought for the sake of combating climate change and creating more "resilient" communities for the next major hurricane (Field Notes, 2013, February 1). Still others argued that the hurricane was only one of several disasters that needed to be addressed in a People's Recovery: "there's the initial disaster, and then there's the long-term disaster that happens after the sorta volunteers leave, after the cameras leave, that is deeply related to the failures and ongoing crisis of capitalism as a system" (Premo, 2013). They saw the practices of solidarity being innovated in the disaster zone as a way of building a movement to address these larger systemic forces that made communities so vulnerable to crisis.

3   Opportunities and Traps of Crisis Mobilisation

While disaster recovery provided activists opportunities to experiment and showcase alternative ways of organisation and care, this activity was significantly shaped and constrained by the context—that of a perceived crisis. Many participants joined the organisation because they saw an urgent need for people to put down their normal activities and step into the field of disaster relief as volunteers (Schuenemann, 2020). It was the idea that in the storm's aftermath, many people were suffering unnecessarily, and the official relief bureaucracies, FEMA and the American Red Cross, would fail to help them (Field Notes, 2013, February 1).

The sense of crisis brought together people with very different goals and values into a common project. For former participants in Occupy Wall Street, this crisis was an opportunity to prove to "themselves" and their larger audience that they can mobilise responses to precarity, to mass need through practices of self-governance, independent of state bureaucracies. This was clearly articulated by one of the initial organisers of Occupy Sandy:

... I think what captivates many of us [in Occupy Sandy] is the idea that... people actually could show the potential for self-governance, the potential for community self-determination... The cavalry isn't coming, it's just us here and we got this, neighbours helping each other, people picking up trash in the street taking care of one another, helping rebuild each other's homes (Liboiron, 2013).

But other participants joined Occupy Sandy with no political goals. They had looked for venues to do humanitarian work and Occupy Sandy was it. What held both the politically minded and the humanitarian-driven participants together was the sense of crisis, the concern of, who will save these people, if not *us?*

Following the logic of crisis response, Occupy Sandy participants focused on their ability to provide effective relief where the officials and trained experts could not. Being effective and practical became a driving principle of their work. Occupy Sandy participants would brag of when FEMA, the Red Cross, and the National Guard each came to them with supplies and resources or asked for guidance (Field Notes, 2013, February 2). National news outlets were praising them on humanitarian grounds: "Where FEMA fell short, Occupy Sandy was there," read a *New York Times* headline. Participants also found meaningful recognition from those in need. One participant stated that to help those who had lost everything was rewarding and addictive: "It feels good. You get that smile when you hand someone a blanket" (Field Notes, 2013, February 2). Horizontal organising, respecting people's autonomy and voice, and an emphasis on direct action—these principles had been valued because they are more inclusive and democratic. Here in Occupy Sandy, they were being appraised under the criteria of crisis response: they are good because they save lives.

The outward recognition of being effective in a crisis and the emotional force that recovery work fostered created new pressures that shaped who participated in Occupy Sandy and how they organised. In a major shift from the Occupy Wall Street, participants made specific demands on the government to determine how federal money should be spent on recovery (Occupy Sandy Policy Working Group and Spokescouncil, 2014). They also collaborated with government agencies and police to achieve their goals. For example, Occupy Sandy participants were gathering large amounts of donated supplies in neighbourhood buildings. Fear of looters convinced some participants to accept offers from the police to guard the buildings overnight. But others saw enlisting the police in the "People's Recovery," or working with City Council members as a betrayal of anti-government principles. Some of them left Occupy Sandy out of frustration (Goodman 2013). The goals of effectiveness in crisis also called into question the values of democratic debate and consensus-building when there was so much urgent work to be done. Many of Occupy Sandy's projects were geographically scattered, making it difficult for many to join in collective decision-making. Some argued that they did not have time for deliberation, while others insisted that people needed to make time for "slower and anti-oppressively cognizant recovery and rebuilding effort" (OccupySandy Coordination Listserve Thread, 2012).

This sense of crisis was so important to Occupy Sandy's activities that as the urgency faded, so too would the organisation. After months of work helped to secure people's access to food, medicine, and shelter, people perceived things as going back to normal. The lack of urgency meant that fewer volunteers and donations were coming into Occupy Sandy projects. This scarcity made it harder to hold together the diverse collection of projects and activists working within the group. They began to compete over the distribution and control of collective resources, including money, supplies, volunteers, "wepay" accounts, websites, social media accounts, physical spaces, as well as who got to speak on behalf of the group. Arguments over what projects are worthy of the scarce resources took on new weight. For example, a few participants had created a lending library of construction tools to make home repairs affordable for residents. Yet some participants argued that the library should be excluded from the pool of shared funding resources because those who ran the library appeared to be straying from their prefigurative values: they had formed a non-profit and were allowing residents unaffected by the hurricane to use their tools for a fee (OccupySandy Coordination Listserve Thread, 2013).

One of the more common fault lines to emerge in these debates was whether Occupy Sandy was falling into a neo-liberal trap, wherein they took on the responsibility of the government, but without its resources. Some participants insisted that they reject a principled anti-state position and avoid exaggerating the capabilities of every-day people in the face of disaster. They expressed embarrassment for their "early days" after the hurricane when they would publicise their accomplishments on Twitter with "#WeGotThis." After months of work, they quickly found themselves in over their heads and insisted that the people should not be left to do the work of government (Field Notes, 2013, February 2).

Another part of the debate was to distinguish between what is "political," or "mutual-aid," and what is "merely charity," and to prioritise the former. As one participant put it, "Charity is easy because it is immediately fulfilling. But the help you are providing is not sustainable, transformative, or empowering for the person you are helping" (Field Notes, 2013, February 1). For this reason, some were uncomfortable with other Occupy Sandy participants working through organisations that also hosted volunteers from FEMA. If FEMA and Occupy Sandy are each sending volunteers to the same projects, then how can Occupy Sandy claim to be doing more than just charity (Nir, 2013)? Others insisted that participants should be serving hurricane victims in whatever they needed, on humanitarian grounds. Still others held that helping people can be transformative, acting as an inspiring example of directly building a better world for future movements.

The evaporation of the sense of crisis also made the need for establishing long-term goals a priority. But Occupy Sandy participants and the communities they were working with did not share a common vision for the future. For example, activists spent months attempting to build a grassroots coalition in the Rockaways, a neighbourhood hit espe-cially hard by the hurricane. Their goal was to force new development projects to address residents' underlying vulnerabilities to future storms: they advocated for affordable housing, local jobs, and storm-resistant designs in new buildings. But they failed to amass enough local partici-pation in the coalition, leaving it too weak to secure commitments from developers or city government. This failure was partially because the resi-dents of the Rockaways were divided on the issue of affordable housing, but also because of tensions between residents and activists. Residents expressed distrust toward some activists, viewing them as outsiders who were driving the agenda. They also expressed frustration with their model of horizontal organising (Bondesson, 2017, pp.147–157). In the early days of the disaster, such divisions were irrelevant because the tasks at hand were so basic and urgent: secure food, water, shelter, electricity, etc. But now that the crisis had passed, the lack of consensus strained the unity of Occupy Sandy and kept it from forging a larger movement.

4   Conclusion

Disaster recovery can be a generative venue for drawing together citizens for collective action because it offers meaningful experiences of agency in moments of perceived crisis. The case of Occupy Sandy demonstrates the productive potential of "mutual aid" practices in such venues, acting as a democratic alternative to prominent administrative models of disaster response that emphasise top-down authority and a return to the pre-disaster status quo. These alternatives included horizontal organising, building power for fighting political battles within the recovery phase, and democratising the agenda of disaster response.

The analysis above reveals several features that might set apart this and similar prefigurative projects from prefigurative efforts in other sectors. First is that people came together because they perceived a crisis, and the logic of that urgent, extemporaneous action shaped both what actions were prioritised and who participated. There was an emphasis on addressing short-term needs and compromising on political commitments to do so quickly and effectively.

Second is that organising to address a crisis strained the longevity of the project. Even though many in the group were motivated to address long-running problems of climate change and capitalism, Occupy Sandy was not the venue through which they would ultimately pursue those goals. The group dissolved after the urgency of that disaster seemed to pass, and those who continued the work of "mutual aid" had to find new venues. Some participants would go on to help set up mutual-aid groups in the aftermath of other disasters, including a tornado in Oklahoma in May of 2012 and flooding in Boulder, Colorado in October 2013. The Mutual Aid Disaster Relief network also traces its origins through Occupy Sandy. As of 2025, they are still active in promoting "mutual aid" throughout the United States with instructional workshops and literature. Moreover, former participants of Occupy Sandy also joined a broader movement for climate justice that would win major legislative victories at the municipal and state level in 2019 (Aldana Cohen, 2020, pp. 14–15). These new laws commit the city and state to ambitious decarbonisation goals and environmental justice initiatives for disadvantaged communities facing environmental hazards.

Third, disasters often have a spectacular character to them, providing a built-in audience for this kind of organising. This spectacle draws in participants from outside of established activist networks. It also gives activists valuable opportunities to showcase their practical innovations and spread their ideological message to a broader audience than they would otherwise have.

Competing Interests The author has no conflicts of interest to declare that are relevant to the content of this chapter.

References

Aldana Cohen, D. (2020). New York City as 'fortress of solitude' after hurricane Sandy: A relational sociology of extreme weather's relationship to climate politics. *Environmental Politics, 30*, 687–707. https://doi.org/10.1080/09644016.2020.1816380

Bondesson, S. (2017). *Vulnerability and power: Social justice organizing in rockaway, new York City, after hurricane Sandy* (pp. 140–157). Department of Government, Uppsala University.

Hewitt, K. (1983). Interpretations of calamity in a technocratic age. In K. Hewitt (Ed.), *Interpretations of calamity: From the viewpoint of human ecology* (pp. 3–30). Allen & Unwin.

Imperiale, A. J., & Vanclay, F. (2016). Experiencing local community resilience in action: Learning from post-disaster communities. *Journal of Rural Studies, 47*, 204–219. https://doi.org/10.1016/j.jrurstud.2016.08.002

Landau, L. (2022). Mutual aid as disaster response in NYC: Hurricane Sandy to COVID-19. *Journal of Extreme Events, 9*(2), 2241001. https://doi.org/10.1142/S2345737622410019

Liboiron, M. (2013, July 24). Superstorm Research Lab. https://superstormresearchlab.org/2013/07/24/interview-with-justin-wedes-occupy-sandy-volunteer/

Luft, R. E. (2009). Beyond disaster exceptionalism: Social movement developments in New Orleans after hurricane Katrina. *American Quarterly, 61*(3), 499–527. http://www.jstor.org/stable/27735005

Nir, S. M. (2013, May 1). Storm effort causes a rift in a shifting occupy movement. *The New York Times.* https://www.nytimes.com/2013/05/01/nyregion/occupy-movements-changing-focus-causes-rift.html

O'Sullivan, T., & Schwartz, R. (2023). Disaster response of short-term emergent citizen volunteer groups: Hurricane Sandy. In S. Feldmann-Jensen, S. Jensen, & J. Slick (Eds.), *Case studies in disaster response* (pp. 103–121). Elsevier.

Premo, M. (2013). *Occupy Sandy operationalised Occupy Wall Street.* More Like People. http://morelikepeople.org/michaelpremointerview/

Rampell, E. (2013, January 7). Occupy Sandy: Volunteers for America's 99%: The cavalry is us. *Rock Cellar Magazine..* http://www.rockcellarmagazine.com/2013/01/07/occupy-sandy-volunteers-for-americas-99/

Rodríguez Soto, I. (2020). Mutual aid and survival as resistance in Puerto Rico. *NACLA Report on the Americas, 52*(3), 303–308. https://doi.org/10.1080/10714839.2020.1809099

Schuenemann, T. (2020). Making and scheduling citizens: Political time and the democratic potential of hurricanes. *New Political Science, 42*(1), 87–108. https://doi.org/10.1080/07393148.2020.1726259

Sitrin, M. (2012, September 14). Occupy Wall Street and the meanings of success. *HuffPost.* http://www.huffingtonpost.com/marina-sitrin/occupy-wall-street-anniversary_b_1884829.html

Stallings, R. A., & Quarantelli, E. L. (1985). Emergent citizen groups and emergency management. *Public Administration Review, 45*(2), 93–100. https://doi.org/10.2307/3135003

Open Access This chapter is licensed under the terms of the Creative Commons Attribution-NonCommercial-NoDerivatives 4.0 International License (http://creativecommons.org/licenses/by-nc-nd/4.0/), which permits any noncommercial use, sharing, distribution and reproduction in any medium or format, as long as you give appropriate credit to the original author(s) and the source, provide a link to the Creative Commons license and indicate if you modified the licensed material. You do not have permission under this license to share adapted material derived from this chapter or parts of it.

The images or other third party material in this chapter are included in the chapter's Creative Commons license, unless indicated otherwise in a credit line to the material. If material is not included in the chapter's Creative Commons license and your intended use is not permitted by statutory regulation or exceeds the permitted use, you will need to obtain permission directly from the copyright holder.

27

Between Situated and Speculative: (Im)possibilities of Prefigurative Politics in Research?

Cheshta Arora ⓘ

1 In Lieu of an Introduction

There's a relationship between the stories we tell about the world and the politics it enables. The word stories in the previous sentence could stand for a host of things. It can refer to frameworks, worldviews, ideas, narratives, angles, anchors, ground, positionality, perception, rumour, lies, facts, fiction, report, etc. Similarly, politics can stand for praxis, action, agency, conflict, co-operation, values, transformation, greed, corruption, violence, rebellion, policy, compromises, struggle, consensus, transition, dissensus, revolution, etc. You may or may not agree with all these connotations that the word story or politics has been made to evoke here. Also, some might prefer facts over rumours or vice versa or consensus-building over violence, while others might have

---

The road to prefiguration is long and winding and unpaved, with several dead ends and no clear signposts.

C. Arora (✉)
Western Norway Research Institute, Sogndal, Norway
e-mail: car@vestforsk.no

© The Author(s) 2026
S. Sareen and S. Juhola (eds.), *Societal Transitions to Sustainability*, Palgrave
Studies in Environmental Transformation, Transition and Accountability,
https://doi.org/10.1007/978-3-032-07395-2_27

a taste for corruption. These preferences or meaning making are not innate but emerge in relation to the kind of stories we come to be a part of and the political that germinates therein.

This relationship between the stories and the politics we come to embody has haunted me for a while now, or at least since 2019. The first time I became aware of this relationship in its strong version was sometime in 2014 when I read a pirated copy of a California-based professor Donna Haraway's "A Cyborg Manifesto" (Haraway, 1991) on my Amazon Kindle, which was a gift from a 24-year-old brother who had already started labouring at the fringes of India's Silicon Valley. These details will soon come to matter.

The text written in 1985 to make sense of "our time, a mythic time, [where] we are all chimeras" was an open invitation to fiddle with irony, blasphemy, perversity that the jhola (or the tote-bag), often fabricated as part of NGO-facilitated women self-help groups funded through micro-credit schemes meant to eradicate poverty through financial empowerment, could not hold. Haraway's cyborg was an open invitation to take pleasure in the "border wars" of nature and culture, machine and organism, human and animal, where the stakes have been the "territories of production, reproduction, and imagination" (p. 4).

The encounter, for the time being, opened the world of an Australian cyberfeminist collective VNS Matrix, "the virus of the new world disorder" that stimulated the senses to seek new pleasures. A socialist feminist professor insisted I read Firestone (2003) along with Haraway to keep myself grounded while others often probed the relevance of such texts for "our context"—that meant, among other things regurgitating stories of neoliberalism, hard-fought women's rights, digital rights, intersectionality, development, middle-class positionalities, violence, and so forth. The metaphors of pleasure, disorder, and messiness had no place in an already chaotic, violence-ridden third world, postcolonial developing economy in the global south. The problem had many names. Amidst the different stories that each of these descriptors laid out in the previous sentence evokes, what could a cyborg do?

Let's break into another section with a question.

2   Can a Womanmachine Dream of a Cyborg?

The above question was first posed to me in another text encountered in 2014 during a writing workshop. The workshop, titled "Critical Review", was designed to teach second year master students (labouring to write their dissertation for a degree) the skill of academic reading and writing or, in its own words, "How do we read (academic) texts? How do we produce critical writing? Is it possible to ask not just what a text means but, also how it means?" (Workshop poster 2014). The keynote speaker had referred to a text by Dina Siddiqui as an example of using gender as an analytical category to foreground questions of contextuality and positionality and their consequences.

To my pleasure, Siddiqi (2000) had finally engaged with Haraway in "our context" only to disappointingly point out that the "womanmachine" of garment factory workers in Bangladesh couldn't dream of the first-world, postmodern cyborgs. Siddiqi (2000) juxtaposed Haraway's cyborg with her own figure of the womanmachine who cannot speak of what Haraway describes as the "intense pleasure in skill, machine skill" (Haraway, 1991: 425, 451 as cited in Siddiqi (2000). Instead, womanmachines "confess to feelings of subordination or subjection to those machines" (p. 15).

A careful reader of Haraway will experience this juxtaposition as a misreading. However, it is the sticky tension of contextualization and space of difference evoked through the juxtaposition that is of interest here, and not how it misreads the question of perverse pleasure in Haraway's oeuvre. The tension takes us back to the oft-repeated relationship between the stories and politics, and at stake are relations of production, reproduction, but most importantly, imagination. Siddiqui's struggle was against those stories that reproduced "third world sweatshop workers" as exemplars of universal subordination of women or victims of western imperialism and capitalism, or poor disenfranchised workers looking for a saviour (2015). Amidst these three stories, Siddiqui struggled for another one which foregrounded the "lived realities and priorities of workers themselves" (2015, p. 172).

These stories, while caught in an ideological conflict with each other, however, enable politics that can only speak about/for the garment workers despite foregrounding the lived realities of workers themselves. The problem inherent in such gestures of speaking on behalf of is quite adequately compressed and expressed in a simple rhetorical question, "Can the subaltern speak? (Spivak, 1988)" where the short answer is no, because "there is no space" for the subaltern. A long answer takes us to realise that within conditions of mediation, translation, and sublation, the subaltern is often lost and overdetermined. The ethical obligation to determine these conditions of mediation and present the subaltern to the available logic of representations then falls upon the researcher.

In 2018, I ended up at IMT Manesar, a newly developed industrial modern township neighbouring Delhi, to understand the activities of an autonomous workers' collective, Faridabad Majdoor Samachar (FMS) (Faridabad Workers' Newspaper). FMS had been distributing an ultra-local monthly newspaper since the 1980s in Faridabad, Okhla, Gurgaon, and more recently in Manesar. The newspaper, archived online, will serve as a significant resource to make sense of the industrial transitions witnessed in the region in the last decades. During this period, the factory floors in Faridabad, one of the biggest industrial zones in North India, producing tractors, motorcycles, mini steel plants, textiles, chemicals, pharmaceuticals, shoes, tyres, paper, printing press, and electric motor factories, underwent rapid automation.

Witnessing these transitions, FMS dabbled with several stories to make sense of the situation. First, among the permanent workers, the Leninist framework advocating union representation to focus on wages and factory conditions was tested and rejected. The demand for union representation became irrelevant amidst the wave of temporary workers and brought forth several instances of self-activity of wage workers. Here, the stakes of the narrative are not whether temporary workers can be mobilised or legally join unions, or what is to be done to organise temporary or contractual workers under union leadership. With the rejection and redundancy of unions, the plot twist rests on recognising the radically different nature of the relationship between transient, contractual, temporary workers, and the factory.

Simply put, the new workers labouring under different contractual regimes didn't care about improving the conditions of this or that factory. The protagonists of the story had moved on, and in the self-activity of the workers hired through contractors, the regime of work/labour itself had come under scrutiny. While in Siddiqui's story, the woman-machine was different from the cyborgs of advanced capitalism, in the FMS's story, a worker was not a worker of this or that factory, this or that country. She was global. She was everywhere—on the factory floor, inside the universities, and in the service economy.

Where the intellectual trend in academia argued for situatedness, i.e. context-specific analysis of conditions, situations, negotiations, contradictions, analysis, of varieties of capitalism—FMS, in its anti-activist, tragic-comic state, implored the need to reflect beyond a situation. This need rested on starkly different ways in which the concept of "situation" is understood. For academics and activists, a situation is a source of data that needs to be understood and analysed to make theoretical or policy recommendations. For tragic-comic-anti-activist activists, as self-described by FMS, to understand a situation is to fall into the trap of politics of negotiations, trade-offs, and compromises as well as demands of sacrifice, martyrdom, awareness, consciousness raising, and social work. Or, to evoke Spivak's logic, it is to get lost and be rendered silent. Or, cheekily put, it is death by discourse. For FMS, the prefigurative emerges from those experimental moments in life that interrupt this logic of negotiations. Such moments become visible only momentarily and refuse to be "scaled up" through an interventionist approach, and only some stories have enough room to keep them alive.

Let's fall back on two vignettes from 2011 and 2019 to unpack this.

It's June 2011. We are in IMT Manesar with FMS witnessing a massive strike in the Maruti Suzuki Factory that will go on till October 2011 (FMS 2024). The strike exemplified one of those rare moments of synergy where permanent workers and workers hired through contractor companies will learn to coordinate on their own without mediators. The synergy that developed without any mediated attempts at organising these otherwise divided groups of workers will bring them to a point where it will be possible to assemble a car in 1 minute instead of 45 seconds, take 16 days of holiday in a year, and significantly increased wages for both permanent and workers hired through contractors. Despite this improvement in working conditions, an insignificant incident in July 2012 culminated in the death of a manager. The incident jolted FMS out of its routine to see things anew. The event presented a question that needed new answers. Why did the anger erupt despite significant improvements in working conditions? Or, what exactly did the workers want? In their reading of the situation in IMT Manesar in 2012, the anger was directed at the wage-labour system itself. The moment was ripe for perverse pleasures and blasphemy.

This reading makes sense in light of the realisation that the contemporary is marked by petering out of various political agents—socialist welfare states, diffusion of left-fronts, a decline of trade union social movements on the left, which results in a general impasse to political mobilisations. What becomes of "unions" in such times? Who decides what is to be done? To whom does one fall back? What does one address, mobilise or evoke—the state for its lack of protectionism, the liberal middle-class for its apparent amnesia, or the workers, and how?

A quick look at most of the labour studies papers published under keywords such as lean manufacturing, Industry 4.0, precarity will most probably tell the following story which can be quickly sketched as follows: Lean manufacturing signifies automation, no permanent employment, increased labour precarity, no unions, flexibilisation, just in time production presents situations of hire and fire, capital is all mighty, workers' rights are eroding, workers need unions, demands of more stability, security, rights, transparency, justice and so on.

FMS's re-reading of the situation, however, interrupts this narrative. Not only does it allow one to exit the trap of infrastructures of representation and delegation (i.e., speaking for /about) that enact rituals of agency, but it also expands the meaning of lived realities to life itself. To understand lived realities is to come face-to-face with the question of what it means to live and who gets to decide. When posed thus, the question of life itself cannot have its proper place in either this or that situation, or this or that worker or this or that identity. The question of life presents itself in moments that are in excess of a situation. Let's continue with the second vignette.

When I began my fieldwork with FMS in 2019, it had already adopted a stylistic device for the newsletter that, at least in its form, enabled a collective reflection on the question of what it means to live and which situations allow us to expand the meaning of life. After the events of 2012, the front page of the newsletter presented a WhatsApp-style fictionalised conversation between two or more people. The events witnessed or heard in conversations with workers, or while distributing the newsletter, often became the source of inspiration for the first page. Several activist-students from the University had often complained that the first page was too stylistic. It made no sense.

During my fieldwork in 2019, the events took another unexpected turn that advanced the story that began in 2011. During my time with FMS in IMT Manesar, contractual workers of the Honda factory hired through labour contractors who were not part of the company-recognised unions, had "de-occupied" the factory for 20 days. This was the first time in the region that the workers hired through labour contractors were at the forefront of leading a strike without any union recognition or mediation. Witnessing the event solidified an important lesson—the intricate relationship between the stories we tell and the politics it enables. A brief detour into the events of November 2019 is necessary to further reiterate what this means.

On 9th November 2019,

> #Tonight, at around 8:30, workers sitting inside the Honda Factory were given biscuits, snacks, bread, and juice.
> # Tomorrow friends, do come from 11 to 12 o'clock to encourage these casual brothers and if you can cook at home then do bring food so that there is no let-up in our brothers' movement... Beedi to smoke...
> (Faridabad Majdoor Samachar, 2019)

Previously, on November 8, three days after the strike, the management had decided to shut the canteen inside the premises which prompted workers hired through contractor companies, de-occupying the factory premises to call upon their friends for food/water/cigarettes allowing other workers, from various other factories spread in the area, to participate in the strike.

This moment of solidarity, exchange, and interaction, however, is turned into a moment of crisis and workers as victims; the de-occupation of the factory is turned into a "hunger strike". The tone of "collective support of each worker" morphs into a state of emergency:–

> November 13th: The narratives of forcibly keeping workers sitting inside the factory hungry to create an emergency has advanced. Inside, three workers have fallen ill. Leaders by beating the drum of "thousands of workers inside the factory on fast unto death" remind of the deceptions in the war epic Mahabharata. (ibid.)
>
> Propaganda that 2500 workers sitting inside the factory have begun fast unto death [but in reality] … the Honda union stopped supply of food. To see to it that no food enters the factory from any side, some from amongst the workers hired through contractor companies and sitting outside the factory were put up as additional guards. (ibid.)

Indeed, after the failed attempts to turn the deoccupation into a "hunger strike", the union began offering milk/bananas to the workers inside. However, "due to an old order from a civil court, assurances of negotiation from the administration, and at the insistence of the supporting unions", the workers are made to leave the factory premises and are relegated to dry barren land, around a km away from the Honda factory. It is only on 20th November, as the workers move out of the factory premises, that a Delhi-based left-leaning media portal, The Wire, reported the demonstration of workers (by then already out of the premises) by using phrases such as "thundering slogans evoking workers' unity" (Kumar, 2019) while on the contrary one of the workers at the site pointed out to us that they felt strong while they were inside as the management appeared anxious, but now they can see their strength receding. The demonstrations, speeches, and rallies on this barren land continued for over a month and eventually fizzled out.

3   Can the Cyborg and Womanmachine Meet?

It's 2020. While presenting my thesis draft to the advisory committee:

> "You have done a very good job with your dissertation, but where are the (women) factory workers in your work?
> "We don't see their conditions of work. A political economic analysis of the situation, their aspirations or desires".
> "You critique regimes of work/labour, delegation and representation, but a lot of women in India are still not part of the formal wage/labour regime; they aspire to get an office job. Where is this story from the field?".

To give a quick overview of my doctoral work, I conducted a multi-sited ethnography of women scientists, software engineers, and factory workers. In relation to each other, these three figures can easily be plotted on a social hierarchy while each navigates its own patterns of inclusion/exclusion and identity formation. In their lived realities, they seldom meet each other but are otherwise deeply entangled. Through each of these, I resonated an experimental statement "I do not want to work" that highlighted frictions in the way each of these figures are understood. I borrowed my understanding of "experimental statement" from Isabelle Stengers (2000), who defines it to distinguish it from theory, where:

An experimental statement can be upsetting, subverting the landscape of knowledges, connecting some regions, disconnecting others, but it defines possibilities available to everyone, constraints everyone must take into account, but which everyone must be able to profit from, if they invent the means. By contrast, a theory requires that the hierarchization of the landscape of knowledges it is proposing be socially ratified...Every theory affirms a social power, a power to judge the value of human practices (Stengers, 2000, pp. 111–112).

The definition also helps redefine the figure of a scientist/researcher "where the scientist submits to a becoming that cannot be reduced to the simple possession of a knowledge" (pg. 90) and "to let oneself become interested" is the prerequisite to this submitting where "those who have accepted to gather around the experimental apparatus (or statement) to recognize its possible relevance, we must first of all say that they have allowed themselves to become interested" (pg. 91). To this notion of a scientist, Stengers (2000) presents a notion of "experimental statement" around which interested scientists gather, to recognize its possible relevance, "who ask themselves if an experimental statement can intervene in their problematic field" but the statement "does not say for whom this difference has to count" (pg. 90). "I do not want to work" emerged as an experimental statement (note: not as an empirical) that could make a difference across all three sites for all three figures—scientists, engineers, and factory workers without situating their lived realities in this or that difference and plotting it on a well-known map of socio-economic hierarchies.

How did a detour help resolve the friction between cyborg and womanmachine with which we started?

Unlike Siddiqui, who rejected any affinity between the womanmachine and the cyborg, the quick detour enacted a different kind of encounter where womanmachine and cyborg can meet, but only when/if the stakes of the debate are relevant for both. They will have to gather around an experimental statement that is not situated but has the speculative strength to make a difference to both their worlds. To prefigure is not about debating means and ends, purity of politics or blasphemy of compromises, or strategies or goals. To prefigure is to present spaces and statements of gathering that interest people, that provoke, that can be discussed outside of existing frames of knowledge or what's practical or makes sense, and most importantly, that make a difference for the 8 billion or more.

Let's change the scene again and go back to Faridabad, May 2014.

"While distributing the paper, we were stopped twice and advised: "Don't distribute the paper here. Workers here are very happy. Are you trying to get factories closed?" That reading, writing, thinking, and exchange can lead to factory closures—where does this thought come from?

Perhaps this fear is a result of messages that circulate between the mobile phones of tailors. Or perhaps this fear emerges because workers on the assembly line are humming!" (Faridabad Majdoor Samachar, 2014)

What can the academic world say about humming on the assembly line? What can the academic world say about a joyful, vibrant life?

Who gets to talk about it? Do we have a method that can help study life? Which kind of stakeholder consultations do we need to do to understand it? Is there a framework? Perhaps a policy brief? Another tool that can facilitate a conversation about life?

Not sure. But we will need an entire 8 billion of us to join in to speculate, do thought experiments that seem bizarre, absurd, and non-sensical. Researchers, especially, will need to unlearn the business of analysis. They will have to stop the double-speak. Rid themselves of irony. Relinquish their concepts and mental maps of how the world works. They will have to relearn the difference between relevant and irrelevant knowledge. They have to acknowledge that things don't make sense and most of what we do is part of the "bullshit jobs" that Graeber (2018) spoke of. The corridor talks are far more interesting than the work we do and present at conferences. They will have to learn to gather rather than present.

The very meaning of joy will have to be redefined. We will also need to find space where we can talk about it without shrinking it into fragments of working life, citizens' life, immigrants' life, women's life, tribal life, past life, future of life, dignified life, protected life, productive life, wasteful life, city life, rural life, biological life, automated life.

Why?

To reduce "life" to fixed situated identities, demands, negotiations, or immediacies of the present, of the here and now that mandate polemics and strengthen the vertical orders (of management, states, global north, global south), and the future that is already available in the present. To prefigure would be to begin from "life" in all its contingencies; it would require relinquishing how things are known. The word "life" itself has to be thought differently. The road to prefiguration is long and winding and unpaved, with several dead ends and no clear signposts.

Ok. But who will fund it?

Competing Interests The authors have no conflicts of interest to declare that are relevant to the content of this chapter.

References

FaridabadMajdoor Samachaar. (2014, May). Some urgent questions have come to shore. Faridabad Majdoor Samachar. https://faridabadmajdoorsamachar.blogspot.com/2014/

Faridabad Majdoor Samachaar. (2019). New Axis, New Terrain, New Milieu. *Faridabad Majdoor Samachar, 378.*

Faridabad Majdoor Samachar. (2024). *मारुति सुजुकी कम्पनी की इन्डस्ट्रियल मॉडल टाउन मानेसर और गुड़गाँव फैक्ट्रियों में मजदूरों की गतिविधियाँ मजदूर समाचार, मई 2007 सेअगस्त 2019 में प्रकाशित.* मजदूर समाचार-कम्युनिस्ट क्रान्ति प्रकाशन. https://faridabadmajdoorsamachar.noblogs.org/books/

Firestone, S. (2003). *The Dialectic of Sex: The Case for Feminist Revolution* (1. ed). Farrar, Straus and Giroux.

Graeber, D. (2018). *Bullshit Jobs: A Theory.* Simon & Schuster.

Haraway, D. J. (with Internet Archive). (1991). *Simians, cyborgs, and women: The reinvention of nature.* Routledge. http://archive.org/details/simianscyborgswo0000hara

Kumar, A. (2019, November 20). Honda Workers' Protest Enters 16th Day, Over 3,000 Camp Outside Manesar Plant. The Wire. https://thewire.in/labour/honda-workers-protest-manesar-haryana

Siddiqi, D. M. (2000). Miracle worker or womanmachine? Tracking (Trans) national realities in Bangladeshi Factories. *Economic and Political Weekly, 35*(21/22), L11–L17.

Siddiqi, D. M. (2009). Do Bangladeshi factory workers need saving? Sisterhood in the post-sweatshop era. *Feminist Review, 91*(1), 154–174. https://doi.org/10.1057/fr.2008.55

Siddiqui, D. (2015). Miracle worker or Womanmachine? *Economic and Political Weekly,* 7–8.

Spivak, G. C. (1988). Can the subaltern speak? *Die Philosophin, 14*(27), 42–58. https://doi.org/10.5840/philosophin200314275

Stengers, I. (2000). *The invention of modern science.* University of Minnesota Press.

Open Access This chapter is licensed under the terms of the Creative Commons Attribution-NonCommercial-NoDerivatives 4.0 International License (http://creativecommons.org/licenses/by-nc-nd/4.0/), which permits any noncommercial use, sharing, distribution and reproduction in any medium or format, as long as you give appropriate credit to the original author(s) and the source, provide a link to the Creative Commons license and indicate if you modified the licensed material. You do not have permission under this license to share adapted material derived from this chapter or parts of it.

The images or other third party material in this chapter are included in the chapter's Creative Commons license, unless indicated otherwise in a credit line to the material. If material is not included in the chapter's Creative Commons license and your intended use is not permitted by statutory regulation or exceeds the permitted use, you will need to obtain permission directly from the copyright holder.

28

Exploring How Metaphors of Change Prefigure Futures in Public Policy, Social Movements, and Community Projects

Johan Holmén, Clara Saglietti, and John Holmberg

1 Introduction

Spaceship earth, doughnut economics, growth addiction, nurturing community ecosystems, transformation pathways, all represent terms we come across in work with sustainability and the environment. These terms are arguably *metaphorical*, that is, they draw meaning from one thing to help us understand another. This chapter builds on conceptual metaphor theory (Lakoff & Johnson, 1980) to discuss how various metaphors employed in sustainability initiatives and movements not only

---

Metaphors are pervasive in everyday life, not just in language but in thought and action—George Lakoff and Mark Johnson, 1980.

J. Holmén (✉)
Department of Engineering Science, University West, Trollhättan, Sweden
e-mail: johan.holmen@hv.se

J. Holmén · C. Saglietti · J. Holmberg
Department of Space, Earth and Environment, Chalmers University of Technology, Gothenburg, Sweden

© The Author(s) 2026
S. Sareen and S. Juhola (eds.), *Societal Transitions to Sustainability*, Palgrave Studies in Environmental Transformation, Transition and Accountability,
https://doi.org/10.1007/978-3-032-07395-2_28

function as linguistic devices, but play a role in the prefiguration of alternative futures, bridging desired ends with thinking, action, and being in the here-and-now.

Conceptual models on systemic transformations tend to position world-view, paradigm, and their influence by metaphor constructs at the deepest leverage of change (Davelaar, 2021). In the face of contemporary and persistent societal challenges lie a natural necessity for deliberate and purposeful change towards desired futures—involving new ways of thinking, doing, being, relating, and imagining in this world (Lotz-Sisitka et al., 2024; Scoones et al., 2020; West et al., 2024). Actors, initiatives, and movements seeking to enact transformations into alternative futures operate on the edge between two worlds: the world as it is, and the world they aspire to create (e.g. Senge, 1990; Seyfang & Smith, 2007; Swilling, 2019). As futures are open-ended (e.g. Horst & Gladwin, 2024), change strategies differ across initiatives, and actions situated, transformative practices might look very different depending on context.

This chapter seeks to elicit, contrast and discuss the metaphors-in-use to guide change in three diverse and distinct deliberate transformation initiatives: the 100 *climate neutral cities mission* by the European Commission; *degrowth* as an academic and social movement, and *transition towns* as a representative community grassroot project. Prefiguration—that is, the achievement of means–ends coherence in thought and action—may play out differently depending on context.

The structure of the chapter is as follows: Sect. 2 provides a brief background into metaphors of change. Section 3 introduces the three selected cases with emphasis on their metaphors-in-use. Section 4 analytically compares the cases and their metaphors. Section 5 integrates observations and claims from the previous sections into a discussion on how metaphors of change might prefigure transformations from a backcasting perspective. Section 6 summarise the main findings and provide conclusions and ways forward.

2   Metaphors of Change

Metaphors are more than just a matter of words or language. They are cognitive devices underpinning the very frames within which we make sense of and engage in the world. Lakoff and Johnson (1980) illustrated in their seminal book how our ordinary conceptual system—in terms of how we think and act—is fundamentally metaphorical in nature. Two classical examples involve *up is good*, and *affection is warmth*, which build on our experiences (e.g. associating the affection from a parent to temperature) and create complex systems of meaning that characterise individual and cultural frames. Moreover, viewing and talking about organisations as machines normalise emphasis on efficiency, control, rational planning, and hierarchical structure. Approaching organisations as organisms rather bring attention to adaptation, environment and interconnections (Morgan, 1980; Scott & Davis, 2006).

Metaphors can work in both the direction of reducing something more complex or abstract into a particular frame and may as well open up and pluralise something initially conceived of as simple or unambiguous. Metaphors that stick tend to be persuasive because they offer an entry point—or shortcut—into grasping abstract or complex phenomena by strategically drawing from something more familiar (Flusberg & Thibodeau, 2023). They transfer meaning from one domain to another, making the unfamiliar more understandable. For example, we often borrow metaphors from computing to conceptualise the brain as an information processing device, or from war to describe political discourse—attacking opponents, winning debates, or shooting down arguments. Such metaphors do more than describe; they uphold particular frames, norms, and rules, shaping not only how we talk about something but also how we act within it (Lakoff & Johnson, 1980).

Work on sustainability and the environment comes with a wealth of metaphors. Some are deliberately chosen and designed, whereas others are more subtle. Kenneth Boulding's (1966) space ship earth is an early example, where the earth is conceptualised as a shared vessel with finite supplies. Doughnut economics (Raworth, 2017) conceptualises humanity's safe and just operating space as the area between two concentric boundaries: an inner social foundation and outer ecological ceiling, represented by the inner and outer edges of a doughnut-shaped diagram. Most would however agree that the planet is not de facto a spaceship, nor are we living in a baked doughnut. What these metaphors do effectively is that they package complex scientific and moral imperatives into graspable concepts that guide thinking and action. They also tell stories of boundaries, limits and constraints (see Leach et al., 2010). More subtle use of metaphor may include the framing of transformation strategies as a matter of *pathways*—inviting imagery and vocabulary of roads, journeys, crossroads, and dead ends.

In this chapter, any metaphor part of a deliberate change effort is referred to as a metaphor of change, to distinguish between the general metaphors part of everyday language, and particular conceptual metaphors used to guide and motivate action in a certain direction.

3   Case Studies

Below, metaphors used to narrate, embody and enact alternative and desired futures in selected cases are surfaced, presented, and discussed. It is also stressed that metaphors naturally become part of a wider discourse, that is, a patterned way of talking about, understanding and structuring reality.

3.1   EU Cities Mission

The EU Cities Mission, formally, the mission to achieve over 100 Climate-Neutral and Smart Cities by 2030 provides a strategic agenda by the European Commission. The initiative is deliberately framed as a mission, a metaphor reminding us of the moonshot but in this case to inspire bold action on climate change as a collective endeavour bringing together cities, sectors and actors around a shared and meaningful objective. As with the moon landing, it was by the public considered unrealistic when announced by John F. Kennedy in 1961, and there was no ready-made plan on how to achieve the mission by the end of the decade. Ursula von der Leyen has described the European Green New Deal, within which the EU Cities Mission is part, as Europe's man-on-the-moon moment (European Commission, Brussels, 2019). The initiative draw from Mariana Mazzucato's concept of mission-oriented innovation (Mazzucato, 2018) and transformative innovation policy (European Commission. Joint Research Centre, 2023).

The metaphor carries an expectation of a transformative achievement of something that has never been done before, which may lead to significant technological development as well as a changed perspective on our place and this planet (Cf. The Blue Marble)—as missions set a tone of a grand challenge and collective purpose. When encouraging cities to join the EU Cities Mission, climate neutrality becomes a society-wide quest requiring innovation, creativity, and "moonshot" courage. To what extent is the choice and use of a mission metaphor central in achieving systemic innovation and a paradigm shift in urban development and ways of living towards climate neutral, smart, and inclusive cities, whose form is not known in advance but emerge through process?

Climate neutrality may at first be understood as giving a sense of balance or harmlessness. But it means net-zero carbon emissions, and much practical effort is technically oriented towards accelerating a transition decarbonising urban energy and transportation systems. The mission blends technological and social imaginaries, with top-down policy, coordination and technological intervention—complemented with mantras such as "by and for the citizens" and "co-creation processes" hinting towards more bottom-up and human-centric change processes. Here, cities become arenas or places where all sectors and actors intersect and interact. By employing accessible and inclusive metaphors such as mission, journey, hubs, contracts, the EU initiative frames an otherwise complex and technicist mix of policies on energy, transportation, buildings, and infrastructures into a compelling mission of transforming cities into sustainability by 2030.

Within the mission umbrella lie initiatives such as pilot cities and experimentation hubs. These position cities as living laboratories where new solutions and policies can be tested, evaluated, and scaled. Cities become pioneers, taking the lead, moving faster than others, exploring unchartered terrain—an extension of the metaphor. The mission also employs climate city contracts, an agreement that each city signs committing to climate neutrality including a co-created plan with local stakeholders. Contracts may be more literal than metaphorical, but they evoke ideas of a social pact or partnership. Journey metaphors are also used in their communication, where the mission is depicted as a collective voyage with milestones—requiring navigation and adaptation given challenges and surprise that arise along the way.

Early evidence on its communicative impact involves cities celebrating their role of being among the 100 chose ones and a showcasing of their mission labels. Critics argue the mission upholds a bold intention while also mobilising mainstream actors—and as incrementalism within existing bureaucracies will not do. What the case does is providing a concrete example of how metaphors help promote ideas, muster, and mobilise social momentum at scale.

3.2   Degrowth Movement

While the EU Cities Mission to seeks inclusivity in terms of framing and actor involvement, they are less critical towards dominant cultures, structures and practices. In that regard, the degrowth movement may be increasingly counter-hegemonic (Hamilton & Ramcilovic-Suominen, 2023) by contesting deeper held assumptions, values, beliefs, discourses—and so metaphors of progress, well-being, and change.

At its core lie the term degrowth (Latouche, 2004) which has been termed a missile word—a jolty term piercing the complacency of pro-growth discourse. By taking the well-established concept of economic growth that is generally seen as positive, and put a negation in front of it, established conventions are challenged. It challenges the central held metaphors that equal up with good and more with better. The movement builds on earlier critique on resource-dependent economic growth on a finite planet (Boulding, 1966; Daly, 1973; Meadows et al., 1972). And rather than being agnostic towards solutions or arguing for decoupling, the movement asks critical questions on forms of whether growth at all is good, for whom, and why—and suggests an alternative development ideal.

Metaphorically, the economic system can be treated like a living organism that can grow—quantitatively and qualitatively—with generally positive connotations. Growth is good, healthy, upward, and even natural. Stagnation or shrinkage means failure, weakness or death. As a result, any alternative to the positive growth story faces a cognitive uphill battle. In their book *The Case for Degrowth*, Kallis et al. (2020) nuance this view and explain growth as possessing cyclical, perpetual, and compounding rhythms. Cyclical processes are also natural and not viewed negatively.

To this come a longer conversation, where degrowth proponents continuously put efforts on explaining that degrowth is not recession, but a planned and purposeful activity to downgrade and down scale life into well-being, harmony, conviviality, sufficiency, and meaning beyond destructive material consumption and accumulation. The danger rather lies in ignoring or negating the role of growth as structuring economic and societal activity and its implications for humanity and the planet (Parrique, 2019), to which critique of capitalism often follow (e.g. Feola, 2019).

Degrowth comes with a host of additional metaphors reframing economy and society within their discourse. A prevalent one is the *addiction* metaphor coined by Serge Latouche, diagnosing the problem of our societies as a term of addiction to growth—like a drug. Or endless expansion of GDP according to Edward Abbey being classified as a *pathological dependency*. Both examples cast the current paradigm as inherently self-destructive and in need of intervention and cure—calling for withdrawal, rehab and reversal into a new everyday life, on a societal level. These metaphors provide a stark contrast to the idea of growth as good, positive, and natural. Other metaphors used include the *stepping out of the growth bubble*—including warning that bubbles may burst, the necessity of *stepping off the growth treadmill*, or *slowing down to a snail's pace*.

Degrowth in contrast to the EU mission project less prevalent in mainstream politics. The movement focuses on critique in the search of a different metric and logic for evaluating progress and success. The degrowth movement may be conceived of as increasingly counter-hegemonic, which by some is argued to be its strength, but by some its weakness. It invites critical conversation on the role of GDP while challenging the big institutions working on global trade, financial stability, development and industrial growth—such as the World Bank, IMF, WTO, OECD, Federal Reserve, the G7, or G20-meetings or World Economic Forum. With equal emphasis, the degrowth movement is also present in sharing and repair circles and other forms of bottom-up engagement outside dominant institutions.

Degrowth metaphors function prefiguratively in the sense that they invite for and embody deviating processes from materialist lifestyles, in search for simplicity as a present virtue and critique of an established development paradigm. In the process of surfacing and questioning deeply held beliefs and conceptual root metaphors (e.g. up is good and more is better), the movement tends to mobilise some groups while risking excluding those who find it too radical. The movement relates to and critiques mainstream growth discourse, to reframe its broader structures, narratives, and assumptions.

The third and last case of Transition Towns is a more grounded grassroots example that instead of bashing, taming, or corroding the existing seeks to escape incumbency and instead focus on escaping and building up safe alternatives on the side (Cf. Wright, 2010). They seek to inspire local action and frugal innovation in the here and now. The transition town movement is a usual suspect in this context and well-researched (e.g. Aiken, 2012; Smith, 2011; Taylor, 2012), and function well to make a point and propose a third alternative in this analysis from a basis of their metaphors-in-use.

3.3   Transition Towns

Transition towns, originating in Totnes, UK in the mid-2000s is a global network, seeking to empower and equip bottom-up grassroots and communities with concrete practices to transition away from fossil fuel dependence toward resilience in the face of climate change and resource depletion.

The concept of transition town is a prefigurative element and arguably metaphorical. It casts the community as being "in or of" transition, an ongoing journey from one state of being to another. Rather than mission, revolution, planned decline, or collapse, it is a guided and phased deliberate purposeful change. Their frontperson, Rob Hopkins, frame the transition movement as being a matter of steps, pathways, stories of travel from a problematic present to a post-carbon future (Hopkins, 2008). By using a simple and concrete language associated with transition, radical change is sought to be feasible and inviting, as a matter of practicing better and hopeful futures in the here-and-now. They maintain optimism that communities have resources and capability to act and can move forward together, also through difficulty.

Resilience is a key concept, borrowed from ecology, characterising the ability of a system to withstand, absorb, and recover from shocks. Resilience is portrayed as a development ideal, rather than degrowth (case 2) or climate neutrality and sustainable development (case 1). They encourage communities to enhance local resilience, by developing local food production, renewable energy, and community skills—to withstand external crises like oil price spikes, economic recession, or climate impacts. The ecological metaphor brings a holistic, natural systems perspective to community development—emphasising diversity, decentralisation, and balance.

Transition towns have a long record of positive visioning work, crafting rich narratives and persuasive images of towns as gardens with tress, fruits, vegetables and animals; local economies as strong webs; and neighbourhoods as villages. Transition towns embody a transition period into futures where we are more connected to each other, nature, and the places we live. Settlements can be net exporters of energy, seasonal and local food, and urban landscapes characterised by permaculture. This paints an imagery of connected places (social bonds), diets in season (living with nature's cycles), and urban landscapes of food production (town as garden).

These images guide practices such as community gardens, tool-sharing libraries and local currencies. Under an umbrella of transition and resilience, transition towns prefigure being and acting towards healing or strengthening the community fabric, like enhancing an ecosystem's biodiversity and restoring soil health. Through embodying and enacting values of community self-reliance and sufficiency, actions are encouraged towards, e.g. planting vegetables and installing solar panels. Change is gradual and led from the bottom-up in terms of what can practically be done, while aiming towards a fundamentally different destination (not just greening cities or shrinking economies), emphasising community resilience to merge individual and collective agency.

Transition town has led to community-led social innovations through frugal means. Repair cafés, oil awareness sessions, seed swaps, celebration events on local culture, and storytelling evenings. These activities function symbolically for the futures they seek to build through meaningful action in the present. Moving beyond critique, pilots, or experiments, transition towns host an assembly of conceptual metaphors and images that help people enact radically different futures involving new socioecological bonds, thereby living post-carbon futures. Success is however not guaranteed and stability as well as scaling is complex and dynamic, building on local–global learning processes, contextual fit, attachment to place, cohesion, available resources, political support, and more (Feola & Nunes, 2014).

4   Contrasting and Comparing Cases and Metaphors

Below, central metaphors of change identified in the three cases are discussed alongside their overall (transformational) ambition and prefigurative efforts, as summarised in Table 1.

The metaphors used by the EU Cities Mission, the Degrowth Movement, and the Transition Towns initiative provide rich insights into how each movement conceptualises transformation and mobilises action. While all three aim to respond to contemporary sustainability challenges and enact systemic change, their metaphorical framings in terms of style and scope diverge, providing increased possibility to compare and contrast their underlying assumptions about agency and change.

The EU Cities Mission employ metaphors of mission, moonshots, piloting and journeys, drawing from technical, goal-oriented and strategic concepts. These metaphors emphasise directionality, coordination, shared ambition, and a positioning of transformation as a collective,

Table 1 Summary of common metaphors-in-use, exemplary overall (transformational) ambition, and types of prefigurative efforts in the three selected cases. *Source* Authors

Case	Common metaphors-in-use	Exemplary overall (transformational) ambition	Types of prefigurative efforts
EU cities mission	Mission and moonshot. Collective journey and endeavour to climate neutrality. Hub, labs, pilot, experimentation and frontrunner cities.	Climate neutrality in 100+ European cities by 2030, establishing these cities as innovation hubs, for other cities to follow suit by 2050. Ancillary benefits include improved air quality, reduced noise pollution, energy-efficient buildings, inclusive and participatory governance.	Climate city contracts: Breaking from business-as-usual by bringing 2050 climate goals to present decision making and development processes. Inspires bold action by invoking historical task, encouraging systemic innovation and risk taking, by and from within existing systems. Aligns actors under shared goal, legitimisd through mission labeling; rallying multi-level governance, researchers, citizens, private sector actors etc. to collaborate and create shared roadmaps.
Degrowth movement	Degrowth as provocative inversion of growth metaphor. Society is addicted, growth as disease. Slowing material throughput – health and balance rather than speed and growth, living well with less.	Overthrowing growth-based model of economic development, societal progress and overall success. Ecological sustainability, social equity, democratic participation well-being beyond GDP; staying within planetary boundaries. Deliberate downscaling of economic production and consumption in rich countries, reimagining prosperity, promoting just and regenerative futures. Slowness as care elaborated via a snail metaphor.	Paradigm critique, exposing current economic model as myth-based and self-destructive. Opens space to think and live economy differently into post-growth societies and planned phasing. Experimentation with post-consumerist lifestyles, adopting simple measures to detox now.

(continued)

Table 1 (continued)

Case	Common metaphors-in-use	Exemplary overall (transformational) ambition	Types of prefigurative efforts
Transition towns	Transition and journey, community moving from oil dependency to sustainable future. Resilience and emphasis on the local and the ground. Towns as ecosystems, resilience-oriented development.	Community resilience and self-sufficiency/reliance: ecosystem. Local economies, circularity, weaving social fabrics, roots and soil, growing local, energy descent, re-localisation. Inner transition, revived culture and caring and regenerational-socio-ecological relationships.	Empowers community agency and adaptation, accessible, constructive, inviting. Motivating ordinary people to act in the here-and-now on an everyday, simple basis, planting the seeds of the future. Global and planetary changes handled through hope, transformation as possible adventure of revival, overcoming fear, sense of belonging, momentum and inclusive participation. Prototype future solutions now, suburbs becoming food forests. Toolboxes for change, co-creation, empowerment.

time-bound, and bureaucracy-led process. The moonshot metaphor positions the policy initiative as a (post-)historic achievement that required institutional alignment, bold experimentation, risk taking, and a mobilisation of deep technical expertise. This framing and associated metaphors align with systemic prefiguration within existing structures via tools such as climate city contracts and innovation hubs, where widened collaboration and experimentation is legitimised but ultimately managed from the top-down.

In contrast, the Degrowth Movement radically subverts dominant metaphors associated to progress. Portraying growth, particularly in rich countries, as form of addiction or disease and illusory objective being a pathological attachment to endless accumulation. Through metaphors of detox, slow down, living well with less, re-purpose, degrowth open a space for radically re-thinking prosperity and well-being. The movement's metaphors are both critical and reconstructive: by both critiquing and dismantling economic orthodoxy they also propose alternative imaginaries grounded in sufficiency and planetary care. Prefigurative efforts, while less apparent than in EU Cities Mission or Transition Towns, are enacted through everyday practices experimenting with ways of post-consumerist living, aiming to embody post-growth futures in the here-and-now similar to transition towns and grassroot communities alike, while also working towards changing dominant institutions and macro-scale economic transformations.

Transition towns occupy a third space, combining metaphorical elements of journeying, ecosystems and local rootedness. Their central metaphor of transition conveys both movement and becoming—shaping communities towards agents of adaptive and regenerative change. Rather than describing cities as economic machines that need growth to sustain, or as sites for innovation, transition towns portray towns as living systems, thriving through diverse local interconnections. The ecological metaphors of roots, seeds, soil, and resilience emphasise interconnection, circularity, and care, while fostering hope and belonging. Unlike the strategic orientation of EU Cities Mission or rupture manifested in Degrowth, Transition Towns increasingly rely on culturally resonant and emotionally supportive metaphorical landscape that encourage practical action and inner transformation at the community level. Prefiguration is expressed in the everyday way of being and co-creating in community gardening, tool-sharing, and re-localised economies from the grassroot level.

EU Cities Mission and Transition Towns share a generally optimistic "it is possible" framing. Missions can succeed—we made it to the moon—and transition towns have a reachable destination evidenced by the fact that several towns and communities already manage. Here, the obstacle is rather a matter of pace and scale given the width of the challenges ahead.

5   Metaphors as Prefigurative Devices in Transformations

As Lakoff and Johnson (1980) demonstrated over 40 years ago, we argue that metaphorical distinctions may have significant implications in societal transformations, as illustrated in the above cases. Also, on a level above the particular cases lie their general type of initiative—policy, critical movement, or community project—which also likely influence their orientation and use of metaphor. They reflect distinct theories of change in terms of their primary orientation: working within the system, against the system, or alongside existing systems.

The EU Cities Mission risk reinforcing technocratic control, a top-down logic and ecological modernisation towards green growth if not critically examined, even as it seeks to stimulate systemic innovation. The Degrowth Movement offer a paradigm critique but may struggle with broader resonance partly due to its critical orientation strengthened by provocative metaphors. Transition Towns, in contrast, encapsulate several messages from the degrowth movement but offer increasingly emotional and practically grounded pathways in, while lacking structural leverage or institutional power.

There are links or traces of the metaphors-in-use in the respective case's transformational ambitions and prefigurative efforts. The metaphors become represented and embedded through being and action, to the extent metaphors possibly can be viewed as tools or devices that influence the transformative impact of action and initiative. When metaphors render futures increasingly imaginable and emotionally resonant, they provide guidance in larger mobilisation efforts including motivating action and commitment. They can also mediate tensions between individual agency and systemic change.

Throughout history, dominant metaphors have structured societal development and stabilised certain worldviews: the machine metaphor of the industrial age enabled standardisation and top-down control; emerging ecological metaphors reframe the economy as embedded in the biosphere. Paradigm and other forms of deep leverage shifts often hinge on the adoption of new metaphors—those that make the invisible visible, challenge assumption, and open up novel trajectories (Davelaar, 2021; Hamilton & Ramcilovic-Suominen, 2023; Inayatullah, 1998). The power of metaphors possibly comes from its capacity to reframe reality and make a new normal–making the unimaginable imaginable, the abnormal normal, or distant close. They lower barriers to collaboration through inclusive language, promoting alignment, and through raising ambition–by making the intangible accessible, underpinning shared stories, and a sense of being part of something bigger together.

For practitioners and changemakers interested in the art of deliberate and purposeful prefiguration of desired futures, backcasting and anticipation provide two possible methodological frameworks that may be useful but remain underdeveloped in terms of their relation to prefiguration and metaphor. Backcasting seeks to articulate desired futures and figure out ways of getting there (Robinson et al., 2011). Originally, the approach was developed as a complement to forecasting, and in particularly principles-based backcasting (Holmberg, 1998) hold several similarities with prefiguration. However, most contemporary backcasting work overlook the power of metaphor and prefigurative possibilities to dissolve temporal boundaries between the present and future. Similarly, anticipation work has advanced understandings of imaginaries, narratives, and temporalities (Fuller, 2018; Muiderman et al., 2020), but largely overlooked the role of metaphor as a constitutive tool in how futures can be imagined, communicated, and enacted.

A well-chosen metaphor can unite diverse actors, inspire commitment, and anchor complex goals in resonant stories. Poorly chosen metaphor might equally backfire and constrain action. Moreover, metaphors are not fixed—they evolve with practice and context and can be refined as new needs and understandings emerge. In this sense, they are both symbolic and strategic. Across the three initiatives explored in this chapter, it was illustrated how metaphors not simply reflect prefigurative practices but help constitute them. Moonshots in the EU Cities Mission generate a sense of urgency and shared ambition, while soil and roots in Transition Towns invite for grounded, regenerative change. The Degrowth movement's metaphors of detox and addiction challenge the dominant economic development paradigm and challenge deeply held root metaphors that up is good and more is better.

As we navigate an era that of systemic change, being mindful of the metaphors that are historically inherited and intentional in those that are created becomes a crucial act. In a spirit of prefiguration, metaphors offer building blocks that bring futures that are not yet there to the here and now. If it is true that "in order to change the world we must change the story", then metaphors are foundational to that change.

Competing Interests   The authors have no conflicts of interest to declare that are relevant to the content of this chapter.

References

Aiken, G. (2012). Community transitions to low carbon futures in the Transition Towns Network (TTN). *Geography Compass, 6*(2), 89–99. https://doi.org/10.1111/j.1749-8198.2011.00475.x

Boulding, K. E. (1966). The economics of the coming spaceship earth. *Environmental Quality Issues in a Growing Economy.*

Daly, H. E. (1973). *Toward a steady-state economy.* W. H. Freeman and Company.

Davelaar, D. (2021). Transformation for sustainability: A deep leverage points approach. *Sustainability Science, 16*(3), 727–747. https://doi.org/10.1007/s11625-020-00872-0

European Commission. Joint Research Centre. (2023). *Capacities for transformative innovation in public administrations and governance systems: Evidence from pioneering policy practice*. Publications Office. https://data.europa.eu/doi/https://doi.org/10.2760/220273

Feola, G. (2019). Capitalism in sustainability transitions research: Time for a critical turn? *Environmental Innovation and Societal Transitions*. https://doi.org/10.1016/j.eist.2019.02.005

Feola, G., & Nunes, R. (2014). Success and failure of grassroots innovations for addressing climate change: The case of the Transition Movement. *Global Environmental Change, 24*, 232–250. https://doi.org/10.1016/j.gloenvcha.2013.11.011

Flusberg, S. J., & Thibodeau, P. H. (2023). Why is mother earth on life support? Metaphors in environmental discourse. *Topics in Cognitive Science, 15*(3), 522–545. https://doi.org/10.1111/tops.12651

Fuller, T. (2018). Anticipation and the normative stance. In R. Poli (Ed.), *Handbook of anticipation* (p. 7). Springer International AG.

Hamilton, R. T. V., & Ramcilovic-Suominen, S. (2023). From hegemony-reinforcing to hegemony-transcending transformations: Horizons of possibility and strategies of escape. *Sustainability Science*. https://doi.org/10.1007/s11625-022-01257-1

Holmberg, J. (1998). Backcasting: A natural step in operationalising sustainable development. *Greener Management International, 23*, 30–51.

Hopkins, R. (2008). *The transition handbook: From oil dependency to local resilience*. Bloomsbury Publishing. https://www.tandfonline.com/doi/full/https://doi.org/10.1080/09581596.2010.507961

Horst, R., & Gladwin, D. (2024). Multiple futures literacies: An interdisciplinary review. *Journal of Curriculum and Pedagogy, 21*(1), 42–64. https://doi.org/10.1080/15505170.2022.2094510

Inayatullah, S. (1998). *Poststructuralism as method*. 15.

Kallis, G., Paulson, S., D'Alisa, G., & Demaria, F. (2020). *The case for degrowth*. Polity Press.

Lakoff, G., & Johnson, M. (1980). *Metaphors we live by*. The University of Chicago Press.

Latouche, S. (2004). Degrowth economics. *Le Monde Diplomatique*.

Leach, M., Scoones, I., & Stirling, A. (2010). *Dynamic sustainabilities: Technology, environment, social justice*. Earthscan.

Lotz-Sisitka, H., Pahl-Wostl, C., Meissner, R., Scholz, G., Cockburn, J., Jalasi, E. M., Stuart-Hill, S., & Palmer, C. (Tally). (2024). Interrelated transformative process dynamics in the face of resource nexus challenges: An invitation towards cross case analysis. *Ecosystems and People, 20*(1), 2297707. https://doi.org/10.1080/26395916.2023.2297707

Mazzucato, M. (2018). Mission-oriented innovation policies: Challenges and opportunities. *Industrial and Corporate Change, 27*(5), 803–815. https://doi.org/10.1093/icc/dty034

Meadows, D. H., Meadows, D. L., Randers, J., & Behrens, W. W. (1972). *The Limits to growth: A report for the Club of Rome's project on the predicament of mankind*. Universe Books.

Morgan, G. (1980). Paradigms, metaphors, and puzzle solving in organization theory. *Administrative Science Quarterly, 25*(4), 605. https://doi.org/10.2307/2392283

Muiderman, K., Gupta, A., Vervoort, J., & Biermann, F. (2020). Four approaches to anticipatory climate governance: Different conceptions of the future and implications for the present. *WIREs Climate Change, 11*(6). https://doi.org/10.1002/wcc.673

Parrique, T. (2019). *The political economy of degrowth*. 872.

Raworth, K. (2017). *Dougnut economics: Seven ways to think like a 21st-century economist*. Chelsea Green Publishing.

Robinson, J., Burch, S., Talwar, S., O'Shea, M., & Walsh, M. (2011). Envisioning sustainability: Recent progress in the use of participatory backcasting approaches for sustainability research. *Technological Forecasting and Social Change, 78*(5), 756–768. https://doi.org/10.1016/j.techfore.2010.12.006

Scoones, I., Stirling, A., Abrol, D., Atela, J., Charli-Joseph, L., Eakin, H., Ely, A., Olsson, P., Pereira, L., Priya, R., van Zwanenberg, P., & Yang, L. (2020). Transformations to sustainability: Combining structural, systemic and enabling approaches. *Current Opinion in Environmental Sustainability*. https://doi.org/10.1016/j.cosust.2019.12.004

Scott, R. W., & Davis, G. F. (2006). *Organizations and organizing: Rational, natural and open systems* (6th ed.). Prentice Hall.

Senge, P. (1990). *The fifth discipline: The art and practice of the learning organization* (1. Currency paperback ed). Currency Doubleday.

Seyfang, G., & Smith, A. (2007). Grassroots innovations for sustainable development: Towards a new research and policy agenda. *Environmental Politics, 16*(4), 584–603. https://doi.org/10.1080/09644010701419121

Smith, A. (2011). The transition town network: A review of current evolutions and renaissance. *Social Movement Studies, 10*(1), 99–105. https://doi.org/10.1080/14742837.2011.545229

Swilling, M. (2019). *The age of sustainability: Just transitions in a complex woSrld*. Routledge.

Taylor, P. J. (2012). Transition towns and world cities: Towards green networks of cities. *Local Environment, 17*(4), 495–508. https://doi.org/10.1080/135 49839.2012.678310

West, S., Haider, L. J., Hertz, T., Garcia, M. M., & Moore, L. (2024). Relational approaches to sustainability transformations: Walking together in a world of many worlds. *Ecosystems and People, 20*(1), 1–27.

Wright, E. O. (2010). *Envisioning real Utopias*. Verso.

Open Access This chapter is licensed under the terms of the Creative Commons Attribution-NonCommercial-NoDerivatives 4.0 International License (http://creativecommons.org/licenses/by-nc-nd/4.0/), which permits any noncommercial use, sharing, distribution and reproduction in any medium or format, as long as you give appropriate credit to the original author(s) and the source, provide a link to the Creative Commons license and indicate if you modified the licensed material. You do not have permission under this license to share adapted material derived from this chapter or parts of it.

The images or other third party material in this chapter are included in the chapter's Creative Commons license, unless indicated otherwise in a credit line to the material. If material is not included in the chapter's Creative Commons license and your intended use is not permitted by statutory regulation or exceeds the permitted use, you will need to obtain permission directly from the copyright holder.

29

Opening at the Close: Societal Transitions to Sustainability

Sirkku Juhola⊙ and Siddharth Sareen⊙

1   Introduction

This volume began with the notion of the necessity of present transformations towards more sustainable futures for all. The sense of urgency linked to the polycrisis directs attention to the scope for present transformations and the extent to which they can fulfil this need. Present transformations are those that are taking place around us. Their limited nature compared to the vast challenges we face related to climate change, biodiversity loss, and global conflict and inequality can make us question to what extent these efforts will suffice and how fast and far they can scale.

S. Juhola
Faculty of Biological and Environmental Sciences, University of Helsinki, Helsinki, Finland
e-mail: sirkku.juhola@helsinki.fi

S. Sareen (✉)
Fridtjof Nansen Institute, Lysaker, Norway
e-mail: ssareen@fni.no

© The Author(s) 2026
S. Sareen and S. Juhola (eds.), *Societal Transitions to Sustainability*, Palgrave Studies in Environmental Transformation, Transition and Accountability, https://doi.org/10.1007/978-3-032-07395-2_29

The enormity of this global and complex challenge places us in a position where we must surrender to the idea that ordinary, day-to-day practices are not enough; rather, drastic, and wide-reaching changes need to take place to make the planet liveable for future generations. This implies that requisite action needs to take place now, undertaken by us, through the various ways and means at our differential disposal. It means surrendering to the task at hand and accepting that the status quo is no longer acceptable. This is a big, difficult concession to make, especially since the status quo suits many of us rather well at the moment, such as fossil fuel incumbents enriching themselves while escalating our planetary peril.

It is this necessary surrender and what it enables that we highlight through the title of this chapter. In the Harry Potter book series, the protagonist wins a golden snitch in his first game of Quidditch, which carries the phrase "I open at the close". There, the "close" refers to Harry submitting to his fate at the wrathful hands of Voldemort, at which point the snitch opens to give Harry the resurrection stone as a helpful tool for his eventual victory. In our case, "opening at the close" refers to the action of being the change that is needed in the world in the face of polycrisis. By examining the concept of prefigurative politics, we understand approaches to present transformations as they unfold to challenge the rigid structures and relations that must be undone, reoriented, and fashioned anew.

At the beginning of this volume, we offered a definition of prefigurative politics as *embodied strategies to render desirable futures with immediacy* (Sareen & Juhola, 2025). This definition, drawing creatively on existing scholarship on prefigurative politics, somewhat broadens the very concept. Indeed, this is the purpose of the definition and the collective effort in this book. Prefigurative politics as a concept, we hold, has the potential to shed light on how and why present transformations come about and why they might not. In fact, each single word of our definition, when examined separately, illustrates the contribution that this volume makes to the understanding of prefigurative politics in present transformation.

To embody means to be an expression of, or to give a tangible or visible form to, an idea or feeling. Our collection of chapters illustrates how different ideas and notions of sustainability transformations are becoming tangible across different thematics and sectors. There are numerous examples of challenging existing systems with regard to matters like sustainable production and consumption across a variety of sectors. What ties these contributions together is a focus on ideas for seeing and doing things differently, and the acknowledgement that this is sorely needed.

These ideas are being enacted in various spaces, groups, and networks. Geographically speaking, they are becoming tangible across the urban and the rural, and from formalised spaces of social organisation such as the medical field to frontier spaces where little formal organisation currently exists. These ideas are emerging through vibrant networks of individuals and groups, who foster them into life and give them meaning.

Strategies, here interpreted loosely as a plan of action, have been a contentious topic in prefigurative politics. Here, we wish to specifically note Monticelli's concern about the role of prefigurative politics and its relationship to (state) power. The role of prefigurative politics and how it is situated in relation to formal institutions is also a theme that features in many of our chapters, which discuss the extent to which prefigurative politics can occur within existing institutions or even be nurtured by them. The accepted notion thus far has been that the very concept of prefigurative politics requires emancipation and alternatives that cannot take place within the structures that already hold power.

However, our contributions highlight how—within these structures and institutions—ways of prefigurative politics such as experimentation and citizen panels are changing the decision-making structures or adding and bending them to further account for issues of sustainability within them. Ultimately, the question is: does prefigurative politics require unmaking existing institutions, as some of our chapters discuss as well, or is remaking for sustainability possible within the existing institutional landscape by reconfiguring key power relations?

This leads us to discuss rendering, the act of making something or of something becoming a particular type of thing. The act of making something, in this case sustainability transformations, does not necessarily mean that that something is permanent or even successful to the point that prefiguration and the intent towards sustainability lead to substantial change. There are questions as to how far prefigurative politics relies on sudden dynamic challenge and how that translates into institutional developments in the longer run. Does prefigurative politics create a sudden disruption and change, which leads a new normal? Also, can prefigurative politics enable change at wider scales, and if so, what are the mechanisms for scaling up and forward?

These issues are also discussed in our chapters, where some of the prefigurative examples have faded away or not led to more permanent changes. In some of the cases discussed in our chapters, rendering is aided, in a sense, by outside funding, such as support from public officials or city departments. To what extent does this affect the act of rendering and the transformation becoming established, or does it in fact hinder it? Is spontaneity and truly bottom-up movement the only way to support sustainability transformations to truly emerge, or is there virtue to sowing seeds of institutionalisation that can be scaled up within larger systems upon recognition of their desirability in wider movements as well as administrative chambers?

The direction, the aim, and objective of these transformations speak directly to the idea of desirable futures. In this volume, we frame this desirability to account for a liveable planet on which societies can flourish within ecological and economic limits, with socially just outcomes. In many cases, this translates to smaller-scale changes that aim towards supporting these broader, global visions. Various examples collected here show the struggle towards not only breaking unsustainable patterns, but also the social struggles of marginalised and unheard communities against injustice in their attempt to ensure that their and their future generations are guaranteed a different future than those of the past.

The final word of our definition, *immediacy*, relays the urgency of the task at hand and directly relates to the need to see transformation emerge and take place across systems along a constrained timeline. Immediacy not only relates to the idea of instant involvement of different kinds of people and groups and organisations, but also to how this sense of urgency can be communicated more widely in communities to engage those who are not directly involved in the action.

Some of our examples of prefigurative politics of transformations address how established communities are facing unemployment and potential hardship as green transitions move through their territories which have previously relied on fossil fuel-based industries. While the accompanying emotions and fears can spark off the need for change, it is also important to pay attention to the everyday instances of inspiration, discovery, and laughter that bring people to the idea of change and lead them to espouse and nurture momentum. This is the type of excitement that captures peoples' attention and fosters further emotions and space for different type of techniques, for example through art-based methods or the use of language and metaphors.

2 Moving Towards Sustainability Through Prefigurative Politics?

Having worked through the definition, we now turn to the individual contributions to highlight the salient insights they contribute in this respect. This volume is loosely divided into four parts that range from the urban as a focal point of prefigurative politics to spaces of conviviality and politics. We also illustrate how prefigurative politics is taking place within sectoral movements, with a notable focus on food system sustainability and food sovereignty. Finally, our fourth part highlights cross-sectoral and transdisciplinary transitions.

In this section, we reflect upon the individual contributions convened in these four parts, drawing out the salient insights to elaborate upon what they collectively contribute to the overarching theme of the book in the conclusion.

The first part comprises chapters that focus on *urban sites of experimentation and contestation*, illustrating how cities as concentrations of people and social life create a fruitful environment for prefiguration.

In Chapter 2, Haugland and von Wirth (2025) address the relationship between prefiguration and urban experimentation. They unpack this nexus by focusing on car-free zoning in downtown Oslo as a manifestation of low-carbon future place-making. Materialising this experiential reality for residents provoked ephemeral contestation that was resolved through adjustments, while the broader push gained momentum through institutionalisation in the politics of urban development and the fabric of the city. The authors zoom out of these dynamics to reflect upon how legitimating present interventions relies upon particular constructions of desirable urban futures, signalling temporal asynchronicity across the infrastructural, cognitive, and normative planes of transformation.

In Chapter 3, Stella (2025) considers the Golf Vinoř Citizen Science Project on Prague's periphery in terms of residents' varied engagement with reclaiming an abandoned golf course. This prefigurative engagement encompasses participatory mapping, biodiversity monitoring, and workshops to rewild the territory, co-produce knowledge and transform land use towards a multifunctional landscape. This locally anchored and participatory approach to land management is at stark odds with a corporatised and widely influential single-use model of recreational land use. The chapter documents challenges of fragmented governance and ownership, as well as a grassroots-powered shift in public perception, towards commoning and participatory governance of land use change.

In Chapter 4, Batterbury, Chakori, and Uxo (2025) examine the prefiguration of lower consumption models for households through examples of community bike workshops and tool libraries located in Melbourne and Brisbane respectively. These interventions advance resource maintenance and reuse, a do-it-yourself approach premised on empowerment and building capabilities, and are anchored in convivial sharing of skills and resources. These prefigurative efforts propagate through what the authors posit as shadow networks with potential to challenge urban discourse and policy. But they also recognise the fragility of these efforts which are contingent on voluntary dedication and knowhow to engender a culture of commoning and salvaging.

In Chapter 5, Grumiau and Hughes (2025) focus on the radical vehicle of democratic assemblies through a novel lens that combines attention to societal innovation and human flourishing for transformation to sustainability. Their approach problematises the very idea of action in prefiguration, by pointing out its generative capacity to birth new points of departure and trouble and reframe the taken-for-granted. They project Chilean citizen assembly participants' refrain "starting by doing" to emphasise how action can inform the evolution of thinking and set in motion processes that may not emerge from the realm of thinking as much as through socio-material embodiment in a collective movement. They thus argue for open-endedness and a shared experiential and grounded understanding of prefigurative politics.

In Chapter 6, McGreevy et al. (2025) argue that societal transformation to post-growth is bound up with community engagement on desirable futures. This makes arenas of exchange vital tools to co-create, experience, and advance prefigurative processes in ways that themselves embody conviviality and offer opportunities for mutual learning. Their chapter compares and draws lessons from three convivial space cases across Paris, Amsterdam, and Kassel, critically examining their modes of prefiguring action and enabling impact that transcends the community scale. Their systematic reflection on points of departure, strategic approaches and systemic impact covers the micro-meso-scale dynamic encompassed by these agri-food sustainability initiatives, bringing out the full chain along which prefiguration must extend to register impact.

Our second part, entitled *spaces of conviviality and politics*, convenes contributions that show how prefigurative politics takes place in shared social spaces by engaging different people, often through different means and emotions.

In Chapter 7, Feola (2025) takes up the matter of deconstruction in prefigurative spaces. He argues that while prefiguration scholars have devoted considerable focus to grassroots initiatives, there has been relative neglect of refusal and resistance as prefiguration, for instance in relation to engagement with capitalist logics. The chapter theorises deconstructive processes in grassroots prefiguration, articulating this as an essential part of prefiguring alternatives to hegemonic phenomena. Beyond refusal, the author identifies deconstructive prefigurative practices that include unlearning, sacrifice, and defamiliarization, and argues that deconstruction plays a crucial role in spaces of prefigurative politics.

In Chapter 8, Firinci Orman (2025) addresses youth climate politics through an Arendtian lens on the politics of novel openings. She understands youth activists as not only envisioning but also embodying and advancing sustainable and democratic futures. The chapter unpacks this through Arendt's political philosophy in relation to natality, political action, and plurality, viewing youth-led climate movements as constituting generative political space with transformative characteristics. The councils, civil disobedience, and everyday acts of sustainability that youth activists mobilise as tools are themselves contributions to building the world anew. The chapter thus posits these ontological acts as constitutive of an urgent and emergent participatory politics.

In Chapter 9, Lode, Basu, and Macharis (2025) consider the prefigurative potential of citizen panels for transformation across five case studies. They draw on a multi-political framework to understand deeper systemic transformations in energy and mobility systems. This approach recognises the inextricable link between such transformations and changes in value structures to prioritise just futures. The authors draw on five citizen panels across Europe where 150 citizens deliberated on these issues, and point out the challenges linked to divergence, disruption, and scalability that surface in relation to transformative change. The chapter emphasises the interplay of prefigurative politics with contentious and institutional dynamics.

In Chapter 10, Eckerle (2025) turns attention to the audience in various acts of carnival that are interpreted as prefigurative politics. He draws on social psychological, anthropological, and philosophical work to reflect upon the experience of prefigurative politics from "without" as an expression of Ernst Bloch's "concrete utopias". The author argues for an explicit combination of the prefigurative and carnivalesque to promote radical alternatives in an experiential manner that effectively generates solidarity. Using the novel lens of carnival, the chapter makes the case for joyful community-building and its subversive potential towards impactful prefigurative politics.

In Chapter 11, Aiken (2025) examines proleptic community environmentalism, namely a move away from established praxis, channelling Kafka's notion of the way away from here. This takes the form of a case study of a zero-waste community store in Luxembourg, Ouni, which is premised on imagining and responding to a world of food scarcity by moving away from wasteful practices. This move births numerous ways of responding to a proleptic event, as the author shows in recounting

Ouni's context, aims, and orientations. Through its framing of prolepsis, the chapter centres intersubjective connection as a means to deconstruct and transcend entrenched socio-material practices.

In Chapter 12, Ravikumar (2025) discusses the politics of organising through the lens of prefiguration in relation to future visioning exercises. The author reflects upon the interstitial space of after-work events organised in a transdisciplinary research project to enable interactions that integrate future visions with organisational processes. She conceptualises the emergent imaginings as prefigurative impulses. These spaces of prefigurative politics, created on the sidelines of everyday practice, convene organisational processes and networks to advance prospects of organisational change for transformation. Lowering hierarchy and offering scope for collective imagination are key aspects of such a convivial space.

In Chapter 13, Brandt and Larsen (2025) turn to inhabitation itself as a form of prefigurative politics and a source of political transformation, using the case of the prominent Christiania community in Copenhagen. From its origin as a radical squat, Christiania has shifted notably to bureaucratic assimilation, while retaining the capacity to symbolise the politics of prefiguration and transformation by enacting alternative relations in governance, housing, and social organisation. Since its legalisation in 2012, the authors argue that a new planning paradigm has brought bureaucratic domination to bear on its construction practices and aesthetic. This moves Christiania away from Lefebvre's notion of a continuously socially produced differential space which engenders radical democratic participation, towards codified and controlled representations of space.

In Chapter 14, Fung and Htoo (2025) focus on the intersection of conservation, peace and self-determination politics in Salween Peace Park in Myanmar. Under protracted conflict in Karen State, this park is a peace and conservation initiative led by the ethnic Karen community to prefigure self-determination over their territory and way of life. This focus is distinct from much prefigurative politics scholarship which is situated in Global North and urban settings. For the Karen, prefigurative politics includes the collective will to sustain generational livelihoods and practices under threat. The authors show how the initiative is itself under threat through proposals for hydropower development and the wider context of state violence.

Our third part on *sectoral movements* explores how prefigurative politics has the potential to catalyse transformation across various sectors, particularly in land use, health, food systems, migration, and legal education.

In Chapter 15, Goncalves (2025) examines movements that aspire to food sovereignty in Brazil. Through the case of Teia dos Povos in Brazil, a network of communities pursuing food sovereignty, the chapter illustrates how these imaginaries resist capitalist, colonial, and individualist paradigms. By reclaiming land, practicing agroecology, and promoting knowledge co-production, Teia dos Povos challenges dominant onto-epistemologies and embodies socio-ecological alternatives. The chapter concludes that the climate crisis demands both imaginative and ontological shifts rooted in collective emancipation.

In Chapter 16, Savasan (2025) views law clinics through the lens of prefigurative politics and asks to what extent prefiguring can be done within existing institutional structures. Environmental Law Clinics can act as catalysts for legal transformation by integrating prefigurative law into their core functions: legal education, legal assistance, and advocacy. Through real-world environmental cases, ELCs train future lawyers, support communities, and promote systemic change. The chapter highlights ELCs' unique potential to model and advance equitable, sustainable legal practices that prefigure desired environmental futures.

In Chapter 17, Rocha-Gomes (2025) examines practices of prefiguring in healthcare and shows who community drive digital therapeutics can be viewed through prefigurative politics. The #WeAreNotWaiting movement represents a grassroots, patient-led initiative that reimagines healthcare through open-source, community-driven digital therapeutics (DTx). Rooted in prefigurative politics, the movement embodies values of openness, mutual aid, and self-determination. It challenges traditional healthcare by blending activism, technology, and evidence generation outside institutional frameworks. The author addresses the extent to which this case provides an avenue towards participatory, patient-powered innovation in digital health.

In Chapter 18, Vaño et al. (2025) view land use in Europe through 12 regions and explores the potential for transformation through prefigurative politics. The authors identify key intervention points, which are needed to achieve this: decentralising power and enhancing multi-actor participation, bridging policy gaps, adapting to external pressures, and strengthening accountability. These systemic actions have the potential to shift entrenched governance models, build inclusive decision-making, and align policy with sustainability. Perhaps rather than a radical overhaul, the chapter advocates gradual transformation, integrating prefigurative practices within existing institutions to foster long-term, equitable change across sectors, and scales.

In Chapter 19, Bertuzzo (2025) explores how human and plant migrations are woven together to shed light on emerging food supply chains and possible cohabitation. Her ethnography introduces us to farms spreading across Italy, thanks to socio-ecological infrastructures and migrant networks. This trend reflects the potential of a more-than-human cooperation between people, plants, and landscapes. Despite minimal institutional support, *sobji* farms offer a promising, decentralised response to climate change and food insecurity. More importantly, they challenge dominant narratives around migration and agriculture, highlighting the role of care, adaptability, and decolonial practices. As transformative, locally embedded innovations, they point to new ways of imagining future food systems and human-nonhuman coexistence.

In Chapter 20, Martinovska Stojcheska et al. (2025) explore transformations in food systems through local initiatives in a post-socialist context. The authors demonstrate how grassroots efforts, like the National Farmers Federation's e-Farm and the First Organic Cooperative, challenge dominant agro-industrial models, by promoting community-based, democratic, and environmentally sustainable practices, despite facing structural and institutional barriers. Using Yates's framework, the chapter shows how experimentation, new social conduct, and diffusion of values can foster systemic transformation. More broadly, these localised efforts embody alternative food systems and contribute to global movements for justice, sustainability, and rural empowerment.

In Chapter 21, the theme of rural transformations continues, with Dragin et al. (2025) discussing place attachment and quality of life and whether they can act as catalysts for sustainable rural transformation. Analysing three dimensions of place attachment—identity, dependence, and bonding—the chapter shows how Croatian youth report higher levels of dependence and bonding, while Serbian youth exhibit stronger identity ties. Improving infrastructure, education, cultural access, and fostering emotional and social connections can enhance rural sustainability. Spa tourism is suggested as a strategic tool for revitalising rural areas and engaging youth in sustainable development.

The fourth part demonstrates how *cross-sectoral and transdisciplinary transitions* can provide a fruitful springboard and testbed for prefigurative politics, which may advance the aspirations towards more sustainable practices.

In Chapter 22, Shokrgozar et al. (2025) exemplify how extractivism, agropastoralism, and the making of frontiers come together in large developmental projects in the Global South, often justified through climate goals, security, or modernisation. The authors demonstrate that these energy projects—such as coal, solar, and wind projects in Sindh, Rajasthan, and Gujarat—enforce state-capital territorialisation, dispossessing agropastoral communities, especially marginalised caste, class, and religious groups. While exclusion is reinforced through infrastructure, caste hierarchies, and securitisation, resistance movements like the Gorano protests and village alliances offer prefigurative politics rooted in pluralism, solidarity, and ecological justice. These counter-majoritarian struggles seek to reimagine land, citizenship, and development, transforming peripheries into spaces for alternative world-making and sustainable futures.

In Chapter 23, Diesner and Scholl (2025) explore to what extent can prefigurative politics be connected with transdisciplinary action approach and experimentation. By embedding structured experimentation and collective learning—particularly within Urban Living Labs (ULLs)—the authors argue that prefiguration can move beyond symbolic acts to real systemic change. They stress the importance of learning processes, scaling strategies (up, out, deep, through), and knowledge diffusion. Using urban food initiatives across European cities, they show that sustained transformation requires strategic coordination, institutional support, and ongoing collaborative learning within and beyond grassroots efforts.

In Chapter 24, Griffani (2025) presents a case from Italy, examining the prefigurative political practices of Mercato Brado, an agroecological network in post-industrial Terni, Italy. Drawing on Ana Cecilia Dinerstein's interpretation of Ernst Bloch's philosophy of hope and Kim Fortun's notion of late industrialism, the author explores how toxic legacies from steel production shape political and ecological imagination. Through the Participated Guarantee System (PGS), producers, and consumers co-created ethical food systems grounded in care, shared meaning, and intentional exposure to toxicity. These practices opposed capitalist abstraction and offered concrete utopias, showing that even on polluted soil, communities can foster reparative, collective futures beyond ecological harm and economic precarity.

In Chapter 25, Milliken (2025) further expands the idea of using diverse methods to engage, particularly the role of art activism in prefigurative politics. Drawing on an example by Agnes Denes' (1982) Wheatfield—A Confrontation that transformed a Manhattan landfill into a wheat field, the chapter explores how creative practices like Denes' can drive food system transformation by engaging with deep "leverage points" such as paradigms, values, and worldviews. Arguing that despite some contradictions, Wheatfield remains a powerful symbol of ecological awareness and political imagination, Milliken shows how participatory art fosters reflection, care, empowerment, and systemic critique.

In Chapter 26, Schuenemann (2025) directs our attention to the role of prefigurative politics in disaster recover, pointing out that it is rarely the state that is the first to respond. Groups like Occupy Sandy, which was rooted in the Occupy Wall Street movement, used horizontal organising and mutual aid to critique traditional disaster responses and challenge systemic inequalities. While Occupy Sandy gained praise for its swift, inclusive actions, internal tensions grew as the sense of crisis faded, leading to fragmentation. Nevertheless, it inspired ongoing mutual aid networks and contributed to broader movements for climate justice and policy reform, revealing both the promise and limits of crisis-driven political mobilisation.

In Chapter 27, Arora (2025) considers the deep entanglement between stories and politics, examining how narratives—ranging from facts to fiction—shape and are shaped by political conditions and identities. Using cyborg feminism, labour struggles in India, and ethnographic insights, she critiques dominant frameworks that reduce complex realities into fixed categories. Through figures like the woman machine and cyborg, Arora challenges representational traps, urging instead for "experimental statements" that provoke collective re-imaginings beyond situated difference. Highlighting moments of worker solidarity and rupture, the piece argues for storytelling that disrupts power, redefines joy, and opens space to think life anew—messy, speculative, and irreducible to policy or theory.

In Chapter 28, Holmen et al. (2025) explore how metaphors play a vital role in sustainability transformations by shaping how people think, act, and imagine the future. Drawing on conceptual metaphor theory, it examines three initiatives—EU Cities Mission, the Degrowth Movement, and Transition Towns—to illustrate how metaphors guide transformative efforts. The EU Mission uses "moonshot" and "journey" metaphors to drive top-down innovation. Degrowth uses disruptive metaphors like "addiction" to critique economic orthodoxy. Transition Towns adopt ecological metaphors of "roots" and "resilience" to encourage grassroots change. These metaphors prefigure alternative futures by making abstract goals tangible, emotionally resonant, and actionable in present-day practices.

3   Concluding Thoughts

We wish to emphasise three cross-cutting insights that run through the volume. These relate to the multi-scalar nature of prefigurative politics, their entanglement with boundary transgression, and the multiple temporalities of present transformation. These are themes that our related work within working group three of the TransformERS network has identified and is drawing out in more detail through other outputs as well.

The systemic aspects of transformation require a multi-scalar understanding to inform and orient the work of prefiguration. Multi-scalarity is evident in several of the chapters when they address the need to scale change, whether upward in size or outward through replication, leading to questions around the spatial scales at which prefigurative politics emerge and are advanced by various actors. Certain kinds of action take place in communities, in cities, in regions, or countries, but all change necessarily happens somewhere, and the local is thus an important site for understanding prefigurative politics.

Boundary transgression takes place both in the cognitive domain of perception and recognition and the normative domain of social value (inner boundary work), as well as partly in the socio-material domain of bridging sectoral boundaries, interest groups, and changing infrastructures (outer boundary work). Transgression indicates unsettling established divides in ways that go against something, but it also results in reconfiguring, in rethinking the need for and purpose of boundaries, and identifying and constituting novel forms of cooperation and co-existence. Aspects of denial, refusal, and resistance are vital to boundary work. Thus, prefiguration is also about which boundaries should *not* be transgressed, and how publics mobilise for this.

Finally, the many spaces of experimentation, conviviality, change efforts and innovation that constitute prefiguration would not amount to much over time without institutionalisation. Where this institutionalisation happens—whether in formal institutions or community practices—is captured by multi-scalarity as well as levels of governance; and what patterns it challenges or brings about is partly addressed as boundary transgression. But *when* institutionalisation happens and for *how long* are also crucial aspects of prefiguration—even present transformation plays across a *now* of different durations. What is important to recognise is that multiple temporalities of change are at play, and prefigurative politics must necessarily engage with several of these to engender enduring change.

An inclusive, full-bodied approach to the prefigurative politics of present transformation entails understanding the phenomenon as *embodied strategies to render desirable futures with immediacy*. This embodiment is multi-scalar, desirable futures come with boundary transgressions, and rendering transformative change with immediacy happens across multiple temporalities to fulfil its strategic prefigurative potential. As we work with diverse collectives and epistemic communities to bring better common futures into being, we are supported by the clarity that comes from mutual engagement and the collaborative work of sensemaking across our varied examples and trends of the prefigurative politics of present transformation.

Acknowledgements We want to warmly thank and acknowledge the contributions of all participants that joined us in Oslo. The level of engagement, supportive, and collaborative space that we shared over two days was palpable and we hope this comes across through these pages.

Competing Interests The authors have no conflicts of interest to declare that are relevant to the content of this chapter.

References

Aiken, G. T. (2025). Community-based proleptic environmentalism: food, energy, and Kafka's 'Weg-von-hier'. (Chapter 11, this volume).

Arora, C. (2025). Between situated and speculative: (Im)possibilities of prefigurative politics in research? (Chapter 27, this volume).

Batterbury, S., Chakori S., & Uxo, C. (2025). Prefiguring consumption reduction in shadow networks: Recirculating resources, knowledge and skills in urban tool libraries and community bike workshops. (Chapter 4, this volume).

Bertuzzo, E. T. (2025). For a more-than-human politics of short food supply chains in times of human-plant migrations. (Chapter 19, this volume).

Brandt, J., & Larsen, J. L. (2025). Inhabitation as prefigurative politics and source of political transformations. (Chapter 13, this volume).

Diesner, D., & Scholl, C. (2025). Towards a transdisciplinary approach to prefiguration experiments for transformative change. (Chapter 23, this volume).

Dragin, A. S., Tešin, A., Ladičorbić, M. M., Zadel, Z., Košić, K., Amezcua-Ogáyar, J. M., Calahorro-López, A., & Surla, T. (2025). Place attachment and quality of life as catalysts for sustainable rural transformation: Addressing depopulation and promoting socioecological resilience. (Chapter 21, this volume).

Eckerle, F. (2025). Laughter, role-play, and immersion: Rediscovering the audience in prefigurative politics through carnival. (Chapter 10, this volume).

Feola, G. (2025). Broadening the understanding of deconstruction in prefigurative social spaces. (Chapter 7, this volume).

Firinci Orman, T. (2025). Prefigurative politics in action: Youth climate activism and Arendt's politics of new beginnings. (Chapter 8, this volume).

Fung, Z., & Htoo, S. (2025). Prefiguring conservation, peace, and self-determination in the Salween Peace Park in Karen State, Myanmar. (Chapter 14, this volume).

Goncalves, J. E. (2025). Climate change as a crisis of recognition: Prefigurative politics of socio-ecological movements for food sovereignty in Brazil. (Chapter 15, this volume).

Griffani, B. (2025). Feral alchemies: The prefigurative politics of staying on toxic Ground. (Chapter 24, this volume).

Grumiau, B., & Hughes, I. (2025). Starting by doing: Ethical orientations for open-ended prefigurative action. (Chapter 5, this volume).

Haugland, B. T., & von Wirth, T. (2025). Acclimatising to the future: Prefiguration and urban experimentation in public policy. (Chapter 2, this volume).

Holmén, J., Saglietti, C., & Holmberg, J. (2025). Exploring the prefigurative power of metaphors of change in public policy, alternative movements, and deep ecology. (Chapter 28, this volume).

Lode, M. L., Basu, S., & Macharis, C. (2025). A multi-political reflection on the transformative potential of citizen panels: Insights from five case studies (Chapter 9, this volume).

Martinovska Stojcheska, A., Kotevska, A., & Tuna, E. (2025). Present transformations in food systems: Local initiatives towards sustainable and just futures in post-socialist context. (Chapter 20, this volume).

McGreevy, S. R., Baibarac-Duignan, C., & Spiegelberg, M. (2025). Prefigurative politics of post-growth futures in the making: learning from transformative practices that leverage (temporary) convivial space. (Chapter 6, this volume).

Milliken, S. (2025). The role of art activism in the prefigurative politics of food system transformation. (Chapter 25, this volume).

Ravikumar, K. (2025). Imagining futures in the present: Conceptualising prefigurative impulses through the politics of organising. (Chapter 12, this volume).

Rocha-Gomes, J. (2025). Prefiguring healthcare: Community-driven digital therapeutics and the politics of patient innovation. (Chapter 17, this volume).

Sareen, S., & Juhola, S. (2025). The Prefigurative politics of present transformation. (Chapter 1, this volume).

Savasan, Z. (2025). The potential role of the law clinics in prefigurative law and politics: Environmental law clinics as a case study. (Chapter 16, this volume).

Schuenemann, T. (2025). Recovery or transformation? The prefigurative politics of hurricane response in the US. (Chapter 26, this volume).

Shokrgozar, S., Ashraf, U., & Singh, D. (2025). Energy at the edge of the state: Extractivism, agropasto-ralism, and the making of frontiers. (Chapter 22, this volume).

Stella, D. (2025) Prefigurative politics in citizen science: The Golf Vinoř project as a case of present transformations. (Chapter 3, this volume).

Vaňo, S., Leventon, J., Mederly, P. (2025). Prefigurative politics of transformations in land use decision-making systems in Europe. (Chapter 18, this volume).

Open Access This chapter is licensed under the terms of the Creative Commons Attribution-NonCommercial-NoDerivatives 4.0 International License (http://creativecommons.org/licenses/by-nc-nd/4.0/), which permits any noncommercial use, sharing, distribution and reproduction in any medium or format, as long as you give appropriate credit to the original author(s) and the source, provide a link to the Creative Commons license and indicate if you modified the licensed material. You do not have permission under this license to share adapted material derived from this chapter or parts of it.

The images or other third party material in this chapter are included in the chapter's Creative Commons license, unless indicated otherwise in a credit line to the material. If material is not included in the chapter's Creative Commons license and your intended use is not permitted by statutory regulation or exceeds the permitted use, you will need to obtain permission directly from the copyright holder.

Index

© The Editor(s) (if applicable) and The Author(s) 2026
S. Sareen and S. Juhola (eds.), *Societal Transitions to Sustainability*, Palgrave Studies in Environmental Transformation, Transition and Accountability, https://doi.org/10.1007/978-3-032-07395-2

467

MIX
Papier aus verantwortungsvollen Quellen
Paper from responsible sources
FSC® C105338

If you have any concerns about our products,
you can contact us on
ProductSafety@springernature.com

In case Publisher is established outside the EU,
the EU authorized representative is:
Springer Nature Customer Service Center GmbH
Europaplatz 3, 69115 Heidelberg, Germany

Printed by Libri Plureos GmbH
in Hamburg, Germany